PING 2019

Modern Trends in Material Engineering

IOP Conference Series: Materials Science and Engineering
Volume 723

Pilsen, Czech Republic
10 - 13 September 2019

Editors:

Ludmila Kucerova
Hana Jirkova
Stepan Jenicek

ISBN: 978-1-7138-2582-1
ISSN: 1757-8981

Printed from e-media with permission by:

Curran Associates, Inc.
57 Morehouse Lane
Red Hook, NY 12571

proceedings.com

Some format issues inherent in the e-media version may also appear in this print version.

Institute of Physics
Dirac House, Temple Back
Bristol BS1 6BE UK

Phone: 44 1 17 929 7481
Fax: 44 1 17 920 0979

techtracking@iop.org

Additional copies of this publication are available from:

Curran Associates, Inc.
57 Morehouse Lane
Red Hook, NY 12571 USA
Phone: 845-758-0400
Fax: 845-758-2633
Email: curran@proceedings.com
Web: www.proceedings.com

TABLE OF CONTENTS

IOP Conf. Series: Materials Science and Engineering **723** (2020) 011001 doi:10.1088/1757-899X/723/1/011001

Preface

We would like to introduce you to the Proceedings of International conference PING 2019 – Modern trends in material engineering, which took place at 10. - 13. 9. 2019 in Pilsen, Czech Republic.

It was already the fourth year of the conference, which started as a workshop of the Regional Technological Institute (RTI) (www.rti.zcu.cz) of the University of West Bohemia in Pilsen, and which was originally intended to present research programs and results of individual laboratories of the Institute. The research program of Research and development of forming technologies organized this conference for the first time in 2018. Research and development of forming technologies have a long tradition not only within RTI but also at the University of West Bohemia and it includes forming of materials, characterisation of the resulting microstructures by various microscopic methods and mechanical testing. PING 2018 was still organized as a local event with guests invited from other universities and industrial companies. Due to the positive feedback of this event, the fourth year was organized in 2019 with similar topics, only with international attendance this time.

The scope of the topics reflected the research interests of the laboratories of RTI, including metal forming, heat and thermo-mechanical treatment, nonferrous metals, mechanical testing, microscopic analysis. The sessions concentrating on modern high strength steels of various generations developed mainly for the automotive industry constituted a very important part of the conference. Dynamically developing field of additive technologies, which has a high potential for future applications, was also frequently presented.

In total, 64 contributions were chosen for presentation at PING 2019, 41 of them in the form of oral presentations and 23 as e-posters. 33 articles were selected after peer review for publication within this Proceedings. These articles cover all important topics presented at the conference and proved an overview of development in the field of modern high strength steels, 3D printing and also nanotechnologies and nanomaterials.

The committee of PING 2019 paid great attention to reviewing of all the papers submitted to the conference and organizing the sessions to enable the participants to enjoy the scientific as well as the social side of the event. We sincerely appreciate your participation and do believe our joint efforts made this conference indeed successful.
At the very end, we would like to invite you to the fifth year of this conference, which will take place from 7. to 10. 9. 2021.

Organizing committee of PING 2019

IOP Conf. Series: Materials Science and Engineering **723** (2020) 011002 doi:10.1088/1757-899X/723/1/011002

Scientific Commitee

Organizing Committee

Program Committee

Contacts

University of West Bohemia
Faculty of Mechanical Engineering – Regional Technological Institute – PING Conference
Univerzitní 8
306 14 Plzeň
Czech Republic
Phone: +420 377 638 762
E-mail:jeniceks@rti.zcu.cz

www.ping.zcu.cz

IOP Conf. Series: Materials Science and Engineering **723** (2020) 011003 doi:10.1088/1757-899X/723/1/011003

International Conference PING 2019 - Modern trends in material engineering

Editors: Doc. Ing. Ludmila Kučerová, Ph.D.
Dr.-Ing. Hana Jirková, Ph.D.
Ing. Štěpán Jeníček

University of West Bohemia
Univerzitní 2732/8, 301 00 Pilsen, Czech Republic, Europe

This conference is organized under the patronage of the Regional Institute of Technology of the University of West Bohemia in Pilsen.

International Conference PING 2019 - Modern trends in material engineering

10–13 September 2019, Primavera Hotel & Congress Center, Pilsen, Czech Republic

Thank you for your attention!
FACULTY OF MECHANICAL

IOP Conf. Series: Materials Science and Engineering **723** (2020) 011005 doi:10.1088/1757-899X/723/1/011005

Peer review statement

All papers published in this volume of *IOP Conference Series: Materials Science and Engineering* have been peer reviewed through processes administered by the proceedings Editors. Reviews were conducted by expert referees to the professional and scientific standards expected of a proceedings journal published by IOP Publishing.

The potential of application of Ni-layer for enhancement of utility properties of freight wagons

Z Andrsova[1a], P Kejzlar[1b], T Bakalova[1c], M Petru[1d], P Podzimek[2]

[1a,b,c,d]Technical University od Liberec, Studentska 1402/2, 461 17 Liberec 1
[2]Funchem s.r.o., Vratislavicka 59/14, 460 06 Liberec 6

[1a]zuzana.andrsova1@tul.cz, [1b]pavel.kejzlar@tul.cz, [1c]totka.bakalova@rul.cz, [1d]michal.petru@tul.cz, [2]podzimek@funchem.cz

Abstract The construction of 4-axled freight wagons used for transportation of bulk materials is limited with respect of minimal angle allowing emptying the bulk cargo and limited lifetime caused by abrasion caused by intensive wear between the wagon wall and transported cargo. A potential of application of protective chemically-deposited Ni-layer was evaluated based on tribological testing.

1 Introduction

For the mass transportation of bulk materials are mostly used freight wagons, it is a profitable and evironmentally friendly model of transportation. However, as in other sectors, there is an increasing emphasis on further cost reductions. Although the railway sector is very conservative, manufacturers of rail vehicles are trying to offer wagons with reduced weight (eg replacement of conventional steel parts with composite elements, often with built-in smart electronics), increased capacity or longer lasting (operational life of the railway wagon may be up to 30 years). Freight wagons (Figure 1) are limited in their construction by the angle of the walls, which is necessary to empty the load. In addition, the discharge flap under the vehicle wall is heavily loaded to wear and its lifetime is several times lower than the whole vehicle. In cooperation with the manufacturer of freight wagons, the surface treatment of flaps with chemical nickel coating was tested. The coating is intended to increase wear resistance, corrosion resistance and sliding properties which alows to modify the construction of the wagon to increase the capacity and extend the lifetime of the flap. [1]

The surface treatment with chemical (electroless) nickel was chosen because of its suitable physical-mechanical properties, but also for ease of application compared to electroplated coatings with comparable parameters. While in electroplating we encounter many limitations resulting from the laws of electrolysis (especially difficult plating of cavities, uneven distribution of layer thicknesses, necessity of using anodes etc.), in the case of chemical deposition, a coating is formed on the surface simply by chemical reduction from the plating bath which is heated to the specified temperature. In the case of chemical nickel plating, in addition to Ni, also elemental phosphorus is deposited in the coating. The resulting Ni_3P alloys are characterized mainly by high abrasion resistance, corrosion resistance, and uniform thickness of the deposited coating. The overall resulting properties of the coating are influenced by the phosphorus content according to which the coatings are divided into three groups – with low content (1-4 wt% P, 675 HV, corrosion resistance of 24 h due to ISO9227), with medium content (5-9 hm. wt% P, 575 HV, corrosion resistence of 200 h due to ISO9227) and with high content (10-12 wt% P, 525 HV, corrosion resistence of 1000 h due to ISO9227). The Ni_3P layers can also be created with the addition of admixtures which further improve their properties (eg. Al_2O_3, SiC, PTFE, graphite etc.), or they can create more complex alloys with B, Co, W etc. Properties of Ni_3P coatings may be further enhanced by heat treatment, when the phosphorus distribution changes and the porosity of the coating is reduced. At the same time, the brittleness is reduced, the hardness is increased and the adhesion is improved by creating a diffusion layer. Usually the heat treatment, so-called hardening of the coatings is performed in the range of temperatures from 200°C (with the dwell in the tens of hours) up to 400°C (with the dwell in hours). If the coatings are cured at the upper limit of the temperature range, they achieve hardnesses up to 1000 HV. The chemical and corrosion resistance of Ni_3P coatings depends primarily on the layer thickness – with the increase in thickness decreases the number of open pores which are completely closed at thicknesses above 30 microns. The mentioned resistance also depends on the phosphorus content. [2-4]

The paper assesses the suitability of chemical Ni application on discharge flaps mainly from the viewpoint of tribological properties - the coefficient of friction and wear rate. Rotational tribology according to ASTM G99-95 "Ball-on-disc" was used to determine these indicators. Its principle consists in placing the non-rotating ball on the surface of the sample (in the shape of a disc). At a certain distance from the center of the sample is placed a "Ball" loaded with a predefined force. The table together with the disc is rotated at a defined speed (rpm), and performs a predetermined number of rotations or a certain path length. The result of the measurement is the dependence of the coefficient of friction on the path length. [5-7]

In addition to tribological properties, the microhardness of the deposited layers and the effect of welding on the layer were verified – wagons are conventionally welded by fusion welds and due to the anticipated thermal degradation of the layer at the weld site, the replacement of the weld by bonded joints is considered.

Figure 1. Illustration image if the freight wagon, greeen highlighted discharge flaps to which surface treatment has been applied.

2 Experimental

2.1 Sample preparation, deposition of layers

For the purposes of testing, two types of samples were prepared from rolled steel sheets S235JR with a thickness of 5 mm, which are used to produce discharge flaps. The sheets were stripped with a phosphoric acid prior to plating to remove the oxide layer. Due to the fact that it is intended to be applied to 6x2 m sheets, in terms of technical possibilities, the surface treatment towards lower initial surface roughness has not been considered yet, therefore, even the specimens were not ground or polished prior to deposition, the initial surface roughness was around Ra 12.5. Three groups of samples were prepared for tribology and hardness measurement (labelled A,B,C, see Figure 2) with circular shape and diameter of 30 mm. Group A was untreated, groups B and C were coated with chemical Ni. Group C was further hardened in an electric induction furnace for 4 hours at 300 ° C. Two groups of samples were prepared for the analysis of welded joints (labelling D, E, see Figure 2) with rectangular shape and size 40x200 mm. Both groups were coated with chemical Ni, group E was further hardened in an electric induction furnace for 4 hours at 300 ° C. The reverse side of the specimens was protected against nickel plating by a barrier to assess the effect of the coating on welding. Samples were subsequently joined by a lap weld joint. Commercially available solutions were used to deposite coatings on all samples (Coventya Enova EF9B and Enova AM100, 5-9% P, 85-90°C/ 40 min., layer thickness of 20 μm).

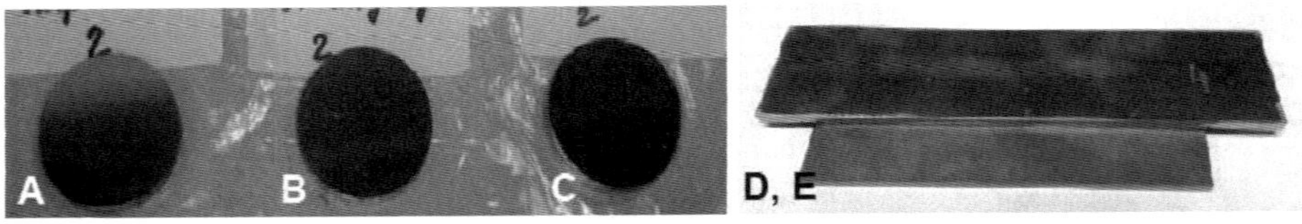

Figure 2. Preview of test samples for tribological testing (A, B, C) and for weld analysis (D, E).

2.2 *Experimental methods*

Rotational ball-on-disc tribology according to ASTM G99-95 was used to evaluate the coefficient of friction. As a counterpart in the friction pair was used ceramic ball of Si_3N_4, resp. Al_2O_3. The choice of the counterpart is based on empirically determined information about the transported materials, which cause the most wear on the flap surface when emptying. The selected ball materials should be as close as possible to their properties. The measurement and evaluation of the coefficient of friction between the samples and the ceramic ball was performed on a tribometer CETR UMT Multi-Specimen Test System. The wear of the disc material from the friction pair was evaluated by mechanical profilometer Dektak XT. The microhardness of layers $HV_{0,1}$ was measured using microhardness tester Zwick 3212 under the load of 1 N for 10 s, using Vickers indenter. Ten indentations were performed on each sample and the mean microhardness was calculated. Un nickel-plated surfaces (corresponding to the flap supporting structure) were welded to the nickel-plated sides of the samples under standard parameters (arc fusion welding). Samples after welding were evaluated visually and a macroscopic examination of the welds was also performed – metallographic samples were prepared and macroscopic and microscopic images were taken. Also detailed samples of the metal layers at the weld site and in the heat affected zone were taken on the same samples.

3 Results and discussion

The aim of the performed experiments was to evaluate the coefficient of friction and comparison of friction properties using different prepared surfaces, as well as to evaluate the degree of wear of the sample after tribology. It can be seen that the coefficient of friction (Table 1) does not change much on contact with the Si_3N_4 ball – as can be seen, for example, from the wear test of the samples (Figure 3), particles shredded from the surface fall between the sample and the counterpart thereby the coefficient of friction increases again. This is is probably due to the roughness of the sample surface. When the surface comes into contact with the Al_2O_3 ball, the friction coefficient of the nickel-plated surface reduces up to 50%. The ball has a higher hardness and there is no crumbling of the particles that can cause an increase in the coefficient of friction. The results of the tribological specimen wear evaluation (Table 2) show that as the coating hardens, wear decreases. Significant is the reduction of wear, especially in hardened surfaces, in case of Si_3N_4 it´s up to 95 %, in case of Al_2O_3 ball even 97 % against the initial state (according to the volume of material removed - wear rate). Microhardness measurements show an increase of Ni_3P coating hardness about 30% after heat treatment (see Table 3). The microscopic

IOP Conf. Series: Materials Science and Engineering **723** (2020) 012001 doi:10.1088/1757-899X/723/1/012001

analysis of the weld showed that the presence of the layer does not affect the quality of the weld – undesirable changes were not evident in the weld microstructure. However, welding will significantly damage the layer at the weld site, in the heat affected zone is then significantly impeded in its integrity (Figure 4). This means that welding of already coated sheets is not permitted – the corrosion resistance of the entire weldment would be fundamentally impaired.

Table 1. Average values of the coefficient of friction [-].

Surface	Base material	Ni – non-hardened	Ni – hardened
counterpart Si_3N_4	**0,857**±0.133	**0,838**±0.234	**0,705**±0.208
counterpart Al_2O_3	**0,828**±0.167	**0,427**±0.251	0,307±0.202

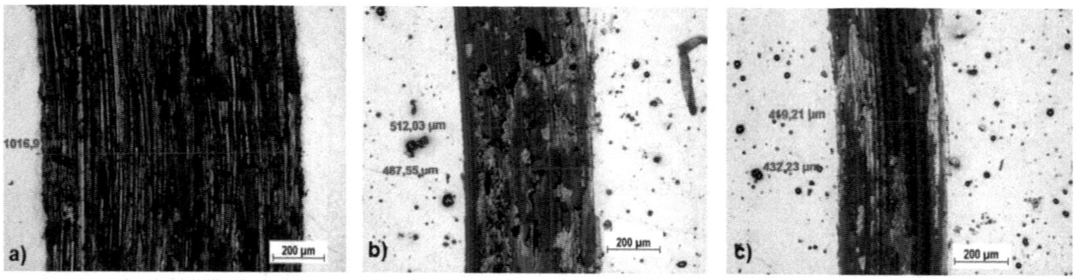

Figure 3. Wear of sample surface after tribology against the ball Si_3N_4 a) initial state, b) non-hardened coating c) hardened coating. LOM, BF, magn. 25x.

Table 2. Measured values of specimen wear after tribology using counterpart Si_3N_4 and Al_2O_3.

Surface	Base material	Ni – non-hardened	Ni – hardened
		counterpart Si_3N_4	
Profile width [µm]	1016.9	499.79	425.72
Profile depth [µm]	6.734	2.827	0.921
Volume of material removed per unit length of profile [μm^3]	6697.8	1233.21	292.08
		counterpart Al_2O_3	
Profile width [µm]	966.6	434.2	212.03
Profile depth [µm]	4.645	1.845	0.523
Volume of material removed per unit length of profile [μm^3]	4310.85	621.09	101.32

Table 3. Measured values of microhardness of layers $HV_{0,1}$.

Sample	Microhardness $HV_{0,1}$ [-]										
non-hardened	594	598	595	594	602	596	592	595	590	598	**595**
hardened	920	923	925	988	924	927	920	921	925	923	**924**

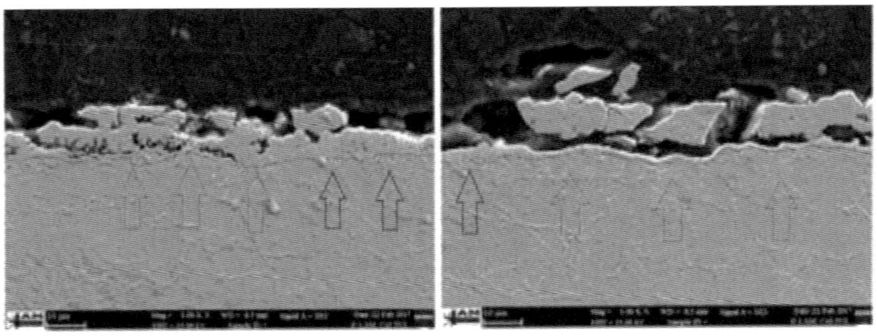

Figure 4. Layer integrity damage at the borderline of weld joint (blue pointers) and the heat affected zone. SEM (red pointers), SE2, magn. 1kx.

4 Conclusion

The results show high potential of application of chemically deposited Ni_3P layers. The samples show an increase in wear resistance of 81-97% against the base material, the upper limit is in the case of heat treatment of the layer (supported by an increase in hardness of about 30%). At the same time, the sliding properties can also be improved by reducing the coefficient of friction when comparing the results of the coated samples with the base material, without pretreating the surface roughness. However, this is conditioned by the characteristics of the counterpart (in practice, a material of lower hardness tends to seize into the substrate) – it would be necessary to achieve a significant reduction in the surface roughness of the surfaces to be coated. The increase of the sliding properties is significant especially when using Al_2O_3 ball, where the coefficient of friction decreased by up to 50%. A significant increase in the lifetime of the discharge flaps can be expected and, due to the achieved coefficient of friction, the possibility of modifying the structure to increase the vehicle capacity. From the analysis of the layers in the area of the weld joints is obvious the need for replacement of the standard fusion weld joint with a joint without thermal interference, eg by bonding, which will require further application development due to the continuous dynamic loading of the joints and the required vehicle service life. Surface pretreatment to reduce surface roughness is possible for parts of this size when using mechanical surface grinding and polishing, but it could cause undesirable increases in production costs and the efficiency of the treatment would have to be considered.

Based on the results of laboratory testing, a pair of flap plates without mechanical pretreatment were experimentally plated (see Figure 5), fixed to the structure of the flap by bonded joint (Letoxit LH 21) and subjected to trial operation of freight wagon that is to verify the properties of the layer in real conditions. In the case of a favorable response, the innovation of the construction of the freight wagon due to higher capacity will be considered (adjustment of discharge angle). The development of the glued joint will be initiated and NSS 480 corrosion tests and, for example, PV1200 shock tests should also be part of the application development of bonded joint technology. At the same time, the corrosion resistance of Ni_3P layer is assumed to be tested for the purpose of this unconventional application.

Figure 5. Application of Ni_3P coating on 6x2m discharge plaps.

Acknowledgements
The result was obtained through the financial support of the Ministry of Education, Youth and Sports of the Czech Republic and the European Union (European Structural and Investment Funds - Operational Programme Research, Development and Education) in the frames of the project "Modular platform for autonomous chassis of specialized electric vehicles for freight and equipment transportation", Reg. No. CZ.02.1.01/0.0/0.0/16_025/0007293.

References

[1] M Pazcek *et al* 2016 A concept of technology for freight wagons modernization *IOP Conf. Ser.: Mater. Sci. Eng.* **161** 012107
[2] P Golias *et al* 2010 Electroless nickel plating *Techpark o.z.: Tribotechnika* 1338-0524
[3] A W Goldenstein, W Rostoker, F Schlossberger and G Gutzeit 1957 Structure of chemically deposited nickel *J. Electrochem. Soc. 104(2): 104-110*
[4] A H Graham, R W Lindsay and H J Read 1965 Structural changes in heated electroless Ni/P alloys. *J. Electrochem. Soc. 112(4): 401-413*
[5] D Jakubeczyova *et al* 2015 Tribological tests of modern coatings *J. Electrochem. Soc. 10): 7803-7810*
[6] P J Blau 2008 Friction Science and Technology:From Concepts to Applications (New York: CRC Press) ISBN 9781420054040 pp 132-174
[7] Bhushan, B. (2002) Introduction to Tribology (New York: John Wiley & Sons, Inc.) pp 23 - 188

Effects of hot forging on the structural condition in HS 6-5-2 high-speed steel

V Průcha[1], S Benešová[1], V Veselý[1], M Kesl[2]

[1]Department of Material Science and Technology University of West Bohemia. Univerzitní 8, 301 00 Plzeň, Czech Republic.

[2]Pilsen Tools s.r.o., Tylova 1/57, 301 00 Plzeň, Czech Republic.

E-mail: vprucha@kmm.zcu.cz

Abstract. Microstructure analysis was performed on rolled bars of high-speed steel after two and three forging cycles, each cycle comprising one upsetting and one drawing out operation. High-speed steels belong to difficult-to-form materials with a narrow forging temperature interval. Forging above the maximum forging temperature may lead to grain coarsening. Below the minimum forging temperature, deformation resistance of the material increases, and the workpiece may fail. Using numerical modelling, special forging dies were designed and effective strain distribution was calculated for an axial cross-section plane in specimens after two and three forging cycles. The purpose of the analysis was to identify the relationship between the amount of effective strain and the shape and size of austenite grain and the volume fraction and density of carbides after forging. The size of prior austenite grains was measured using the linear intercept method which is based on the Snyder-Graff method. Grain shapes were characterized in terms of circularity, which is the difference between the shape in question and a circle. With increasing amount of strain, the grains in the material became finer, as undissolved carbides impeded grain growth. In as-received rolled condition, the austenite grain size was G9. After three forging cycles, it was smaller, G11 (the higher the number, the smaller the grains). Circularity characterizes the complexity of a grain shape. Micrographs of carbide particles were taken using a scanning electron microscope and examined with NIS Elements image analysis software. The majority of carbides were sized between 0.2 and 2 μm. The carbides which are less than 1 μm in size do not shrink in response to increasing strain and their quantity does not change appreciably. Carbides with a size of 1-2 μm show a different behaviour. In the central region of specimens, where strain is the largest, their amounts are much larger than in less-worked regions.

1. Introduction

Although steel is traditionally worked by forging, lack of information persists in terms of the relationship between the microstructural conditions of high-speed steels (HSS) and the amount of strain in a forged workpiece. High-speed steels are characterized by the presence of various carbides [1]. In addition to chromium carbides, they contain complex carbides of molybdenum, tungsten and vanadium. Some carbides dissolve during hot forming (e.g. Fe_3C and $Cr_{23}C_6$), whereas others are stable (M_6C, M_7C_3, M_2C, MC and M_4C_3) and become crushed, fragmented and shifted around. Undissolved carbides prevent grain growth during hot forming and contribute to microstructure refinement [2].

High-speed steels are air-hardenable but they may suffer quench cracks when cooled rapidly from finish forging temperatures. Carbides provide good cutting performance and abrasion resistance. Toughness and resistance to tempering are given by the condition of the matrix after forming and subsequent quenching and tempering [3]. This paper explores the relationship between the amount of deformation introduced into high-speed molybdenum steel HS 6-5-2, DIN 1.3343 (equivalent ČSN 419830 grade) by alternate hot upsetting and elongation out and the prior austenite grain size and shapes. Chemical composition of the steel is shown in table 1.

Table 1. Chemical composition of HS 6-5-2 steel.

Wt. [%]	C	Mn	Si	P	S	Cr	Mo	W	V
	0.80-0.90	max. 0.45	max. 0.45	max. 0.035	max. 0.035	3.80-4.60	4.50-5.50	5.50-7.00	1.50-2.20

Alternate upsetting and elongation out is a sequence of forging operations which successively cause the material to flow in all three orthogonal directions in each cycle, the directions of the flow being opposite in consecutive cycles. As a result, these repetitive cycles ultimately bring the individual locations within the workpiece more or less back to their initial locations. High-speed steels are difficult-to-form materials with a narrow range of forging temperatures from 1150°C to 900°C [4, 5]. In this range, their flow stress may reach 350 MPa, the same as the ultimate strength of ordinary steels at room temperature.

2. Experiments

The specimens were 1.3 kg slugs from rolled bars 50 mm in diameter and 75 mm in height in soft annealed condition. Forging was performed in CKW 6000 programmable laboratory forging press of 1 MN capacity, which offers programmable forging speeds. The specimens were heated in a two-chamber electrical resistance furnace. One chamber was kept at 700°C. The other chamber maintained the upper forging temperature, 1150°C. The chamber at 700°C was used for pre-heating and reheating. This material is relatively difficult to form and the temperature of specimens decreases during forging. This places demands on the speed of processing and handling operations. During the forging process, the forging time of a batch of eight specimens was monitored to ensure that it does not exceed 15 minutes. Hence, the time available for forging a single piece was no more than two minutes. Since forging was performed manually, it is not possible to determine the exact process time. Between individual operations, the specimens were placed in the furnace at 700°C. Before each forging operation, the specimens were transferred to the furnace at 1150°C. After the specimens had been placed in the furnace, the temperature inside the chamber decreased. Once the furnace had returned to the upper forging temperature, the specimens were held for five minutes to ensure that they heat through before forging. After all forging operations, the specimens were left to cool in the furnace to approx. 60°C for several dozen hours to prevent quenching and to obtain material which is easy to cut and machine.

Metallographic sections through the specimens were prepared and photographed using Zeiss Vert.A1 optical microscope. The prior austenite grain boundaries were difficult to reveal due to the complexity of the matrix in this high-speed steel. The matrix of perlite and ferrite was effectively etched with Nital (100 ml C_2H_5OH + 3 ml HNO_3). The complex matrix of martensite, bainite and pearlite resulting from transformations after cooling in furnace was etched with a reagent specified in ČSN EN ISO 643 (42 0462) standard. This etchant consisted of a saturated aqueous solution of picric acid and sodium alkyl sulfonate mixed at 5:1 ratio and heated to 60°C. Neither of the above etching reagents revealed prior austenite grains in the rolled and annealed sample (as-received condition). This obstacle was removed by oil quenching the sample from 1220°C and subsequent etching with Nital. Magnifications for grain size measurement are prescribed by ISO 643:2012 standard. At least 50 particles must be measured. In each case, measurements were taken on five micrographs. The same measurement was performed for determining the shape factor of grains. For analysis of carbides was also used PHILIPS 30XL ESEM scanning electron microscope, the BSE mode, magnification 2000×. Carbides were revealed by etching with Vilella-Bain reagent (100 ml C_2H_5OH + 5 ml HCl + 1 g $C_6H_3N_3O_7$). The amount of carbides in the matrix was evaluated using image analysis software NIS – Elements AR v. 4.40. Carbide particle were also counted within intervals of their equivalent diameter. These intervals were 0.1 μm for particles smaller than 1 μm, and 1 μm for those larger than 1 μm. The ISO 16 232-7 standard deals with particle sizes which can be analysed. Depending on current image calibration, it defines objects five-fold larger than the image calibration factor as acceptable for analysis [6]. With images captured at 2000× magnification, the image calibration factor was 0.04 μm/pixel. Therefore the smallest particles to be considered were 200 nm in size [7] which is sufficient in the present case. 15 micrographs from each location were used for the measurement. The aggregate area measured was 34800 μm^2.

3. Results and discussion

3.1 Numerical modelling and boundary conditions

Effective strain fields in specimens subjected to two and three forging cycles were computed using FE simulation DEFORM software [8] according to equation:

$$\varepsilon_{ef} = \sqrt{2/3}\left[(\varepsilon_1-\varepsilon_2)^2+(\varepsilon_2-\varepsilon_3)^2+(\varepsilon_3-\varepsilon_1)^2\right]^{1/2} \tag{1}$$

Where ε_1, ε_2 and ε_3 denote principal strains. The amount of effective strain in the input stock was considered zero. The calculation therefore only covered the strain introduced by the forging cycle. Materials data, i.e. flow stress curves, thermal conductivity, specific heat capacity and the friction and heat transfer equations were adopted from the literature [9, 10]. Based on these simulations, three pairs of forging dies were designed which were intended to prevent spreading of the forged specimen. When the specimen spreading is restricted by the die walls, tensile stresses are reduced. These stresses normally occur when the specimen buckles, and lead to cracking along the circumference of the disc being upset [11]. Effective strain distribution on a central cross-section plane according to numerical modelling of two and three cycles is shown in figure 1. The locations in which microstructure was examined and which received the largest and smallest deformation were denoted as P1 and P2, respectively. Location P2 is in a peripheral region of the specimen, with low strains when compared to the specimen centre, despite repeated forging operations. Effective strains in P1 were ε_{ef} = 8.1 for two forging cycles and ε_{ef} = 9.9 for three cycles. In location P2, the values were ε_{ef} = 1.6 for two forging cycles and ε_{ef} = 3.2 for three forging cycles, as illustrated in figure 1.

a) b)

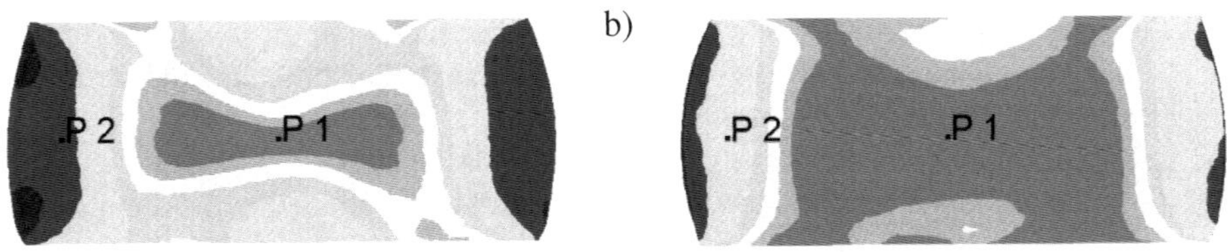

2 cycles		
location	P1	P2
ε_{ef}	8.1	1.6

3 cycles		
location	P1	P2
ε_{ef}	9.9	3.2

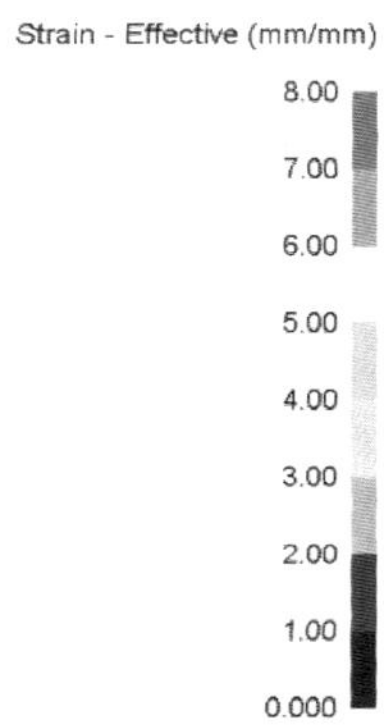

Figure 1. Effective strain distribution on a longitudinal cross section through the specimen centre, as found by numerical simulations of a) two forging cycles and b) three forging cycles.

IOP Conf. Series: Materials Science and Engineering **723** (2020) 012002 doi:10.1088/1757-899X/723/1/012002

3.2 *Evaluation of prior austenite grains*

The prior austenite grain size in HS 6-5-2 DIN 1.3343 steel (ČSN EN 41 9830) was determined using the linear intercept method according to ČSN EN ISO 643 "Steels - Micrographic determination of the apparent grain size", which is based on the Snyder-Graff method [12]. The measurement involves superposing line segments in three directions onto a micrograph. Grain size is indicated with letter G and a number. Higher numbers after the letter G indicate that the grains are finer. For the sake of comparison, measurements were also taken on as-received rolled and annealed specimens. The results are listed in table 2. The microstructure grew finer with increasing amount of deformation. However, refinement only became noticeable when the material undergoes effective strain higher than approximately 3 ($\varepsilon_{ef} \geq 3$). In the centre of the heavily-forged specimen ($\varepsilon_{ef} \approx 10$), the grain size was G11.

Table 2. Prior austenite grain size.

Specimen	Initial condition	Two forging cycles, point P2 (peripheral region)	Three forging cycles, point P2 (peripheral region)	Three forging cycles, point P1 (centre)
Effective strain ε_{ef}	-	1.6	3.2	9.9
Grain size - ČSN EN 41 9830	G9	G9	G9-10	G11
Grain size [μm]	12.8 ± 1.1	13.7 ± 1.2	12.2 ± 0.5	6.5 ± 0.4

3.3 *Evaluation of circularity of austenite grains*

The other grain characteristic assessed in this work was the shape factor. It is a dimensionless quantity which indicates the difference between the shape of an element and an ideal user-defined shape. In this case, the ideal shape was a circle. The calculation followed the equation:

$$f = \frac{4\pi A}{L^2} \qquad (2)$$

Where L denotes the circumference of the particle under measurement and A is the area of the particle. The interval of shape factor values is 0-1. The lower the shape factor is (circularity in this case), the greater is the virtual distance of the particle's shape from the ideal shape [13].

Results of measurement are plotted and relevant micrographs in as-received and forged conditions are shown in figures 2 - 5. In as-received material, most grains exhibit circularity in the range of 0.65–0.75. Strong carbide banding [3] was present and had impact on the final shapes of grains after annealing and quenching. The purpose of these heat treating operations was to reveal austenite grains (figure 2). In the peripheral region of the specimen which was worked with two forging cycles, where $\varepsilon_{ef} = 1.6$, most grains exhibited circularities of 0.7–0.75. Along with them, there were more grains with circularities between 0.5 and 0.55; and the histogram broadened because significantly non-circular grains had emerged. This trend grew stronger with increasing amounts of deformation. As a result, in the peripheral region of the specimen after three forging cycles (where $\varepsilon_{ef} = 3.2$), the largest percentage of grains showed a circularity of 0.55 to 0.7. The variance of grain circularity values also became larger. The central region with strain level $\varepsilon_{ef} = 9.9$ in the specimen which had been worked with three forging cycles contained a large number of considerably non-circular grains. Each of the five circularity classes between 0.4 and 0.7 comprised almost equal number of grains. At high strains, very non-circular grains formed in the microstructure. These were rarely found in the as-received material.

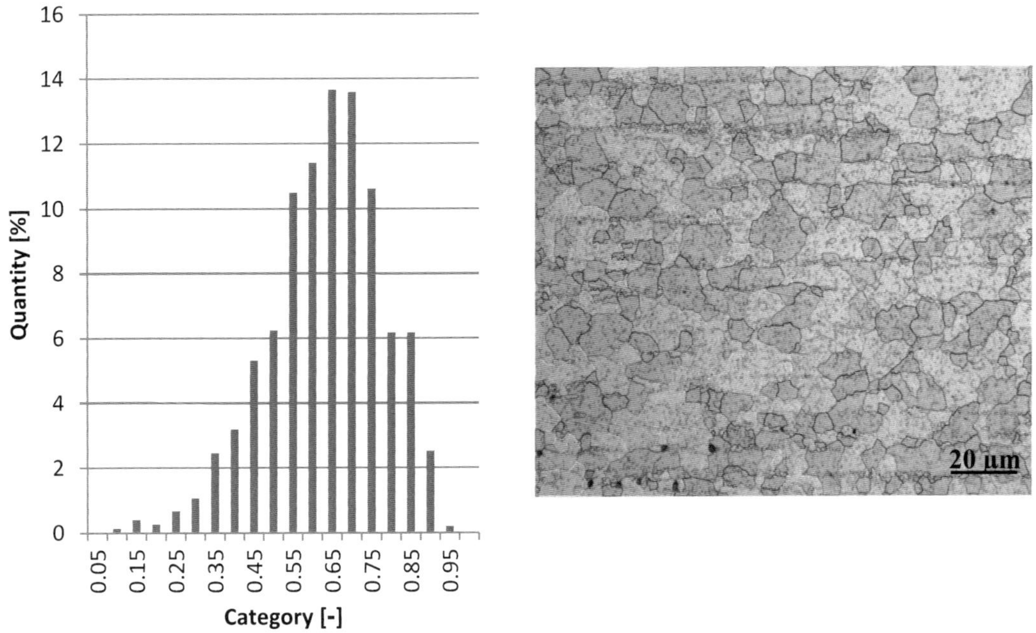

Figure 2. Circularity of prior austenite grains- as-received rolled condition, longitudinal cross-section.

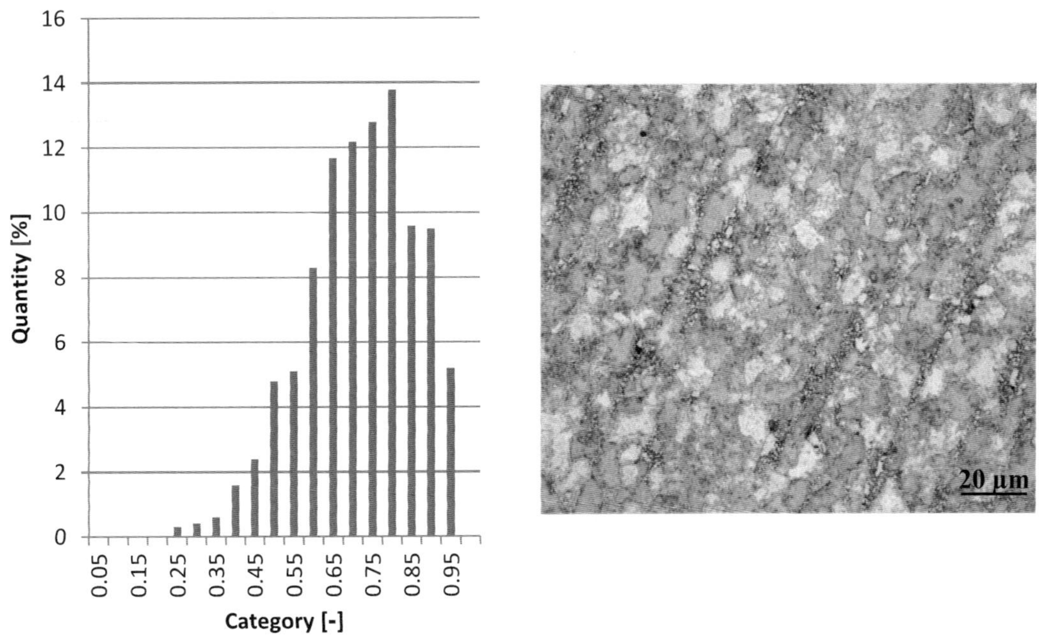

Figure 3. After 2 forging cycles, location P3 (periphery), ε_{ef}= 1.6.

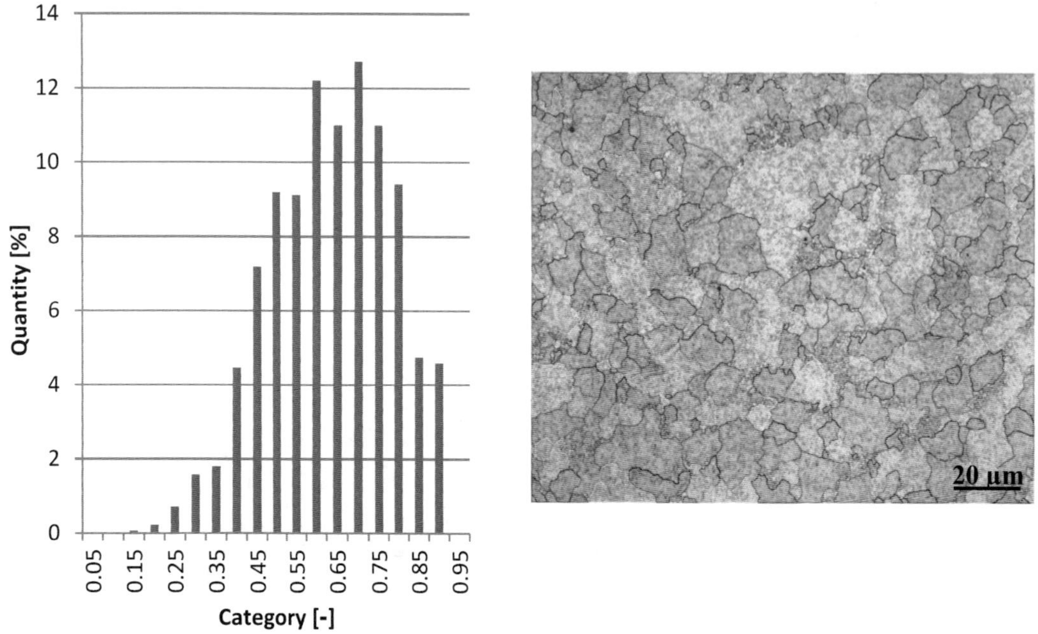

Figure 4. Circularity of prior austenite grains - 3 forging cycles, location P2 (periphery), ε_{ef} = 3.2.

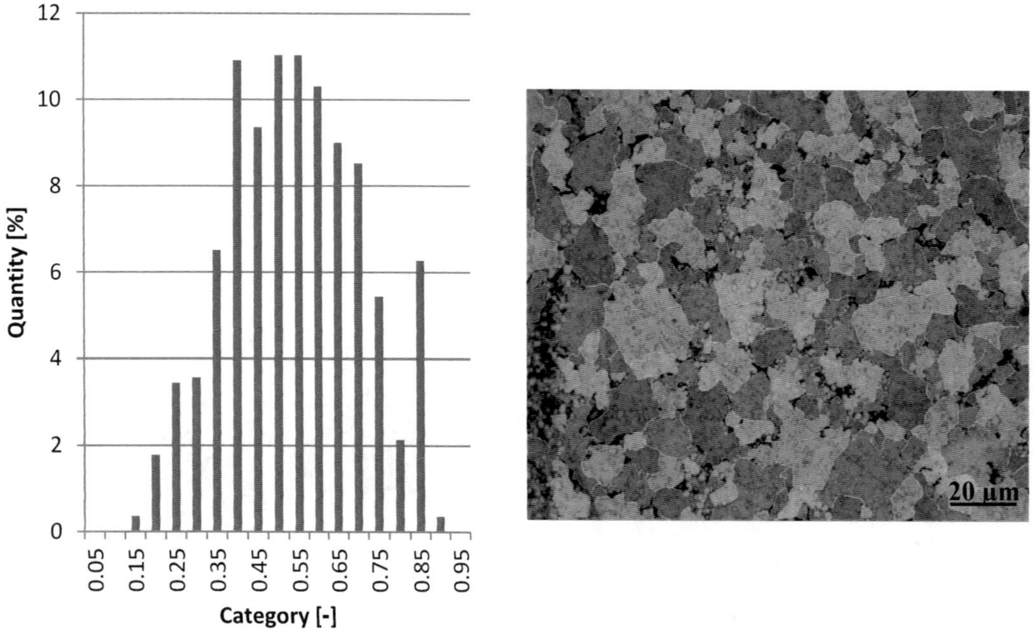

Figure 5. After three forging cycles, point P1 (centre), ε_{ef} = 9.9.

3.4 The size and frequency of carbides

Figure 6 shows the distribution of carbides in the location with the smallest deformation, $\varepsilon_{ef} = 1.6$ (left) and the location of the largest deformation $\varepsilon_{ef} = 9.9$ (right). They clearly show that in the specimen which was subjected to two forging cycles, the peripheral region ($\varepsilon_{ef} = 1.6$) contain large clusters of carbides, which are not present in the central region with high deformation of the specimen forged in three cycles ($\varepsilon_{ef} = 9.9$).

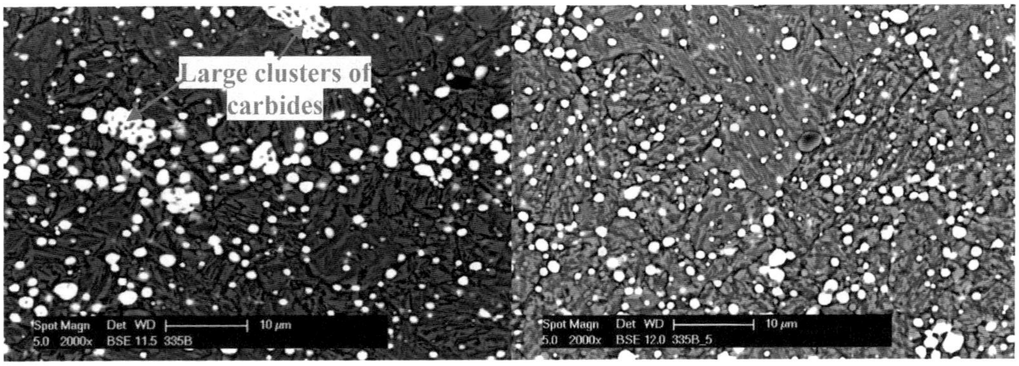

Figure 6. Distribution of carbides in martensitic matrix (in selected specimens): left – two forging cycles, region P2 (peripheral region of the specimen - $\varepsilon_{ef} = 1.6$); right - three forging cycles, region P1 (specimen centre - $\varepsilon_{ef} = 9.9$).

Figure 7 shows a histogram of equivalent diameters of carbides in specimens forged in two cycles. The location on the left is P2 (periphery of the specimen - $\varepsilon_{ef} = 1.6$), in which the fraction of the carbide phase was found to be 15.2 ± 4.5%. The location on the right is P1 (centre of the specimen - $\varepsilon_{ef} = 8.1$), in which the fraction of the carbide phase was found to be 18.9 ± 3.4%.

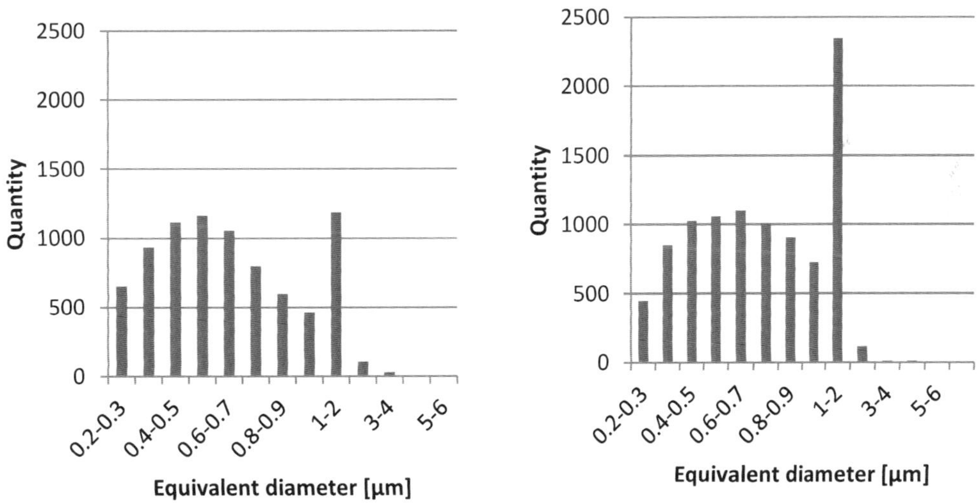

Figure 7. Histograms of equivalent diameters of carbide particles for samples with 2 forging cycles. Left: region P2 (periphery - $\varepsilon_{ef} = 1.6$); right: region P1 (centre - $\varepsilon_{ef} = 8.1$).

Figure 8 shows a histogram of the equivalent diameters of carbides in specimens forged in three cycles. The location on the left is P2 (periphery of the specimen - $\varepsilon_{ef} = 3.2$), in which the fraction of the carbide phase was found to be 14.6 ± 4.7 %. The location on the right is P1 (centre of the specimen - $\varepsilon_{ef} = 9.9$), in which the fraction of the carbide phase was found to be 17.7 ± 4.1 %.

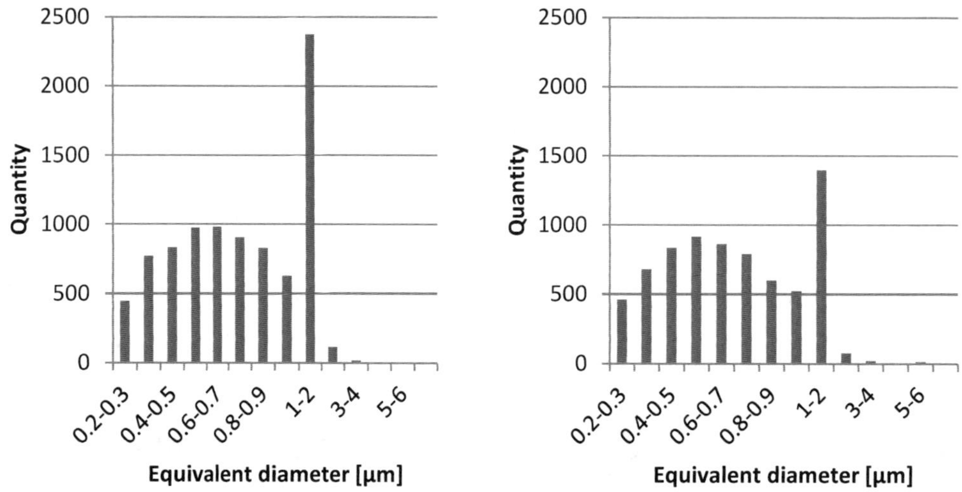

Figure 8. Histograms for samples with 3 forging cycles. Left: region P2 (periphery - $\varepsilon_{ef} = 3.2$); right: region P1 (centre - $\varepsilon_{ef} = 9.9$).

Carbides between 2 and 5 µm are only found in small amounts. In both cases, they are not likely to have a substantial effect on material properties, owing to their low counts.

The amounts of carbides 1-2 µm in size vary. In the highly deformed centres of specimens forged in two and three cycles, their counts are nearly identical. After conversion, they are approx. 660/10000 µm^2. In the peripheral region of the specimen after three cycles, their density was a mere 400/10000 µm^2. In the peripheral region of the specimen processed with two cycles, it was no more than 345/10000 µm^2.

The fractions and distributions of carbides smaller than 1 µm, i.e. sized 0.2 to 1 µm, are very similar in all the specimens forged. Their maximum is in the 0.4–0.8 µm range, where their counts are at 900–1100 in the area under examination.

4. Conclusion

The goal of this experimental work was to assess the impact of forging on prior austenite grains and on the volume fraction, sizes and distribution of carbides in relation to the effective strain field in high-speed steel HS 6-5-2 (according to ČSN 41 9830). The first location is the centre with the largest strain $\varepsilon_{ef} = 8.1$ (after two forging cycles) and $\varepsilon_{ef} = 9.9$ (after three forging cycles). Another location, the least worked one, was at the periphery of the specimen, with a strain of $\varepsilon_{ef} = 1.6$ after two forging cycles and $\varepsilon_{ef} = 3.2$ after three cycles.

The grains were found to shrink with increasing strain: from G9 in the as-received rolled condition to G11 in the heavily-forged specimen. The size of austenite grains influenced the size of grains of other phases. Post-forging heat treatment produced complex, very fine and highly-stable microstructures. Grain refinement led to higher grain boundary densities. The boundaries became obstacles to dislocation movement, which was reflected in material strengthening, specifically higher yield strength (Re). With increasing deformation, the numbers of strongly non-circular grains were rising. In addition, the amount of lattice defects, including the spatial ones, one of which are grain boundaries, increased with the amount of strain. Circularity of austenite grains was decreasing, which means that the total area of grain boundaries expanded.

It was found that larger carbides break up and become finer during upsetting of the rolled material. However after breaking and refining as the forging process continues and strain accumulates, their size remains unchanged, and only their positions change. They prevent austenite grain growth and affect the complexity of the grain shape.

In addition, the area fraction of carbides in peripheral regions was lower than in the centre, due to the smaller proportion of carbides sized 1–2 μm. Forging temperatures under 1150 °C are known [3] to cause dissolution of cementite, Fe_3C, and $Cr_{23}C_6$, as well as other carbides, some of which may be complex. Variation in the quantities of carbides 1-2 μm in size appears to be related to their precipitation which occurs as temperature decreases in the course of and after forging. The quantity of these carbides increases with the amount of strain. The reason is that nucleation of the carbide phase aims to reduce the energy of grain boundaries [14, 15]. The fine-grained austenitic grain with low circularity is characterized by large bounders with high energy. Carbide precipitation is therefore a favourable process in terms of energy.

Acknowledgements
This paper is one of the results of a project under the EPSILON programme sponsored by the Technology Agency of the Czech Republic in support of applied research and experimental development: TH02020161: Development of high-purity high-alloy steel with ultrafine microstructure and an extreme amount of mechanical working for special applications.

References

[1] Davis J R et al. 1998 Tool materials ASM specialty handbook *ASM international*
[2] Kříž A and Průcha V 2019 Condition of carbides in high-speed steels after multidirectional forging *Defect and Diffusion Forum* Vol. **395** pp. 1-15
[3] Fremunt P, Krejčík J and Podrábský T 1994 Nástrojové oceli (Dům techniky Brno)
[4] European Steel and Alloy Grades/Numbers Searchable Database [online] [Accessed on 25/06/ 2019] *Kharkiv Polytechnic Institute* Aviable from: http://www.steelnumber.com/en/steel_composition_eu.php?name_id=1002Materiálový list.
[5] Material Sheet [online] [Accessed on 25/06/2019] *Bohdan Bolzano* Available from: https://www.bolzano.cz/cz/technicka-podpora/technicka-prirucka/nastrojove-oceli/nastrojove-oceli-rychlorezne/materialove-listy/materialovy-list-oceli-hs-6-5-2-5
[6] Průcha V, Kříž A and Veselý V 2018 Aspects of evaluation of carbide phase in high-speed steels *In Kovárenství (*Brno: Svaz Kováren ČR *z.s.*) No. **65**, pp. 59-63 ISSN: 1213-9289
[7] ISO 16232-7:2007 Road vehicles-Cleanliness of components of fluid circuits: Part 7: Particle sizing and counting by microscopic analysis *Technical Committee ISO/TC 22*
[8] Scientific Forming Technologies Corporation [online] © 2019 Available from: http://deform.com [Accessed on 25/06/2019]
[9] Spittel T and Spittel M. 2009 Ferrous alloys, Volume 2, *Springer*, ISBN 978-3-540-44758-0
[10] Altan T Ngaile G and SHEN G 2005 Cold and hot forging, fundamentals and Applications. *ASM International*
[11] Benešová S and Kesl M 2019 Treatment of high quality forged high-speed steel with ulta-fine structure pro specials tools *Kovárenství*, No. **68**, Svaz kováren České republiky z.s. (Czech Forging Association).
[12] ČSN EN ISO 643 (42 0462) 2014 Ocel - Mikrografické stanovení velikosti zrn. OPRAVA 1. (Praha: Úřad pro technickou normalizaci, metrologii a státní zkušebnictví)
[13] Wojnar L, Kurzydlowski K J and Szala J 2004 *Quantitative Image Analysis*, Metallography and Microstructures, Vol 9, *ASM Handbook,(* ASM International), pp. 403–427
[14] Hyspecká L 1983 Teorie tepelného zpracování (VŠB Ostrava)
[15] Smallman R E and Ngan A H W 2007 Physical metalurgy and Advanced Materials *Elsevier Ltd.*

Influence of heat treatment on properties of SD251-PH1 composite produced by additive SLM technology

[1*]David Bricín, [2]Zdeněk Jansa, [1]Josef Somr, [1]Andrea Elmanová, [1]Antonín Kříž

[1]Department of Material Science and Technology University of West Bohemia in Pilsen. Univerzitni 8 301 00. Pilsen. Czech Republic
[2]New Technologies Research Centre. University of West Bohemia in Pilsen. Univerzitni 8 301 00. Pilsen. Czech Republic

*bricda@kmm.zcu.cz

Abstract. This study evaluates how heat treatment affects the structure and properties of the engineered composite material SD251-PH1. This material was formed by mixing two powder blends in a weight ratio of 90 wt. % WC-Co powder SD251 with 10 wt. % of a PH1 precipitation hardening powder of the stainless steel. Samples were prepared from this mixture on an SLM additive device. The printed samples were then divided into groups. Some of them were left in the initial state and the rest were used for heat treatment based on the standard AMS 5659. Light and electron microscopy metallographic analysis together with X-ray diffraction analysis were then used to evaluate structural and phase changes in the prototype samples. The main attention was paid to changes in the phase composition of the samples. In addition, changes in the size and shape of the pores and tungsten carbide (WC) grains were studied. The metallographic analysis was supplemented by an evaluation of the changes in their mechanical properties and wear resistance. Vickers hardness measurements and the ball on disc test were used for this purpose. The experiments showed that the selected heat treatment processes precipitated new structural phases which differed from the original sample structure in terms of shape, size, chemical composition, and mechanical properties.

Key words: SLM technology. Heat treatment. WC-Co. Porosity. Phase transformation

1 Introduction

The processing of sintered carbides by additive manufacturing, known as SLM (Selective Laser Melting), is a complex process. The processing of this material, as stated in previously published studies [1-7], produces different types of pores [8] and imperfections in the structure, see Figure 1. The volume of these defects in the material structure depends on the parameters used for processing the powder mixture, its residual moisture, the chemical composition of the powder particles, their topography, surface defects and shape.

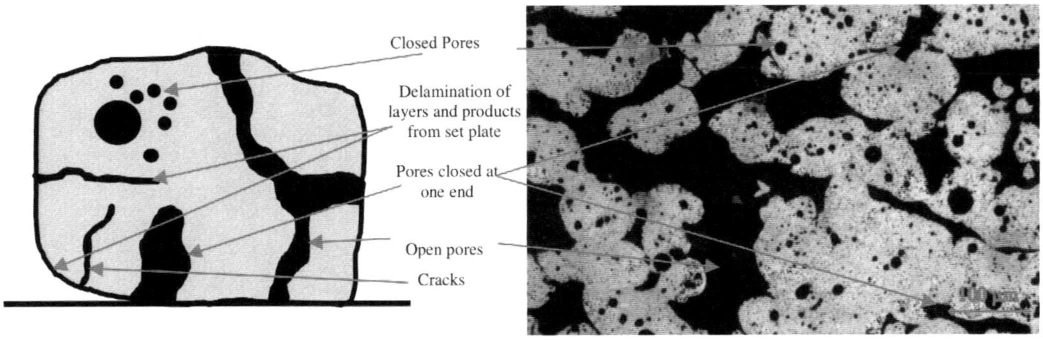

Figure 1. Defects that can be detected in the volume of the builds.

In one previously published study [9], the effect of the SLM process parameters on the properties of the prints made from the SD251 powder blend was evaluated. A high level of porosity and a high degree of delamination of the samples from the set plate was observed in the builds. To reduce the porosity in the builds, WC-Co powder was mixed with PH1precipitation hardening stainless steel powder in this study. One of the aims of this blending was to achieve an increase in the uniformity of the powder layer on the set plate. The unevenness of the powder layer causes changes in the distribution of the temperature field during the interaction of the laser with the powder bed. As a result, different degrees of porosity, different mechanical properties and chemical compositions are observed in different areas of the builds [7, 10]. Furthermore, the addition of secondary powder results in an increase in the proportion of binder in the powder mixture. This should lead, as in the study by R.S. Khmyrov [11], to a crack-free printout structure. Cracks are defects which are very difficult to remove by additional processing. The third reason for adding the steel powder was that precipitates are formed during subsequent heat treatment. All changes which occur during heat treatment could then increase the hardness and abrasion resistance of the formed composite material. This case study aims to assess what structural changes could occur in the composite material thus formed after heat treatment. These changes were evaluated using light microscopy and electron scanning microscopy with EDX analysis of the chemical composition, X-ray analysis of phase composition and analysis of changes in mechanical properties of structural phases using Vickers hardness analysis and analysis of tribological properties by the ball on disc method.

2 Experimental material and processing

2.1 Experimental material

The experimental composite material was formed by mixing two types of powder: SD251 (Figure 2), and PH1 (Figure 3), in a 9:1 weight ratio. The chemical composition of the materials is shown in Table 1. The powder was mixed by mechanical mixing in an unlubricated container for one hour. The aim of this mixing was to obtain a uniform distribution of the PH1 powder in the SD251 matrix. After the powder mixture was prepared, it was processed using an SLM-type additive device. The processing parameters used are summarized in Table 1.

IOP Conf. Series: Materials Science and Engineering **723** (2020) 012003 doi:10.1088/1757-899X/723/1/012003

Figure 2. Shape of SD251 powder particles

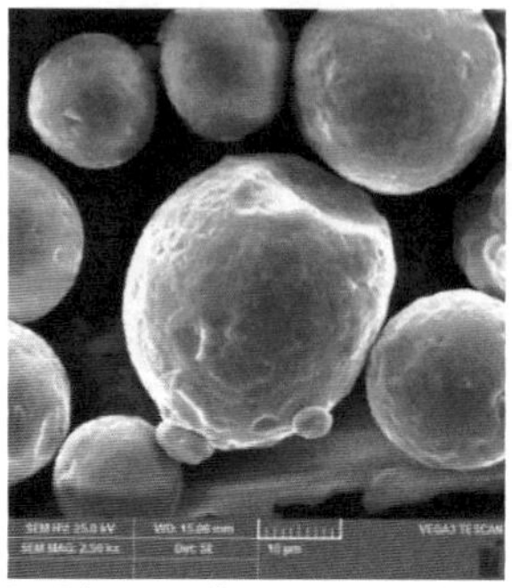

Figure 3. Shape of PH1 powder particles

Table 1. Chemical compositions of powder mixtures and their processing parameters in the SLM additive device.

WC (Wt. %)	Co (Wt. %)	Fe (Wt. %)	Cr (Wt. %)	Ni (Wt. %)	Cu (Wt. %)
87.6±0.61	12.4±0.61	78.6±0.41	14±0.05	3.9±0.23	3.6±0.2
Power of laser (W)	Thickness of powder layer (mm)	Speed of laser spot (mm/s)		Energy density Ev (j/mm3)	
250	0.04	900		58	

After processing the composite powder on the additive device, the builds were divided into two groups. The first group was left in the state after SLM. The second group was used for heat treatment.

2.2 Method of heat treatment of the experimental material

The heat treatment was carried out according to AMS 5659 Rev. S [12]. It consisted of dissolving annealing, fast cooling, and aging. The dissolving annealing was done to form a homogeneous solid solution in the volume of the prototype samples.

A homogeneous solid solution was not formed in the entire volume of the prototype samples, as is the case with some types of aluminium alloys. This is because the elements from which the composite is composed have different solubilities, with some elements having absolute solubility and some partially limited solubility [13]. The dissolving annealing was followed by rapid cooling at supercritical speed by immersing the samples in water at 20 ° C. This was done to create a supersaturated solid solution in the prototype structure. The last step of the heat treatment was the aging of the samples at elevated temperature. The aim of this step was to form precipitates from the solid solution. The heat treatment parameters used are summarized in Table 2. The heat treatment process was carried out in a single-chamber furnace where the temperature was sensed using a K-type thermocouple which measured with a tolerance of ± 1.5 ^{0}C.

Table 2. Parameters used for the heat treatment of the composite samples. Allowed deviations from the standard times and heat treatment temperatures stated in ASM 5659 are shown in brackets [12].

	Temperature (^{0}C)	Time (min.)	Cooling environment
Dissolving annealing	1038 (±14)	60 (+15)	Water
Variant 482 aging	482 (±9)	60 (+15)	Air
Variant 496 aging	496 (±9)	240 (+30)	Air
Variant 552 aging	552 (±9)	240(+ 30)	Air

2.3 Method of evaluation of prototype samples

Cross sections of all the samples were analysed. This consisted of hot-dipping prototype specimens into a PolyFast moulding material, followed by grinding and polishing on a Tegramin 20 semi-automatic sander. Samples were then analysed using a Philips XL30ESEM electron scanning microscope equipped with an EDX chemical composition analyser. Chemical analysis was done at a magnification of 500x at area of 0,0625 mm^2 at five different locations on each sample.

A CarlZeiss Observer Z1m light microscope and a Panalytical X'pert PRO X-ray diffractometer with a CuKα source of X-ray radiation with a wavelength $\lambda K\alpha 1 = 0.15405980$ nm were used to analyse the structure of the samples. Because the samples are bulk, symmetric geometry was used. With this symmetry were observed the lattice planes (hkl) which are parallel to the sample surface. The planes fulfil Bragg's law (λ=2dsinθ) for each measured 2ϑ angle. XRD patterns were recorded from 30 to 100 degrees in 2ϑ scales. An ultra-fast high-resolution PIXcel detector was used for collecting the diffracted beams. The detection limit for the measurement is about 1 % for each detected phase. The structure of the samples was enhanced by etching using Murakami and Adler etchants. This was supplemented by analysis of the tribological properties using the ball on disc method. Tracks created by the Al_2O_3 ball were than evaluated according to ASTM G99 [14]. Hardness analysis of HV0.05 was also done using a semi-automatic DuraScan hardness tester. Indentations were made at five different places on each sample.

3 Results of experiment and discussion

3.1 Metallographic analysis results

The structure of the builds made of SD251 powder consisted of irregular pores and cracks, see Figure 4 (l). The builds made of composite SD251-PH1 consist of pores with circular and irregular shapes, see Figure 4 (r).

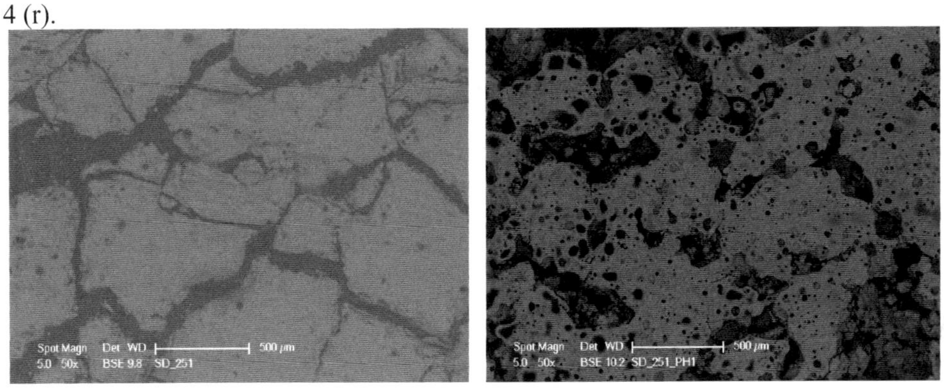

Figure 4. Structure of printed prototype samples. Left image shows build made of SD251 Powder Mixture. Right image shows build made of SD251-PH1 Powder Mixture.

It can be seen from the above that by increasing the proportion of binder by adding steel powder, the development of cracks in the volume of the prototype specimens can be reduced and their degree of delamination reduced.

From the point of view of the proportion of pores in the structure of prototype samples, it was found that the porosity of the samples increased during the heat treatment of the samples. The degree of porosity became higher as the temperature and time of processing increased. In addition to the increase in the porosity, the size of the pores increased and their circularity decreased, see Figure 5.

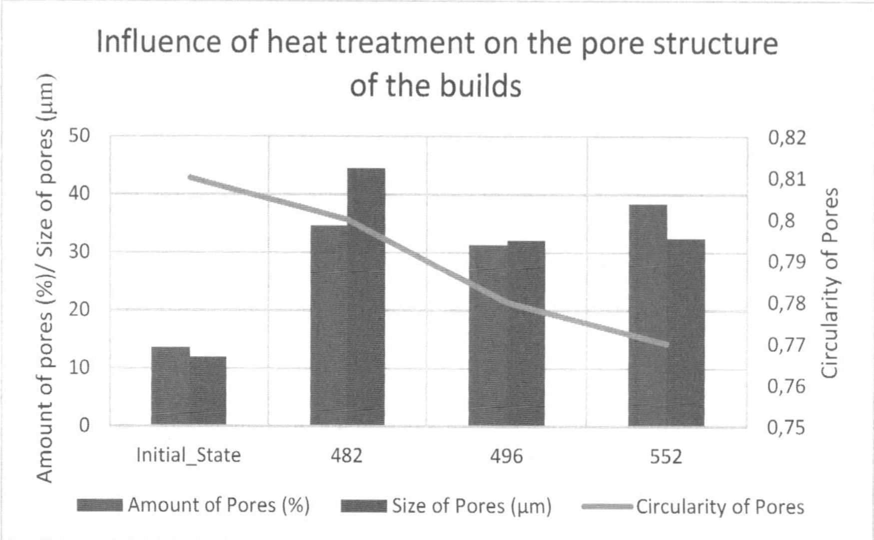

Figure 5. Effect of heat treatment on porosity of builds.

After analysing the porosity of the printed samples, their chemical composition was analysed by EDX element analysis, see Table 3.

Table 3. Comparison of the chemical compositions of builds in the initial state and after heat treatment.

Heat treatment	Cr (Wt. %)	Fe (Wt. %)	Co (Wt. %)	Ni (Wt. %)	Cu (Wt. %)	W (Wt. %)
Initial_state	1.5±0.27	9.8±5.25	8.9±0.62	1.4±0.19	1.4±1	77±6.1
482	2.2±0.23	9.4±0.37	9.2±0.1	1.4±0.14	0.5±0.25	77.4±0.79
496	1.8±0.28	8.6±1.96	9.2±0.2	1.3±0.16	0.7±0.18	78.5±2.15
552	1.9±0.52	9.8±1.84	9.7±0.65	1.6±0.16	1.1±0.38	76±3.3

It can be seen from the table above that there were no significant changes in chemical composition during heat treatment.
However, the metallographic analysis showed structural changes induced by thermal processing, see Figures 6-9 below. It can be seen that the initial triangular to rectangular shapes of the WC grains turn to cigar shaped structures after heat treatment.

Figure 6. Build structure in original state without heat treatment. Magnification 2000x.

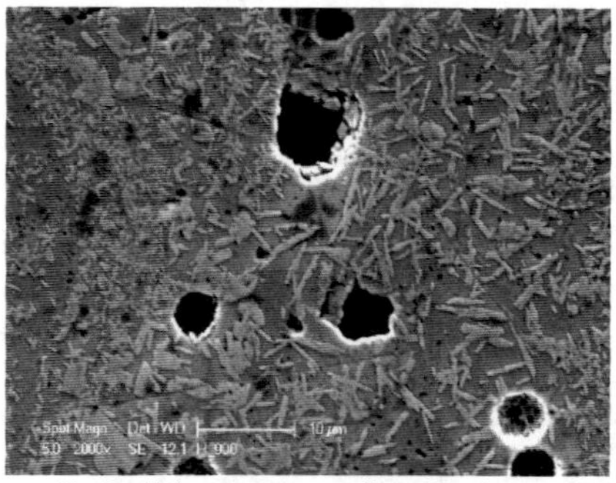

Figure 7. Build structure after Variant_482 heat treatment. Magnification 2000x.

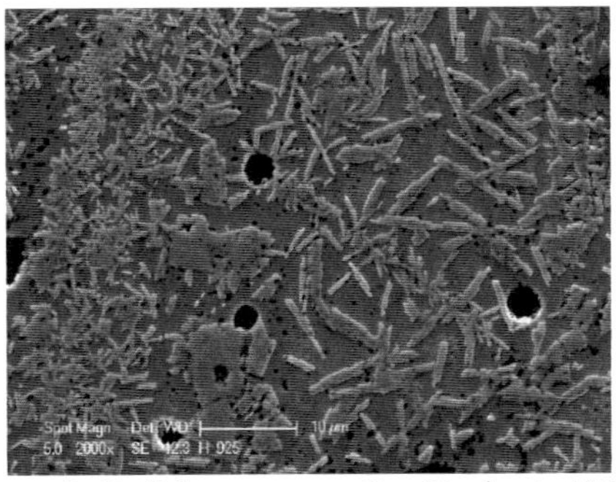

Figure 8. Build structure after Variant_496 heat treatment. Magnification 2000x.

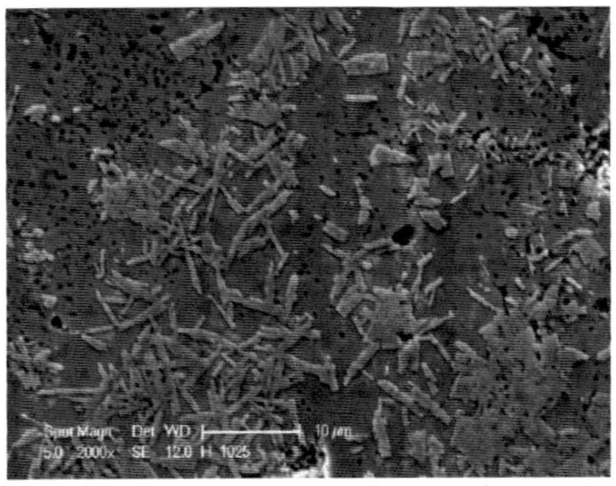

Figure 9. Build structure after Variant_552 heat treatment. Magnification 2000x.

Coarsening of the precipitated grains of WC can be seen in the images above. This change occurs due to the increasing temperature of the aging process. The metallographic analysis was supplemented by a hardness analysis of the observed structural phases, see Table 4.

Table 4. Average values of microhardness HV0.05 of the structural phases of the builds.

Heat Treatment	Average hardness of WC grains	Average hardness of matrix	Average hardness of precipitate phase
Initial State	1860±270	1277±285	-
Variant_482	822±380	1089±165	1016±170
Variant_496	1221±283	947±112	1506±169
Variant_552	1088±57	649±125	1340±185

The results from the hardness measurements show that the newly precipitated phase probably has a changed coherence of its lattice due to the diffusion-activated displacement of atoms during the heat treatment of the builds. The process of diffusion, especially of interstitial elements, also influences the final hardness of the carbide phase and the matrix of the builds, when concentration gradients are gradually balanced in the volume of the evaluated samples.

3.2 X-ray diffraction analysis results

X-ray diffraction analysis confirmed changes in the phase composition of samples before and after heat treatment, see Figure 10.

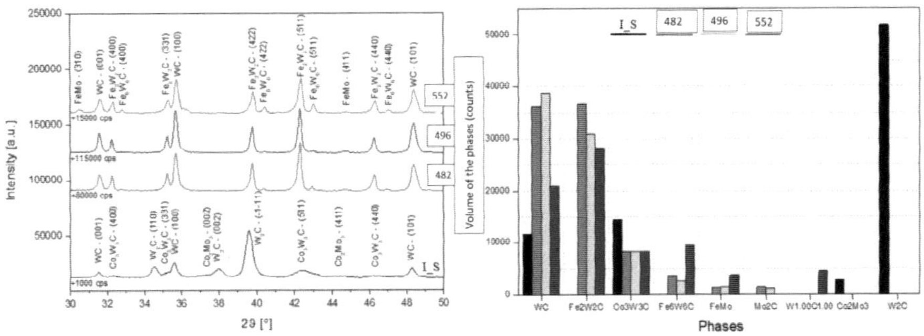

Figure 10. X-ray diffraction analysis and calculated volume fraction of phases in prototype samples.

From the graph above (Figure 11), it can be seen that the initial structure of the builds consisted of the WC, W2C, Co2Mo3, Co3W3C and the Fe-C phase. The Fe-C phase cannot be seen in the image due to its low peak intensity. Therefore it was not possible to count its right volume for the initial state builds. After heat treatment, the structural phases of WC, Fe_2W_2C, Fe_6W_6C, Co_3W_3C, FeMo and Mo_2C were recorded in the volume of the builds.

The difference in the volumes of the phases occurred due to the diffusion flow of the interstitial and substitution elements in the volume of the builds. Diffusion of the carbon especially affected the amount and type of phases in the volume of the builds after their aging.

The less stable phase of W2C decayed. The released carbon then interacted with other elements to form new structural phases, see Figure 11.

3.3 Tribological properties

The aim of this analysis was to assess the effect of the structural changes on the frictional properties of the heat treated builds. The results in Figure 11 show that the heat treatment reduces the wear rate.

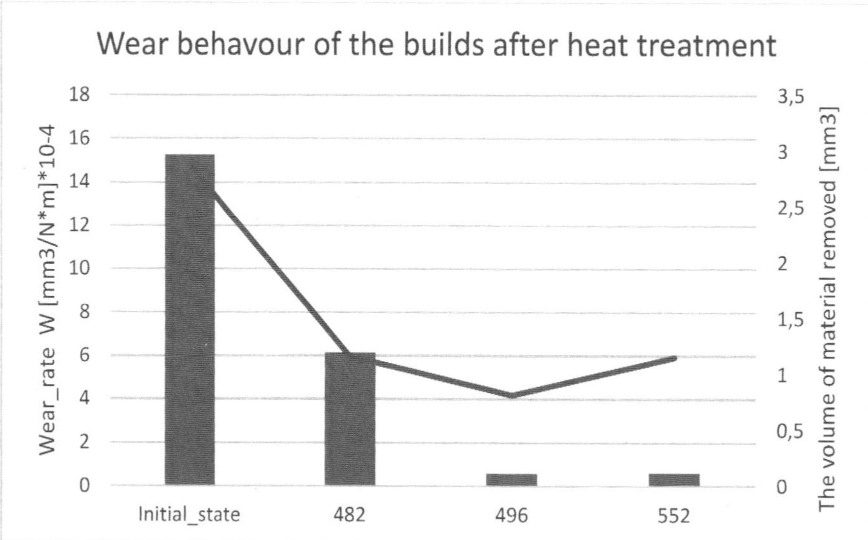

Figure 11. Analysis of tribological properties of builds before and after heat treatment.

These results also show a re-increase in the degree of wear of the samples which were aged according to variant C of the used aging process.. This is probably associated with the recorded increase in the volume and size of the pores detected in the volume of the prints after heat treatment.

4 Conclusion

Heat treatment of builds made from composite material SD251-PH1 had a relatively high effect on the homogeneity of the material structure. This was confirmed by X-ray diffraction analysis, metallographic analysys and subsequent ball on disc tribological testing, where the heat treated samples have lower wear rates than the builds without heat treatment.

During heat treatment, the development of newly precipitating phases was recorded by metallographic analysis, and was subsequently confirmed by X-ray diffraction analysis. The newly detected phases included Fe_2W_2C, Fe_6W_6C, FeMo, and Mo_2C. They were not detected in the heat treated prints due to the diffusion phenomena of the W2C phase, which was transformed into a WC phase. Heat treatment had a negative impact on the development of the porosity of the builds. Porosity increased by up to 24 percent compared to builds without heat treatment. This phenomenon occurred in samples that were aged at the highest temperature. In addition to their volumetric increase, the pores changed shape, with a gradual decrease in their circularity, which in turn has a negative impact on the fracture toughness of the material. In addition, the newly precipitating structural phases gradually coarsened and the coherence of the lattices of the precipitates changed. This was confirmed by measuring a micro-hardness of HV0.05 for the original structural phases and precipitates.

The highly porous material thus prepared can be used in the future as a filter material for liquid particulate media. However, for its possible application, it is necessary to conduct further experiments aimed at evaluating its utility properties.

Next research will also focus on looking for other alternative composite powder blends for SLM processes and other additive techniques. This new research will evaluate the quality of feedstocks and also the influence of the parameters of the manufacturing process on the properties of the final builds, focusing on mechanical behaviour, chemical composition changes and phase changes.

Acknowledgments

This work was supported by CEDAMNF project (CZ.02.1.01/0.0/0.0/15_003/0000358), co-funded by the ERDF as part of the OP RDE programme of the Ministry of Education, Youth and Sports (Czech Republic) and by the project Zéta TJ01000218 'Production of a machine tool using additive SLM technology' of TAČR, Technology Agency of the Czech Republic. The build used in this study was made in cooperation with the Technical University of Liberec, the Institute for Nanomaterials and New Technologies and New Technologies Research Centre at the University of West Bohemia in Pilsen.

References

[1] Uhlmann E Bergmann A and Gridin W 2015 Investigation on Additive Manufacturing of Tungsten Carbide-cobalt by Selective Laser Melting *Proc. CIRP* 35 pp 8-15.

[2] Domashenkov A Borbély A and Smurov I 2016 Structural modifications of WC/Co nanophased and conventional powders processed by selective laser melting *Mat. and Man. Proc.* DOI: 10.1080/10426914.2016.1176195 ISSN 1042-6914 Aviable from: https://www.tandfonline.com/doi/full/10.1080/10426914.2016.1176195

[3] Kumar S 2009 Manufacturing of WC–Co moulds using SLS machine *J. of Material Processing Tech.* DOI: 10.1016/j.jmatprotec.2008.08.037 Aviable from: http://linkinghub.elsevier.com/retrieve/pii/S0924013608006705

[4] Gu D Shen Y Dai P and Yang M 2006 Microstructure and property of sub-micro WC-10 %Co particulate reinforced Cu matrix composites prepared by selective laser sintering DOI: 10.1016/S1003-6326(06)60061-7 Aviable from: http://linkinghub.elsevier.com/retrieve/pii/S1003632606600617

[5] Gu D and Meiners W 2010 Microstructure characteristics and formation mechanisms of in situ WC cemented carbide based hardmetals prepared by Selective Laser Melting *Mat. Scien. and Eng.* DOI: 10.1016/j.msea.2010.08.075 Aviable from http://linkinghub.elsevier.com/retrieve/pii/S092150931000986X

[6] Kumar S and Czekanski A 2017 Optimization of parameters for SLS of WC-Co *Rap. Prot. J.* DOI: 10.1108/RPJ-10-2016-0168. Aviable from: http://www.emeraldinsight.com/doi/10.1108/RPJ-10-2016-0168

[7] Bricín D Špirit Z and Kříž A 2018 Metallographic Analysis of the Suitability of a WC-Co Powder Blend for Selective Laser Melting Technology DOI: 10.4028/www.scientific.net/MSF.919.3. Dostupné z: https://www.scientific.net/MSF.919.3

[8] Zdravkov B Čermák J Šebera M and Janků 2007 Pore classification in the characterization of porous materials: A perspective *O. Chem.* DOI: 10.2478/s11532-007-0017-9 ISSN 2391-5420. Aviable from: http://www.degruyter.com/view/j/chem.2007.5.issue-2/s11532-007-0017-9/s11532-007-0017-9.xml

[9] Bricín D Elmanová A and Kříž A 2019 The effect of selective laser melting technology on the Development of the structure of samples made from WC-Co powder *Mat., met. and tech. conf. 2019*

[10] Bricín D and Kříž A 2019 Processability of WC-Co powder mixtures using SLM additive technology *MM Sc. J.* DOI: 10.17973/MMSJ.2019_06_2018115 Aviable from: http://www.mmscience.eu/2018115
[11] Khmyrov R S Safronov V A and Gusarov A V 2016 Obtaining Crack-free WC-Co Alloys by Selective Laser Melting *Phys. Proc.* DOI: 10.1016/j.phpro.2016.08.091. Aviable from http://linkinghub.elsevier.com/retrieve/pii/S1875389216301985
[12] AMS 5659 2019 Available from: https://www.aksteel.com/sites/default/files/2018-11/15-5-ph-stainless.pdf [Accessed 22 Jul. 2019]. Rev. S *: Heat Treatment of Steel Raw Materials - SAE International*. ASM 5659
[13] Baker H 1999 Alloy phase diagrams *ASM Int.* Materials Park Ohio
[14] ASTM G99 2016 Standard Test Method for Wear Testing with a Pin-on-Disk Apparatus. [online] *Astm.org*. Available at: https://www.astm.org/DATABASE.CART/HISTORICAL/G99-05R16.htm [Accessed 28 Aug. 2019].

IOP Conf. Series: Materials Science and Engineering **723** (2020) 012004 doi:10.1088/1757-899X/723/1/012004

Influence of chemical composition and parameters of heat treatment on the mechanical properties and microstructure of TRIP steels

Dagmar Bublíková[1,2], Hana Jirková[1,2], Štěpán Jeníček[1,3], Jiří Vrtáček[1,4]

[1]University of West Bohemia, RTI-Regional Technological Institute, Univerzitní 22, CZ – 306 14 Pilsen, Czech Republic

[1]dagmar.bublikova@seznam.cz, [2]hstankova@rti.zcu.cz, [3]jeniceks@rti.zcu.cz, [4]vrtacekj@rti.zcu.cz

Abstract. Good mechanical properties of steels, in which appropriate heat treatment can produce mixed hardening structures, make them candidates for a broad range of applications, namely in the automotive industry. TRIP steels (in which transformation-induced plasticity operates) with a carbon content of approximately 0.2 % are one of such steels. Heat treatment of these steels comprises two stages. The first involve heating to the intercritical region between the A_1 and A_3. It is followed by cooling to a bainitic transformation temperature and holding. The resulting mixed microstructure consist of ferrite, bainite and retained austenite. Thanks to the presence of ferrite and retained austenite, the ultimate strength and elongation can reach 1500 MPa and 25-40 %, respectively. The experiments presented in this paper were performed on two steels which chemistries were specially adjusted to support formation of TRIP microstructure. The main difference between them was the level of chromium. Intercritical annealing was carried out on both steels. Aspects of interest included mainly the effect of the cooling rate above the bainitic transformation temperature and the holding time on mechanical properties and final microstructure. The heat treatment led to microstructures ferrite, bainite and retained austenite. The strength was under 1100 MPa and elongation reached 28 %.

1 Introduction

Advances in engineering place ever stricter demands on mechanical properties of advanced high-strength steels. The automotive industry in particular requires high strength along with high ductility in structural and safety components of sheet metal. These requirements are met by multiphase steels, in which retained austenite contributes to final mechanical properties [1]. Among them there are TRIP steels, materials which benefit from transformation-induced plasticity. The phenomenon is based on transformation of retained austenite to strain-induced martensite while the steel is subjected to plastic deformation. As a result, these materials are capable of absorbing large amounts of deformation energy [2].

Their heat treatment involves temperatures between the A_1 and A_3, cooling to the isothermal holding temperature, isothermal holding in the area of the bainitic transformation. Their microstructure then consists of polygonal ferrite, bainite and retained austenite (Figure 1). Mechanical properties are governed by chemical compositions and heat treatment procedure. In these steels, ultimate strengths of about 1500 MPa can be achieved in combination with an elongation of 25–40 % [3].

Microstructure plays an important role in the mechanical properties of TRIP steels. Ferrite phase has the strongest effect in this regard. It is also the one which makes up the largest portion of the microstructure. When the amount of bainite grows, yield strength increases as well. Yield strength increases with cooling rate in the intercritical annealing process. Retained austenite is another phase which affects plasticity. Important aspects include the morphology, amount and stability of retained austenite. In TRIP steels, retained austenite forms either granular particles or needles within bainite sheaves [4]. Stability of retained austenite is controlled by factors which include the size of austenite islands. The latest studies show that austenite islands higher than 1 μm are unstable. By contrast, those which size is around 1 μm require appreciable deformation force in order to transform to martensite. Consequently, retained austenite in them remains stable under cold deformation [5].

The primary focus of this paper is the effect of increased chromium levels on mechanical properties and microstructure of TRIP steels. In ordinary TRIP steels, the chromium level is not more than 0.4 wt. %. The chromium level in the experimental steel was increased to 0,64 %.

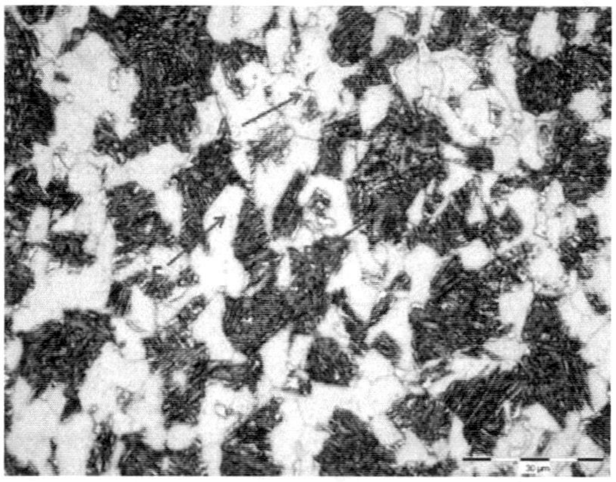

Figure 1. Microstructure of TRIP steel produced by intercritical annealing [6].

2 Experimental programme

Two steels were designed for these experiments: CMnSiCr and CMnSiNb steels with special alloying for achieving the desired microstructure and mechanical properties (Table 1). Their transformation temperatures, A_1, A_3, M_s and M_f, were calculated from their chemical compositions using JMatPro software (Figure 2, Figure 3). The crucial addition is carbon which substantially governs all transformations, the final microstructure and mechanical properties. Both steels contained about 0.2% carbon. The manganese content was 1.5 %. The element stabilizes austenite, depresses the M_s and M_f, enhances the solubility of carbon in austenite and retards pearlite formation [7]. The purpose of silicon as an alloying addition was to prevent carbides from forming, and promote migration of carbide from super-saturated martensite into austenite. The element also supports formation of proeutectoid ferrite during cooling [8]. Another alloying addition important to TRIP steels is niobium because even a minimum amount substantially alters mechanical properties. Both steels contained 0.06 % Nb. Niobium has a strong impact on austenitizing, recrystallization and enrichment of austenite with carbon. It has effects on the type and amount of retained austenite. Niobium-free steels contain globular retained austenite. Those alloyed with niobium contain retained austenite in lath or interlath form [9]. CMnSiCr grade had a higher level of chromium (0.64 %), the purpose of which was to improve hardenability and provide solid solution strengthening.

As-received materials were 50-kg ingots. Tops and bottoms of the ingots were removed and the ingots were cleaned. Then they were homogenized at 1100 °C/6 hours and cut up. The segments were heated to 1150 °C and forged into bars 20 mm in diameter. They were reheated between individual reductions. The forged bars were normalized at 950 °C/2 hours and left to cool in a furnace. Specimens to be heat treated in a thermomechanical simulator were then made from the bars. Metallographic observation was carried out after heat treatment and tensile testing performed on miniature specimens with a gauge length of 5 mm [10].

Table 1. Chemical compositions of steels [wt. %].

	C	Mn	Si	P	S	Cu	Cr	Ni	Al	Mo	Nb	M_s	M_f
CMnSiCr	0.2	1.5	1.8	0.008	0.003	0.11	0.64	0.098	0.009	0.03	0.06	426	317
CMnSiNb	0.2	1.5	1.8	0.005	0.003	0.06	0.008	0.07	0.006	0.02	0.06	360	246

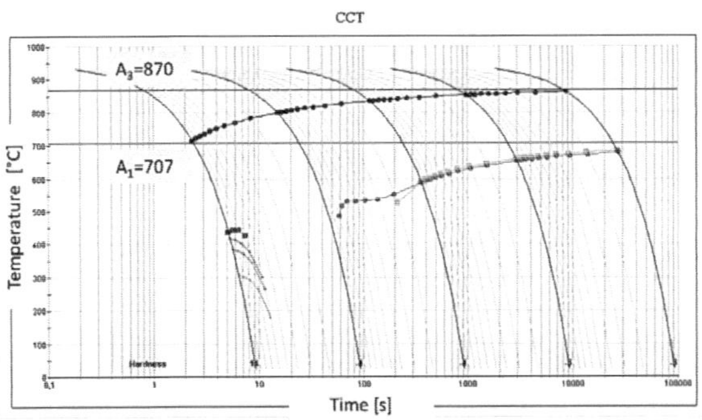

Figure 2. CMnSiCr – CCT diagram.

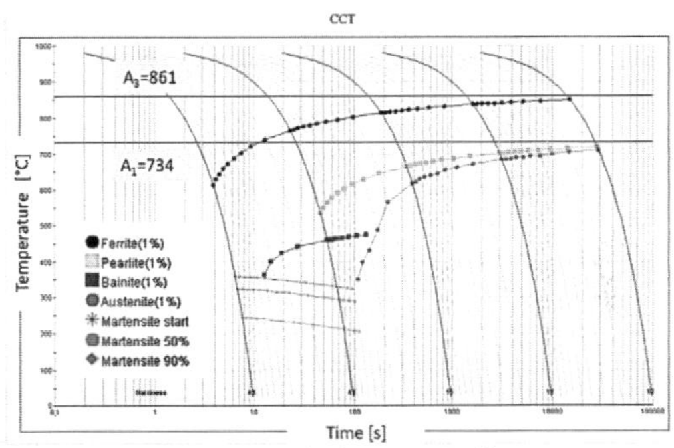

Figure 3. CMnSiNb – CCT diagram.

The first step in designing experimental heat treatment routes was finding a suitable austenitizing temperature for both materials, T6 and T13. To this end, results of calculation with JMatPro and actual heat treatment were compared. Specimens were heated in a furnace to several different temperatures between 800-1200 °C spaced at 50 °C, held for 30 seconds and quenched with water. In the next step, intercritical annealing was designed on the basis of known transformation temperatures. To ensure that the exact temperature profile is maintained, the experiments were carried out in a thermomechanical simulator. Using this machine, real-world heat treatment can be closely replicated on small amounts of material in laboratory conditions [11]. The intercritical annealing sequence comprised five stages (Figure 4), (Table 2). First, the material was heated to a full-austenitization temperature between the A_{c1} and A_{c3}. Based on the experiments, the austenitizing temperature was set at 950°C. The specimens were held at the austenitizing temperature for 100 seconds. The rates of cooling to the bainitic transformation temperature were selected on the basis of CCT diagrams. For CMnSiCr, they were 10 °C/s and 16 °C/s. Since CMnSiNb grade contained no chromium, the pearlite nose shifted to the left. Faster cooling at 30°C/s was used in order to avoid the nose (Figure 3). 425 °C was chosen as the temperature at which bainite would form and retained austenite would become stable. The holding time at this temperature is an important factor governing the amount of bainite and the stability of retained austenite. The holding times were 100 s and 600 s.

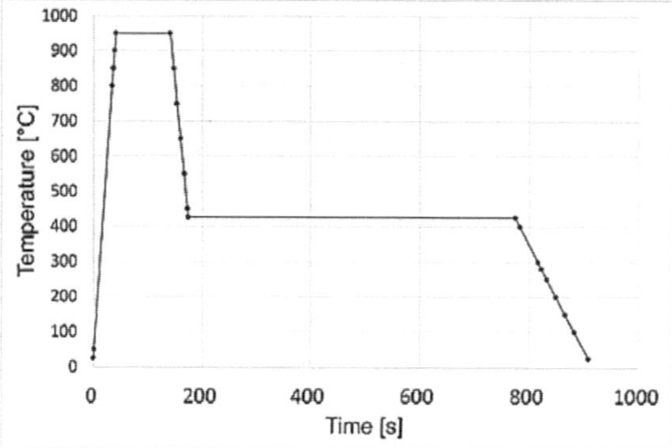

Figure 4. Intercritical annealing diagram.

Table 2. Heat treatment sequences and mechanical properties.

Sequence number/steel	T_A [°C]/t_A [s]	Cooling rate [°C/s]	T_B [°C]	t_B [s]	HV10 [-]	R_m [MPa]	A_{5mm} [%]	RA [%]
1/CMnSiCr	950/100	10	425	100	346	1080	27	12
2/ CMnSiCr	950/100	10	425	600	335	1059	30	13
3/ CMnSiCr	950/100	16	425	100	361	1098	27	12
4/ CMnSiCr	950/100	16	425	600	391	1099	28	12
3/CMnSiNb	950/100	16	425	100	283	905	34	10
4/CMnSiNb	950/100	16	425	600	272	889	39	12
5/CMnSiNb	950/100	30	425	100	301	927	34	10
6/CMnSiNb	950/100	30	425	600	274	873	40	12

3 Results and discussion

First, a suitable heating temperature was required. When the steels were heated to the lowest temperature from the interval, 800 °C, they developed large amounts of ferrite. In line with that, their hardness was low: 301–373 HV10 (Figure 5, Figure 8). Once they were heated to 950 °C ferrite was eliminated, with the resulting microstructure consisting mainly of martensite. Its hardness was 484–503 HV10 (Figure 6, Figure 9). Further increments in the heating temperature, all the way to 1100 °C, brought no additional changes to the microstructure or increase in hardness (Figure 7, Figure 10). Finally, the optimal heating temperature for the experiments was identified as 950 °C.

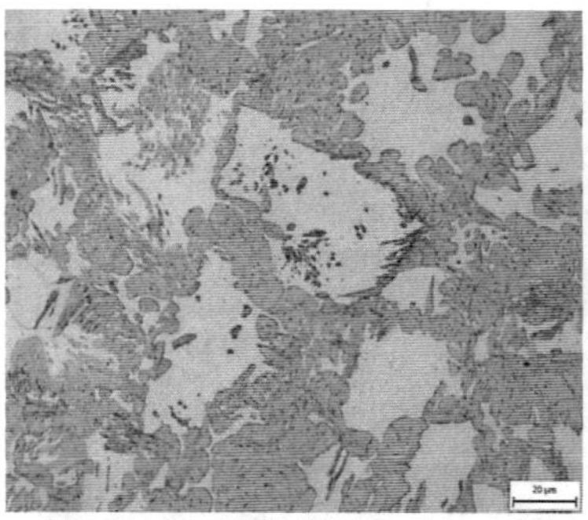

Figure 5. CMnSiCr – heated to 800°C, holding time 30 minutes, 373 HV10.

Figure 6. CMnSiCr – heated to 950°C, holding time 30 minutes, 503 HV10.

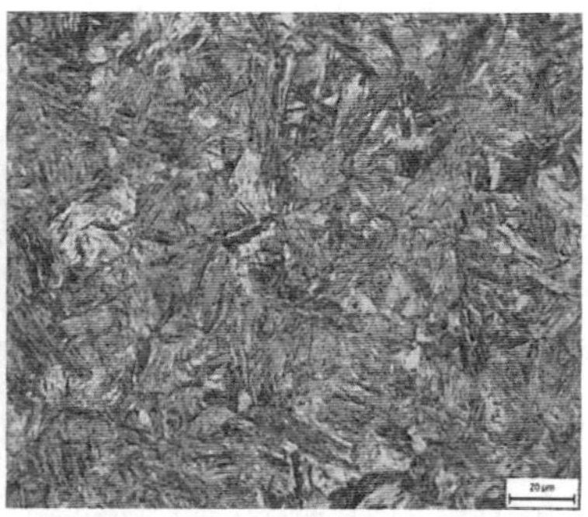

Figure 7. CMnSiCr – heated to 1000°C, holding time 30 minutes, 511 HV10.

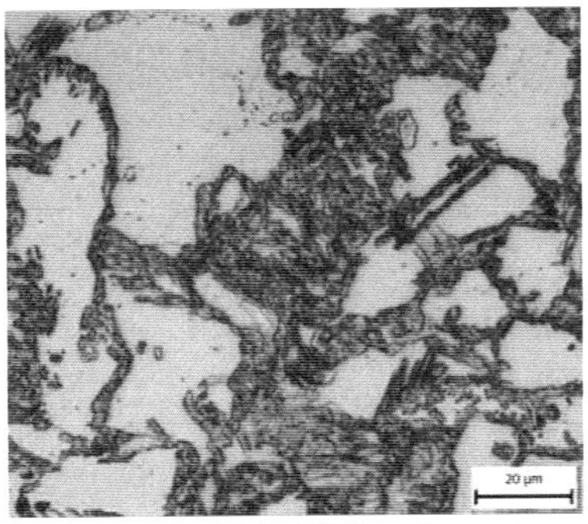

Figure 8. CMnSiNb – heated to 800°C, holding time 30 minutes, 301 HV10.

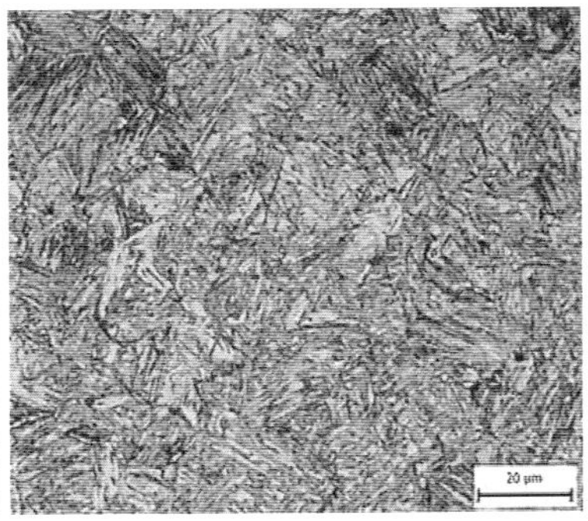

Figure 9. CMnSiNb – heated to 950°C, holding time 30 minutes, 484 HV10.

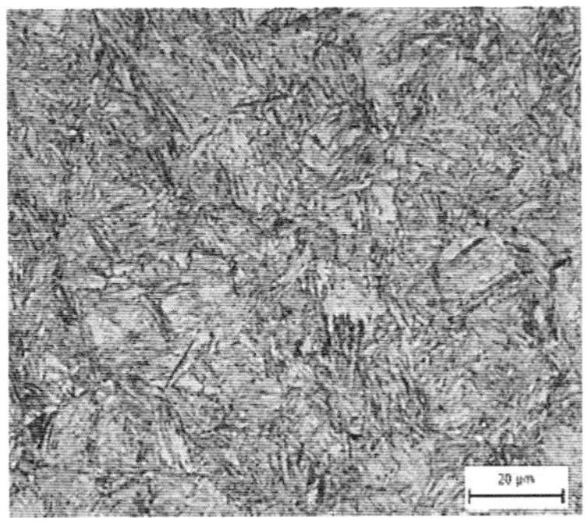

Figure 10. CMnSiNb – heated to 1000°C, holding time 30 minutes, 489 HV10.

Intercritical annealing sequences produced mixtures of ferrite, bainite and retained austenite in both materials (Figure 11 - Figure 20). The amount of ferrite was dictated by the rate of cooling to the bainitic transformation temperature. The time at this temperature, in turn, dictated the volume fraction of retained austenite. The shorter holding time (100 s) failed to secure adequate saturation of austenite with carbon. In the course of cooling from the bainitic transformation temperature to room temperature, retained austenite decomposed into martensite. It was evidenced by the resulting high hardness, lower elongation and a smaller amount of austenite. The longer hold at the bainitic transformation temperature (600 seconds) led to lower hardness, higher elongation and a larger amount of retained austenite (Table 2).

In CMnSiCr steel, the sequence with cooling at 10 °C/s produced a bainitic microstructure with a small amount of martensite, some grains of free ferrite and 12-13 % of retained austenite (Figure 11, a, b). The range of hardness values was 334-346 HV10. The next sequence, with a longer holding time at 425 °C, i.e. not 100 seconds but 600 seconds, led to a slightly lower ultimate strength, i.e. 1059 MPa instead of 1080 MPa, but improved elongation from 27 % to 30 %. In both sequences (which involved different holding times at the bainitic transformation temperature), higher rate of cooling (16 °C/s) from the heating temperature to the isothermal holding temperature led to a lower amount of proeutectoid ferrite and a higher amount of bainite (Figure 12, a, b). This was reflected in higher hardness: 360 -391 HV10. The ultimate strength and elongation were 1099 MPa and 28 %, respectively. X-ray diffraction analysis found 12 % retained austenite for both holding times at 425 °C. The presence of retained austenite was demonstrated and its distribution studied after special colour etching (Figure 13, a, b).

After cooling at the same rate (16 °C/s) CMnSiNb steel developed a ferritic-bainitic microstructure with 12 % retained austenite. The amount of bainite was considerably smaller than in the chromium-alloyed steel (Figure 14, a, b). After holding for 100 s at 425 °C, hardness was 272 HV10, and ultimate strength and elongation reached 889 MPa and 39 %, respectively. Longer holding time produced coarser bainite blocks and a larger amount of bainite. It was reflected in the slightly higher hardness and ultimate strength than in the previous case: 283 HV10 and 905 MPa, respectively, and an elongation of 34 %. The retained austenite volume fraction rose slightly to 12 % (Table 2). Faster cooling (30 °C/s) from the heating temperature to the isothermal holding temperature resulted in a larger bainite fraction and smaller amount of ferrite (Figure 15, a, b). Retained austenite volume fraction was 10-12 %. After the shorter holding time, 100 s, ultimate strength and elongation were 927 MPa and 34 %, respectively. Longer holding time, 600 s, led to lower strength, 873 MPa, and elongation of 40 %. Owing to the absence of chromium, this steel developed lower hardness and ultimate strength (Table 2).

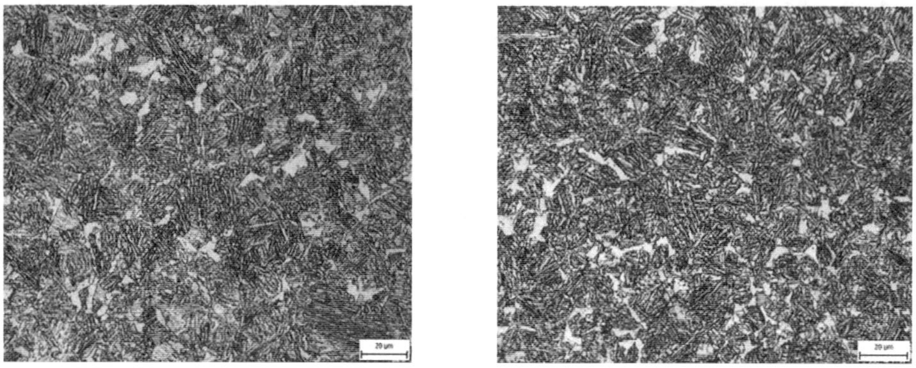

a) sequence 1, 100 s b) sequence 2, 600 s

Figure 11. CMnSiCr, cooling rate 10 °C/s, TB 425 °C, bainitic structure with some amounts of martensite, ferrite and retained austenite.

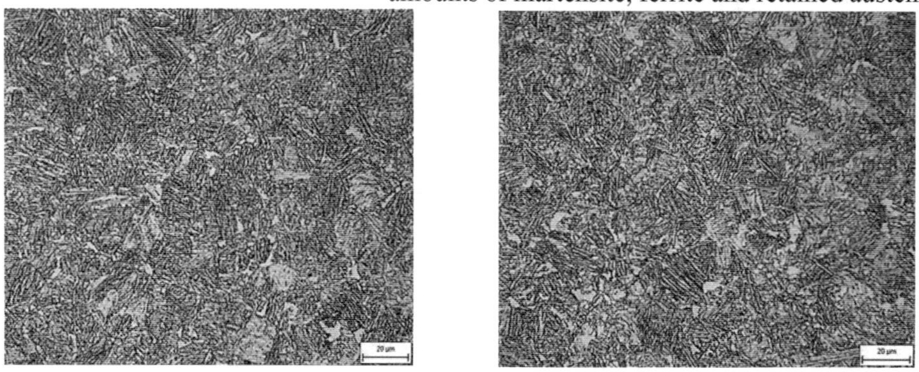

a) sequence 3, 100 s b) sequence 4, 600 s

Figure 12. CMnSiCr, cooling rate 16 °C/s, TB 425 °C, bainitic structure with some amounts of martensite, ferrite and retained austenite.

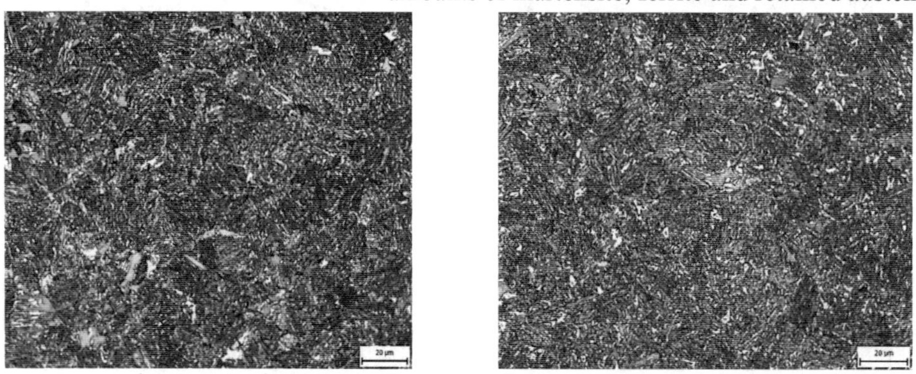

a) sequence 3, 100 s b) sequence 4, 600 s

Figure 13. CMnSiCr, cooling rate 16 °C/s, TB 425 °C/, colour etch made to reveal retained austenite.

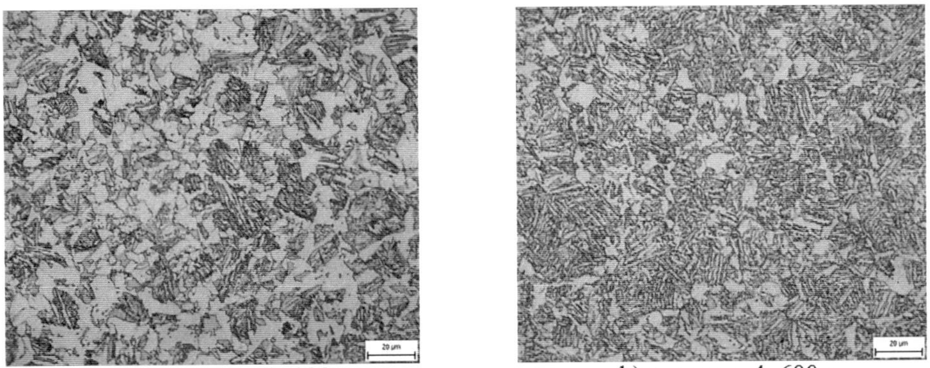

a) sequence 3, 100 s   b) sequence 4, 600 s

Figure 14. CMnSiNb, cooling rate 16 °C/s, TB 425 °C, mixture of bainite and austenite with a certain amount of ferrite.

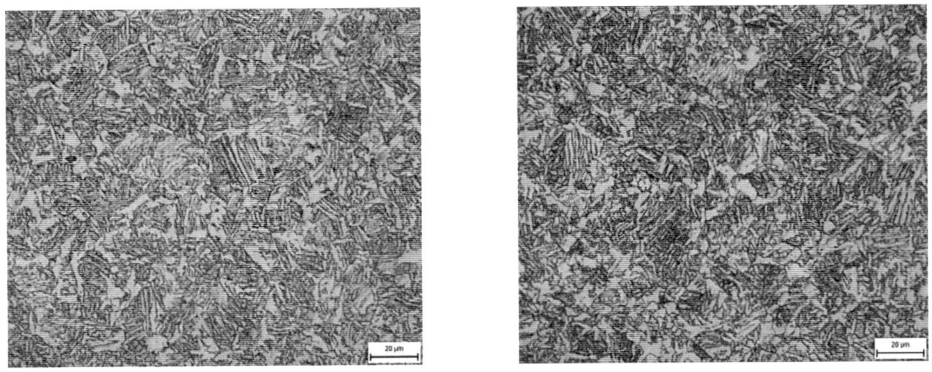

a) sequence 5, 100 s   b) sequence 6, 600 s

Figure 15. CMnSiNb, cooling rate 30 °C/s, TB 425 °C, mixture of bainite and austenite with some retained austenite.

4 Conclusion

All the heat treatment sequences applied to CMnSiCr produced a bainitic microstructure with a small amount of martensite, grains of free ferrite and 12-13 % retained austenite. Thanks the chromium content were hardness values ranged between 346 and 391 HV10, ultimate strengths were 1059-1099 MPa, and elongation reached 27-30 %. In CMnSiNb a ferritic-bainitic structure with 10-12 % retained austenite was obtained. Owing to the absence of chromium, this steel developed lower hardness and ultimate strength. Hardness values ranged between 272 and 301 HV10, ultimate strengths were 873-927 MPa, and elongation reached 34-40 %.

The amount of ferrite, and thus mechanical properties, showed dependence on the rate of cooling to the holding temperature. The holding time at this temperature had a notable effect on stability of retained austenite more in the steel CMnSiNb. With shorter times (100 seconds), austenite failed to saturate adequately with carbon, and eventually decomposed into martensite. The volume fraction of retained austenite was 10 %. After the longer hold (600 s) austenite was saturated appropriately and remained stable among bainite needles. The volume fraction of retained austenite was 13 %. The amount of retained austenite was demonstrated using colour etching.

Acknowledgements

This paper includes results achieved within the project SGS-2019-019 Research of Modern AHS Steels and Innovative Processing for their Manufacturing. The project is subsidised from specific resources of the state budget for research and development.

IOP Conf. Series: Materials Science and Engineering **723** (2020) 012004 doi:10.1088/1757-899X/723/1/012004

References

[1] Xiong Z P, Kostryzhev A G, Saleh A A and Chen L Pereloma E V 2016 Microstructures and mechanical properties of TRIP steel produced by strip casting simulated in the laboratory *Mat. Sc. & En. A* **664** pp 26–42

[2] Qian Z, Lihe Q, Junt T Jiangying and M Fucheng Z 2013 Inconsistent effects of mechanical stability of retained austenite on ductility and toughness of transformation-induced plasticity steels *Mat. Sc. & En. A* **578** pp 370–376

[3] Shi W, Li L and Zhou Y A 2002 Effect of Mn-Mn TRIP steels Proccedings of International Conference on TRIP Aided High Strenght Ferrous Aoys (Belgium)

[4] Basuki A Aernoudt E 1999 Influence of rolling of TRIP steel in the intercritical region on the statility of re-tained austenite *J. of Mat. Process. Tech.* pp 89-90 34-43

[5] Timokhina I B, Hodgson P D Pereloma E V 2010 Morphology and stability of retained austenite in thermomechanically processed C-Mn-Si (-Nb) TRIP steels Proceedings of Euromat (Italy)

[6] Kříž A 2010 Kovové materiály Studijní materiály ZČU

[7] Bleck W 2002 Using the TRIP efekt – the down of a promising group of cod formable steelos Proccedings of International Conference on TRIP – Aided High Strenght Ferrous Alloys (Belgium)

[8] Baik S CH, Kim S and Jin S O 2001 Effect of alloying elements on mechanical propeties and phase transformation of cold rolled TRIP steel steels ISIJ International **41** no 3 pp 290-297

[9] Park S H, Choo W Y, Kim N J and Ko J H 1996 Effects of hot rolling conditions on the microstructure and tensile properties of Nb-bearing TRIP steels International Symposium on Hot Workability and Light Alloys Composites TMS of CIM (Motrioll Quebek Canada) pp 493

[10] Kučerová L, Jeníček Š and Mach J 2018 Effect o sample geometry on tensile properties of chromium alloyed middle carbon steel In Conference Proceedings METAL - 27th International Conference on Metallurgy and Materials (Ostrava: TANGER Czech Republic EU ISBN 978-80-87294-84-0) pp 761-765

[11] Káňa J, Vorel I and Ronešová A 2015 Simulator of Thermomechanical Treatment of Metals Daaam International (Vienna: ISBN 978-3-902734-07-5 ISSN 1726-9679) pp 0513-05018

The effect of application of the plaster as a mould material on the microstructure and properties of AlSi9 aluminium alloy

T Bucki[1], M Sidorko[1] and D Bolibruchová[2]

[1] Kielce University of Technology, Kielce, Poland, EU
[2] University of Žilina, Žilina, Slovakia, EU
E-mail: tbucki@tu.kielce.pl

Abstract. The study involved examination of the effect of mould material on the microstructure and mechanical properties of AlSi9 aluminium alloy castings. The castings were fabricated in: plaster-based moulds, dehydrated plaster-based moulds and steel mould. The structure of castings from plaster mould not subjected to dehydration process was characterized by a significant porosity. This results indicated that as-cast plaster mould is not suitable for casting of AlSi9. Dehydrating of the mould led to significant reduction in the number of pores in the casting, while no pores were observed in the structure of steel mould castings. Microstructure of dehydrated plaster mould castings and steel mould castings was analysed in a Scanning Electron Microscope. Both types of examined castings were characterized by typical microstructure containing a eutectic (α-Al and Si) with dendrites of α-Al and multicomponent phases locally distributed in the matrix. Mould material affected the size of microstructural constituents of observed specimens. Castings fabricated in steel mould showed considerable refinement of the eutectic and dendrites of α-Al in comparison with dehydrated plaster castings. In addition to differences in microstructure, the castings from dehydrated plaster mould were also characterized by lower strength and smaller elongation than steel mould castings.

1 Introduction

Aluminium and aluminium alloys are applied in many branches of the industry where the combination of good mechanical properties and low density is required. The most popular aluminium cast alloys are the alloys containing silicon. Their advantages, in addition to good strength and corrosion resistance, are very good casting properties that enable cheap and efficient production of castings. The Al-Si alloys can be produced by such methods as die casting, sand mould casting and metal mould casting [1-8].

The literature data indicate that non-ferrous castings can be also manufactured by plaster mould casting method [9-11]. This technique seems to be relatively cheap and efficient due to use of widespread raw materials and possibility of casting by gravity method. The liquid consistency of the moulding medium enables reproduction of complex shapes and fabrication of castings by investment casting method.

The disadvantage of using plaster-based moulds is their chemical composition that can negatively affect castings. The plaster hydrates by incorporating water molecules in the structure. That leads to dehydration at elevated temperature and consequently deteriorate the mould properties. The contact of molten aluminium with water is undesirable, not only due to violent evaporation of water in contact with liquid metal but also because of the formation of porosity in castings,

and consequently, decrease of their mechanical properties. The disadvantage of plaster moulds is also their low cooling rate and low gas permeability.
Heating of plaster-based material in high temperature causes detaching water molecules. Dehydration leads to dimensional instability, negatively affecting the properties of mould. The literature data indicate that drying plaster at temperature of about 750 °C for a long period of time leads to formation of an anhydrite II, which can be applied as mould material [12,13]. The research [14] shows a positive effect of additives such materials as Al_2O_3, SiO_2, SiC and talc to plaster-based mould, due to improvement of dimensional stability and strength of the mould at elevated temperatures.
The aim of presented study was to examine the effect of application of the plaster-based moulds on the properties of AlSi9 aluminium alloy. Liquid AlSi9 was poured into three types of mould, i.e. as-cast plaster mould, plaster mould subjected to dehydration in 750 °C for 3 h and steel mould. The microstructure and uniaxial tensile strength of the castings were analysed and the effect of mould material and dehydration of plaster mould on the properties of AlSi9 castings were studied.

2 Experimental procedure

AlSi9 aluminium alloy (9.62 % Si, 0.39 % Mg, 0.38 % Fe, 0.13 % Cu, 0.26 % Mn) was selected as the cast material. The process involved heating of AlSi9 to 670 °C in resistance furnace equipped with a graphite crucible and pouring of molten alloy into the moulds being at room temperature. Three types of the mould were used in the experiment, i.e. as-cast plaster mould (not subjected to dehydration process), plaster mould heated in 750 °C for 3 h (to dehydrate and transform the plaster into anhydrite II) and steel mould. The plaster moulds were fabricated by mixing of plaster mortar with aluminium oxide powder and water (ratio: 1 g : 1 g : 1.5 ml, respectively). Figure 1 presents the cross-section and dimensions of the mould. Steel mould with the same internal dimensions and wall thickness of 10 mm was machined from a steel bar.

Specimens for microstructure analysis and uniaxial tensile test were machined from the castings according to schematic diagram showed in figure 1b. The specimens for microstructure observations were prepared from lower and upper part of castings using standard metallographic procedures. The polishing was performed using 0.3 μm Al_2O_3 suspension and 0.05 μm colloidal silica. The mechanical properties of AlSi9 castings were measured by uniaxial tensile test of cylindrical specimens with a diameter of 12 mm on Zwick/Roell universal testing machine. The displacement rate was 1 mm/min. The microstructure was analysed on a Nikon ECLIPSE MA200 Optical Microscope (OM) and a JEOL JSM-7100F Scanning Electron Microscope equipped with an energy dispersive X-ray spectroscopy (SEM/EDS).

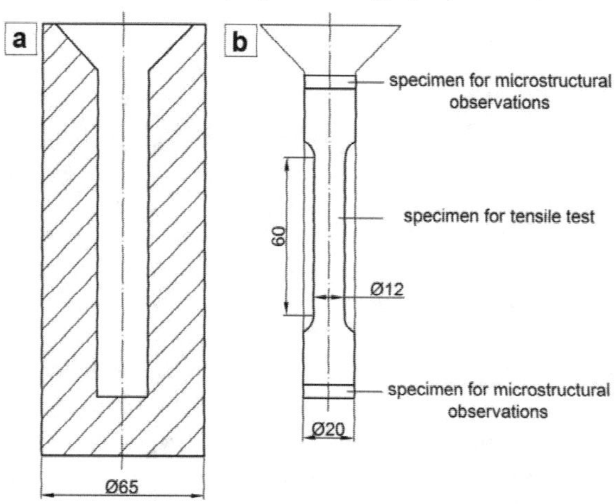

Figure 1. Schematic diagram of the plaster mould (a) and machined specimens (b).

3 Results and discussion

Figure 2 shows low magnification images of the AlSi9 casting fabricated in as-cast plaster mould. Microstructure was observed in upper (figure 2a) and lower part (figure 2b) of specimen. The results present a significant porosity distributed in the structure of casting, both in its upper and lower parts. The porosity suggests that as-cast plaster mould is not suitable for casting of AlSi9 alloy.

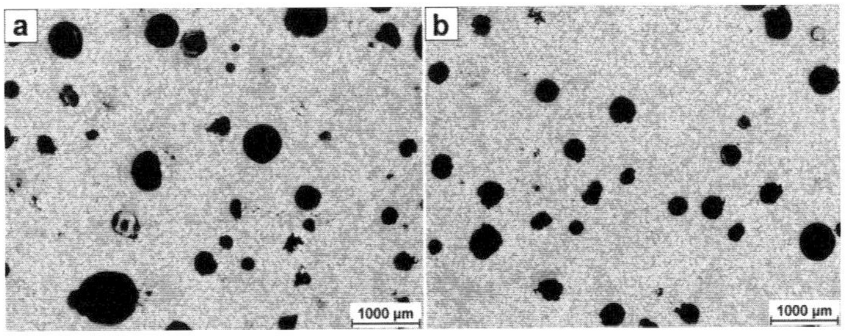

Figure 2. Low magnification OM images of the microstructure of AlSi9 casting fabricated in as-cast plaster mould observed at: upper part (a) and lower part of the casting (b).

Figure 3 presents the microstructure of the AlSi9 casting fabricated in dehydrated plaster mould. Microstructure was observed in upper and lower parts of specimen, in their middle region and close to the surface. From the observations and Al-Si phase diagram [15], it is evident that analysed AlSi9 alloy is characterised by typical microstructure containing white dendrites of an α-Al distributed in a eutectic matrix (α-Al and Si). In the microstructure, grey particles characteristic in shape were also observed. Similar composition of the structure was observed throughout the entire cross-section, but the size of individual precipitates varied. The structure analysis throughout characteristic parts of casting (figures 3a-d) shows that when moving from the surface to middle region of the sample, both in the lower and upper parts of casting, the increase of the size of eutectic precipitations and white dendrites is noticeable. The observation also indicates that drying of plaster mould leads to significant reduction in the number of pores in the casting compared to as-cast plaster mould casting. In this case the porosity was observed only locally, mainly in the middle region of the casting, both in its upper and lower parts (see figures 3b and 3d).

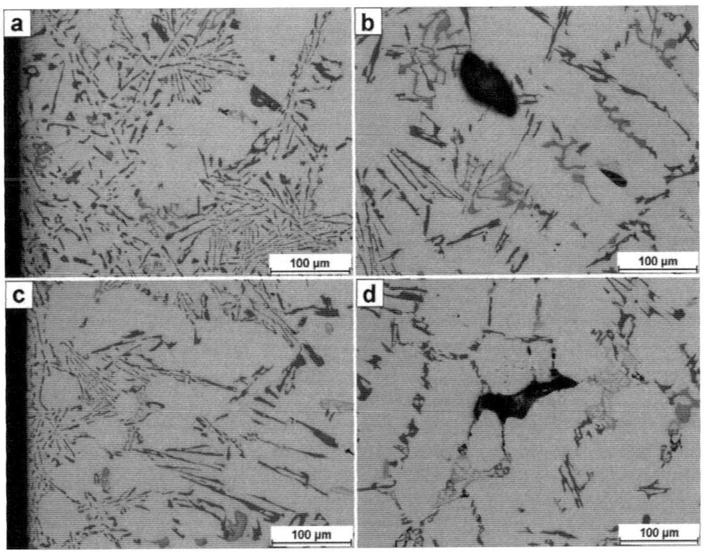

Figure 3. OM images of the microstructure of AlSi9 casting fabricated in dehydrated plaster mould observed close to: surface of upper part (a), middle region of upper part (b), surface of lower part (c) and middle region of lower part of the casting (d).

Figures 4a-d show the OM micrographs of AlSi9 alloy cast to steel mould, observed at characteristic regions of upper and lower parts of casting. The results show that AlSi9 cast to steel mould is characterized by similar structural composition to casting from dehydrated plaster mould. However, a significant difference in the size of structural constituents is noticeable. Casting fabricated in steel mould shows considerable refinement of the eutectic and dendrites of α-Al. In steel mould casting, the trend of microstructure refinement when going from the middle region to surface and from the upper to lower part of specimen is also visible. No noticeable porosity was observed in analysed specimens.

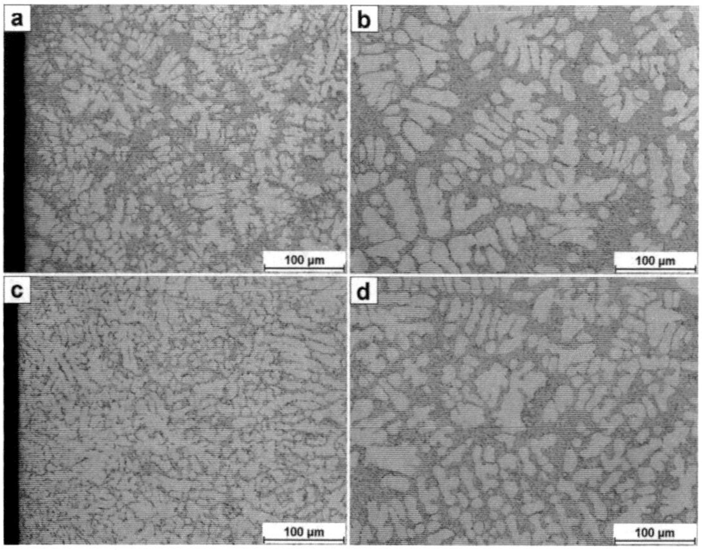

Figure 4. OM images of the microstructure of AlSi9 casting fabricated in steel mould observed close to: surface of upper part (a), middle region of upper part (b), surface of lower part (c) and middle region of lower part of the casting (d).

Figure 5a illustrates the microstructure of AlSi9 casting from dehydrated plaster mould observed in SEM. EDS quantitative analysis was performed for regions marked in the figure to confirm the phase composition. The results are provided in table 1. The composition of white phase marked as 1 and dark particles marked as 2 indicated that two-phase matrix is a eutectic composed of α-Al and Si. The chemical composition of white dendrites distributed in the eutectic matrix (point 3) also corresponded to the composition of α-Al.
The results of EDS analysis carried out in characteristic areas of steel mould casting (presented in figure 5b) are also summarized in table 1. The chemical composition of regions 6 and 7 clearly indicated a eutectic (α-Al and Si). The dendrites marked as 8 were identified as α-Al.
The observations show that grey precipitations, characteristic in shape, can be distinguished in the microstructure of both analysed specimens. The results of EDS analysis for particles marked as 4 in figure 5a and 9 in figure 5b suggest Al-Mn-Fe multicomponent phase, while the composition of particles marked as 5 in figure 5a and 10 in figure 5b indicates the phase containing Al, Si and Fe. The literature data [16,17] indicate that this types of intermetallic phases are typical for Al-Si alloys. The content of Fe in alloy leads to occurrence of Al_5FeSi phase, negatively affecting the properties of the alloy, while the addition of Mn causes the formation of Mn-rich multicomponent intermetallic phases.
The uniaxial tensile test was carried out to examine the mechanical properties of the castings. The castings fabricated in as-cast plaster mould were disqualified from the tests due to the porosity. Test was carried out for 3 specimens machined from dehydrated plaster mould castings and 3 specimens from castings fabricated in steel mould. The results consisting of yield strength,

ultimate tensile strength and elongation are summarized in table 2. The results show that average yield strength and tensile strength of castings fabricated in dehydrated plaster mould was 108.9 MPa and 159.5 MPa, respectively, while the elongation was about 1 %. The measurements for steel mould castings show higher values of mechanical properties. The average yield strength raised to 115.8 MPa, but the highest improvement was observed for ultimate tensile strength (226.0 MPa) and elongation (5.3 %).

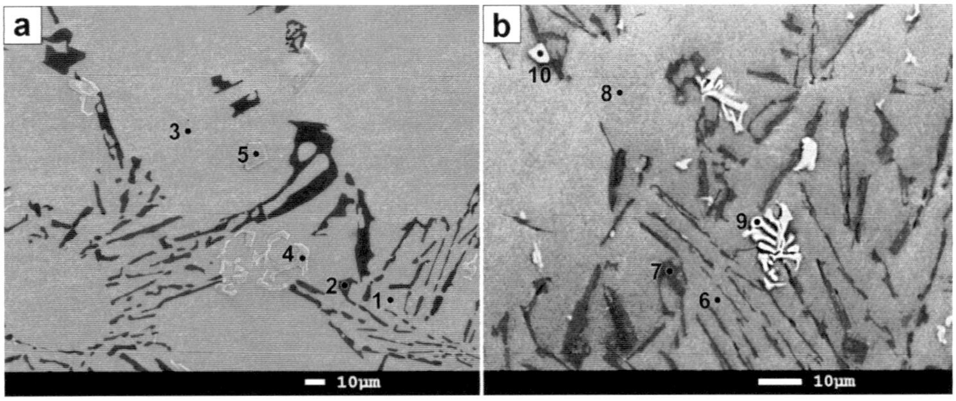

Figure 5. SEM image of the microstructure of AlSi9 casting fabricated in dehydrated plaster mould (a) and steel mould (b).

Table 1. Results of EDS point analysis performed in dehydrated plaster mould casting and steel mould casting specimens in points marked in figures 5 and 6 (at.%).

Point	Al	Si	Mn	Fe
1	98.15	1.85	-	-
2	0.43	99.57	-	-
3	98.24	1.76	-	-
4	70.76	12.00	6.27	10.97
5	64.81	19.57	1.83	13.79
6	97.71	2.29	-	-
7	0.54	99.46	-	-
8	97.95	2.05	-	-
9	73.09	11.05	6.31	9.55
10	72.33	15.76	-	11.91

Table 2. Results of Uniaxial Tensile Test.

		Yield strength $R_{0.2}$ (MPa)	Ultimate tensile strength (MPa)	Elongation (%)
Dehydrated plaster mould castings	1	104.7	151.3	1.4
	2	110.1	166.6	0.8
	3	111.8	160.7	0.7
	Average	**108.9**	**159.5**	**1.0**
Steel mould castings	1	117.0	225.5	5.5
	2	112.2	221.4	4.7
	3	118.3	231.1	5.6
	Average	**115.8**	**226.0**	**5.3**

IOP Conf. Series: Materials Science and Engineering **723** (2020) 012005 doi:10.1088/1757-899X/723/1/012005

The results show that plaster-based mould is suitable for casting of AlSi9 alloy but it requires long time drying of the mould at high temperature, and, despite these expensive treatment, the mechanical properties of castings are lower compared to castings fabricated in steel mould. The lower values of mechanical characteristics of castings fabricated in plaster moulds may be caused by their coarse-grained structure as well as the pores locally distributed in the casting.

4 Conclusions

The AlSi9 castings were fabricated in three types of mould, i.e.: as-cast plaster mould, dehydrated plaster mould and steel mould, to study the effect of mould material on the microstructure and mechanical properties of the casting.

The results show that as-cast plaster mould is not suitable for casting of AlSi9 alloy, due to a significant porosity distributed over entire cross-section of the castings. Dehydrating of plaster mould in 750 °C for 3 h led to significant reduction in the number of pores in the casting compared to as-cast plaster mould casting. The porosity was locally observed there, mainly in the middle region of casting. In the castings fabricated in steel moulds no noticeable porosity was observed.

Microstructural analysis of AlSi9 cast to dehydrated plaster mould and steel mould indicated that, for both mould materials, the structure of casting was composed of a eutectic (α-Al and Si) and dendrites of a solid solution of Si in Al with precipitation of multicomponent phases locally distributed in the structure. Mould material affected the size of microstructural constituents of observed specimens. Casting fabricated in steel mould showed considerable refinement of the eutectic and dendrites of α-Al in comparison with dehydrated plaster casting.

The mould material affected the mechanical properties of castings. Castings fabricated in dehydrated plaster moulds had lower strength and smaller elongation than steel mould castings. These differences may be caused by the different size of structural constituents and by the porosity observed in castings.

References

[1] Hatch J E 1984 ALUMINUM. Properties and Physical Metallurgy (Metals Park, Ohio: AMERICAN SOCIETY FOR METALS)
[2] Niu X P, Hu B H, Pinwill I and Li H 2000 *J. Mater. Process. Technol.* **105(1-2)** pp 119–27
[3] Wang L, Makhlouf M and Apelian D 2013 *Int. Mater. Rev.* **40(6)** pp 221–38
[4] Mae H, Teng X, Bai Y and Wierzbicki T 2008 *Int. J. Solids Struct.* **45** pp 1430–44
[5] Pezda J 2008 *Arch. Foundry Eng.* **8(1)** pp 273–6
[6] Bucki T and Mola R 2017 METAL 2017: 26th International Conference on Metallurgy and Materials (Brno) (Ostrava: TANGER Ltd) pp 1664–8
[7] Mola R and Bucki T 2018 *Arch. Foundry Eng.* **18(1)** pp 71–6
[8] Brůna M, Bolibruchová D and Pastirčák R 2017 *Procedia Eng.* **5(1)** pp 488–95
[9] Pastirčák R, Sládek A and Kucharčíková E 2015 *Arch. Foundry Eng.* **15(2)** pp 91–4
[10] Władysiak R and Pawlak M 2009 *Arch. Foundry Eng.* **9(4)** pp 225–32
[11] Ramesh C, Valliappan M and Prasanna S C 2015 *Int. J. New Technol. Sci. Eng.* **2(1)** pp 1–6
[12] Ghazi Wakili K, Koebel M, Kaufmann J, Derluyn H, Hofer M, Glaettli T and Modregger P 2012 HEFAT2012 9th International Conference on Heat Transfer, Fluid Mechanics and Thermodynamics (Malta)
[13] Pawlak M and Niedźwiedzki Z 2008 *Arch. Foundry Eng.* **8(3)** pp 145–8
[14] Wiktorski M, Pawlak M and Niedźwiedzki Z 2004 *Archives of Foundry* **14** pp 225–32
[15] Murray J L and McAlister A J 1984 The Al-Si (Aluminum-Silicon) System, National Bureau of Standards, Bulletin of Alloy Phase Diagrams vol 5(1) (New York: Springer US) pp 74–84
[16] Mola R, Bucki T and Gwoździk M 2019 *JOM* **71(6)** pp 2078–86
[17] Bolibruchová D, Podprocká R, Pastirčák R and Major-Gabryś K 2018 *Arch. Metall. Mater.* 63(4) pp 1883–8

Small punch testing of Fe-Al based alloys with Ti and Nb additions

Petr Dymáček[1,2*], Ferdinand Dobeš[1], Milan Jarý[1], Yvonna Jirásková[1], Naděžda Pizúrová[1,2], Martin Friák[1,2]

[1]Institute of Physics of Materials of Czech Academy of Sciences, Žižkova 22, 616 62 Brno, Czech Republic

[2]CEITEC-IPM, Žižkova 22, 616 62 Brno, Czech Republic

E-mail: pdymacek@ipm.cz

Abstract. Three Fe-Al-based alloys, a binary with 22 at. % of Al, a ternary with 22 at. % of Al and 7 at. % of Ti and a quaternary with 22 at. % of Al, 4 at. % of Ti and 4 at. % of Nb prepared by arc melting to small button type ingots were studied by small punch test and small punch creep test in order to obtain the high temperature tensile and creep properties. Evaluation of the results shows a significantly improved strength at high temperatures and creep resistance of the ternary and quaternary alloys compared to the binary alloy. The observation of the punched discs fracture surfaces related to the initial microstructure also helps to better understanding of the deformation and fracture behavior of these alloys at high temperatures.

1 Introduction

The iron aluminides represent a group of materials with unique properties such as very good corrosion and chemical resistance, attractive strength to weight ratio and low cost if compared to more expensive Ni-based alloys or superalloys. A successful example of recent replacement of engine part originally produced from Ni-based superalloy can be found in [1]. The iron aluminides have also interesting magnetic properties. Recent investigations devoted to the structural and physical properties of the parent Fe-20 at. % Al alloy modified by Ti and Nb at the expense of Fe point to different microstructures of Ti+Nb alloyed samples compared to samples with only Ti addition [2]. Moreover, a huge increase in saturation magnetization of the sample with 4 at. % Ti + 4 at. % Nb compared to sample with 8 at. % Ti (87.5%) is unresolved and represents an interesting topic for the next experimental and theoretical investigations [3]. Aim of this study is to test and compare the high temperature mechanical properties (tensile and creep) of Fe-Al alloys with 3 different compositions using the cost-effective small scale testing. A detailed comparison of the first two alloys with the data found in the literature was done in [4].

2 Materials and methods

The studied alloys were prepared from high-purity elements (Fe – 99.95%, Ti, Nb – 99.8%, and Al – 99.95%) using arc melting in MAM-1 furnace (Buehler GmbH) into a button-type ingots. They were four times re-melted to guarantee a good chemical homogeneity. The chemical composition of the three studied alloys is given in Table 1.

Table 1. Intended and analyzed compositions of the studied Fe-Al alloys (at. %).

Alloy	Composition	Fe	Al	Ti	Nb
Ti-0	intended	78	22	0	0
	analyzed	77.1±0.2	22.9±0.2		
Ti-7	intended	71	22	7	0
	analyzed	70.2±0.2	22.6±0.2	7.2±0.1	
Ti-4-Nb-4	intended	70	22	4	4
	analyzed	69.5±0.2	22.5±0.2	4.1±0.1	3.9±0.1

Small punch tests were performed in two loading modes: (i) small punch creep (SPC) test at constant force under which the punch penetrates through the specimen and the time dependence of the punch displacement is measured; (ii) small punch test (SPT) at constant rate of punch displacement and the necessary force is measured. Both tests are run up to the rupture of the specimen. A constant load cantilever creep machine was adapted for testing. During the test, punch is pushed against a specimen supported by a 4-mm diameter receiving die (lower die). Punch tip consists of a precise ceramic ball made of FRIALIT F99.7, 2.5 mm in diameter. The disc specimen of diameter D_S = 8 mm and thickness h_0 = 0.5 mm is clamped by an upper die. Central deflection is measured as the difference in the positions of the punch and lower die, using a linear variable differential transformer and is continuously recorded with a PC. The force acting on the specimen is measured using the load cell. All tests are performed in a protective atmosphere of purified argon. Temperature is measured by thermocouple placed in the receiving die in the proximity of specimen. The technique is described in more detail elsewhere [5, 6].

The specimens for small punch testing were prepared by machining cylinders 8 mm in diameter. The cylinders were subsequently cut to slices 1.2 mm thick using electro discharge machining. The slices were ground carefully from both sides equally and finally polished to 1200 grit. The final thickness h_0 = 0.500 ± 0.005 mm was measured by using a micrometer with a resolution of 1 μm.

The microstructural and fractographic observations were done using a scanning electron microscope Tescan Lyra3 XMU. Morphology studies and phase analyses were carried out with the transmission electron microscope JEM 2100F (Schottky cathode, using a 200 kV electron beam). TEM foils of as prepared alloys were prepared by grinding down to a grit size of P1200 and finished using an ion milling device (PIPS).

3 Results and discussion

3.1 Microstructure

The polycrystalline microstructures of the samples were checked by scanning electron microscopy (SEM) and analyzed using transmission electron microscopy (TEM). The morphologies of the alloys in the "as prepared" state were substantially different. The polished cross section microstructures of three studied alloys in the as prepared state are shown in Fig. 1. The irregular larger grains are observed at the Ti-0 sample in Fig. 1a. A finer-grained microstructure was formed in the Ti-7 alloy as seen in Fig. 1b. The Ti-4-Nb-4 alloy microstructure in Fig. 1c shows a matrix formed by primary dendrites with eutectic similar to the Fe-Al-based alloy in [1]. The chemical analysis showed a low concentration of Nb in the matrix and high in the eutectic. TEM images of the sample Ti-0 showed nanocrystalline morphology (Fig 2a). TEM analysis of the sample Ti-7 revealed cuboids of the off-stoichiometric Heusler Fe_2TiAl phase (crystal lattice $L2_1$) with the side lengths of 50 nm embedded in the bcc Fe-Al-(Ti) matrix (crystal lattice A2) (Fig 2b). TEM analysis of the sample Ti-4-Nb-4 shown in Fig. 2c from the dendrites that are Nb depleted (dark areas in SEM in Fig. 1c) using the selective electron diffraction also points to a superstructure but needs to be further evaluated. It seems that addition of some alloying elements like Ti can lead to more regular substructure within the polycrystalline grains that might resemble the superstructure of the Ni-based superalloys.

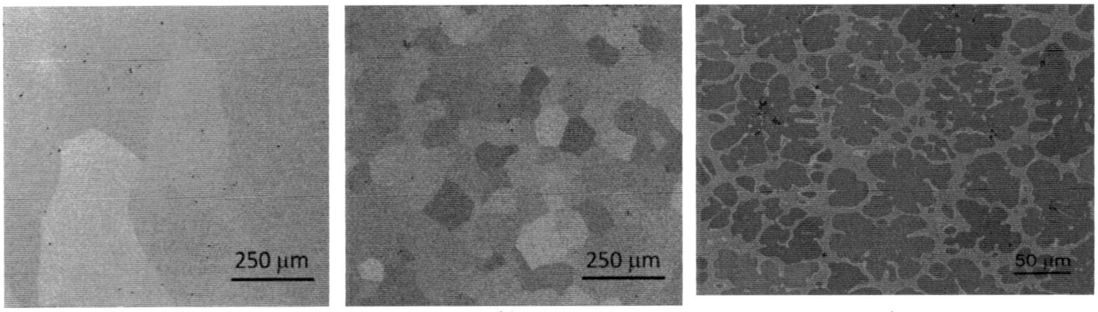

a) b) c)

Fig. 1. SEM micrographs of "as prepared" state of samples SEM back-scattered electrons a) Ti-0, b) Ti-7, c) Ti-4-Nb-4.

a) b) c)

Fig. 2. TEM micrographs of "as prepared" state of samples: a) a nanocrystalline subgrain morphology of the alloy Ti-0 [4]; b) cuboid regions with the side lengths of ~50 nm of the alloy Ti-7 [4] c) morphology of the Nb depleted area of the alloy Ti-4-Nb-4 (dark areas in SEM in Fig. 1c).

3.2 Small punch test

Dependence of force F on punch displacement v measured by small punch tests at the temperature of 600 °C is shown in Fig. 3. A notable yielding and higher plasticity of the Ti-0 alloy is obvious in comparison with the Ti-7 and Ti-4-Nb-4 alloys. The value of maximum recorded force can be used as an indicative parameter for the strength of material tested in small punch at constant deflection rate. The yield and ultimate tensile strength can be estimated using the correlations proposed by García et al. [7] or Purmenský and Matocha [8] listed in Table 2:

Table 2. Empirical correlation formulas for determination of strength from SPT.

Yield tensile strength R_e (MPa)		Ultimate tensile strength R_m (MPa)		Authors
$R_e^{SPT} = 0.346 \left(\frac{F_e}{h_0^2} \right)$	(1)	$R_m^{SPT} = 0.277 \left(\frac{F_m}{u_m \cdot h_0} \right)$	(2)	García et al. [7]
$R_e^{SPT} = 1.67 F_e - 5.6$	(3)	$R_m^{SPT} = 0.35 \left(\frac{F_m}{u_m \cdot h_0} \right) + 23$	(4)	Purmenský and Matocha [8]

where F_m is the maximum force, u_m is the deflection at F_m and h_0 is the initial disc thickness. In this case of lower ductility alloys for simplification we consider the punch displacement equal to disc deflection. The value of $R_m^{SPT} = 154$ MPa is obtained for the alloy Ti-0 by Eq. 2. This is in good agreement with the $R_m = 165$ MPa obtained by a single mini tensile test at the same temperature [4]. The values of the ultimate tensile strength that can be obtained from the indicative data of Ti-7 and Ti-4-Nb-4 curves in Fig. 3 by the same correlation (Eq. 2) is equal to 979 MPa and 1088 MPa respectively. Table 3 shows a significant increase in the static strength of Ti-7 and Ti-4-Nb-4 compared to the Ti-0 alloy.

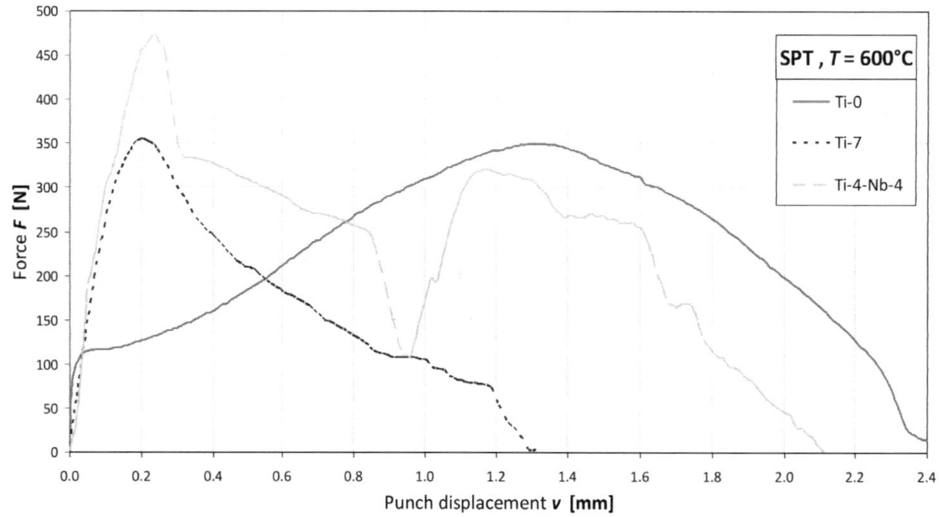

Fig. 3. SPT curves of studied alloys at 600 °C.

Table 3. Strength determined by means of empirical correlation formulas.

Alloy	R_e^{SPT} (MPa)	R_m^{SPT} (MPa)	Authors of formulas
Ti-0	169	154	García et al. [7]
	186	217	Purmenský and Matocha [8]
Ti-7	463	979	García et al. [7]
	556	1260	Purmenský and Matocha [8]
Ti-4-Nb-4	464	1088	García et al. [7]
	559	1397	Purmenský and Matocha [8]

3.3 Small punch creep test

The SPC tests were performed at 600 °C for all three alloys and also at 700 °C and 800 °C for Ti-7 and 800 °C for Ti-4-Nb-4 alloy. The time to rupture dependence on force is shown in Fig. 4a and the minimum deflection rate in Fig. 4b. The highest creep resistance has the Ti-4-Nb-4 followed by the Ti-7 followed by the Ti-0 alloy. The rate and rupture time comparison shows that the creep performance of the Ti-4-Nb-4 at 800 °C and Ti-7 alloy at 700 °C is similar as for Ti-0 alloy at 600 °C. Interesting is the comparison of the minimum deflection rates during a single test of Ti-7 alloy at 600 °C with stepped loading from 150 to 300 N (open squares in Fig. 4b) with the multiple tests with instant loading 225-300 N (filled squares in Fig. 4b). It suggests that instant high loading causes early cracking of the disc, possibly already in the loading phase. In contrary small load at the beginning of the test and small load increments prevent this early specimen cracking and result in the lower deflection rates. This effect is notable at load 250 N and higher. The SPC force and deflection rate can be converted to equivalent stress and creep strain rate according to the methods described in [9-12] and compared with other creep resistant materials; this is however out of scope of this paper.

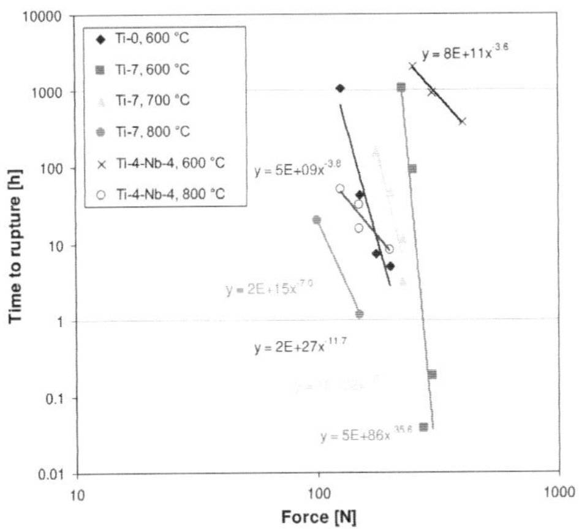

a)

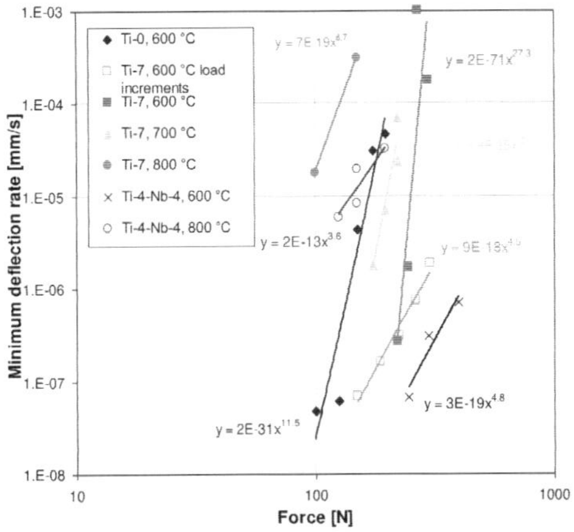

b)

Fig. 4. SPC results a) time to rupture vs. force, b) minimum deflection rate vs. force.

3.4 Fractographic analysis

The fractographic analysis of ruptured SPC discs is shown in Fig. 5. The Ti-7 and Ti-4-Nb-4 alloys rupture at grain boundaries. Most notable it is for Ti-7 alloy in Figs. 5b and 5e; this specimen cracked even in the clamped area. The fracture of Ti-0 alloy is mixed due to large grains that are often lager than the specimen thickness 0.5 mm. The grain boundaries become well visible on the deformed sombrero-like specimen in Fig. 5d. The fracture surfaces in Figs. 5e and 5f correspond well with the microstructures.

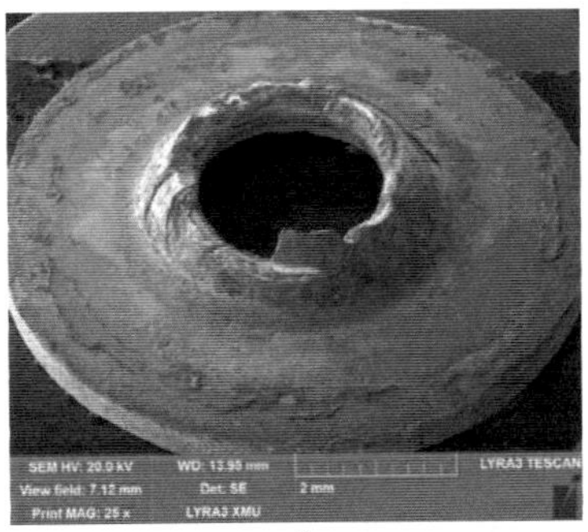

a)

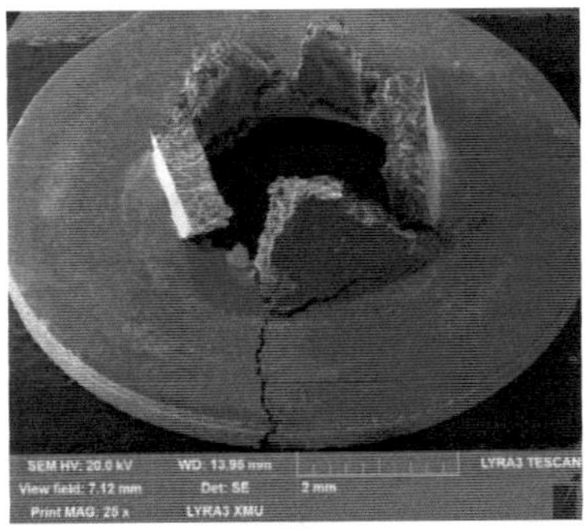

b)

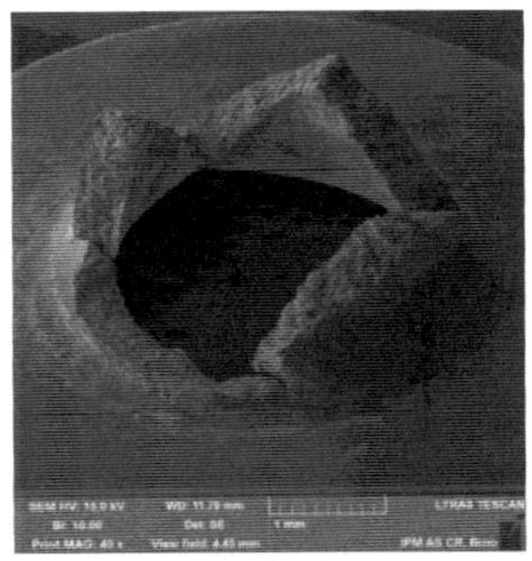

c)

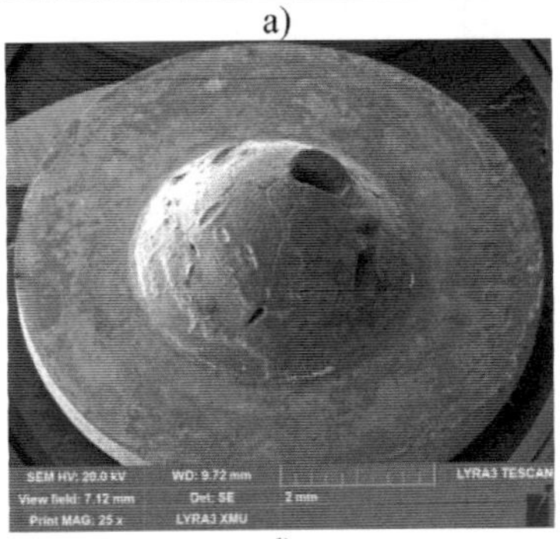

d)

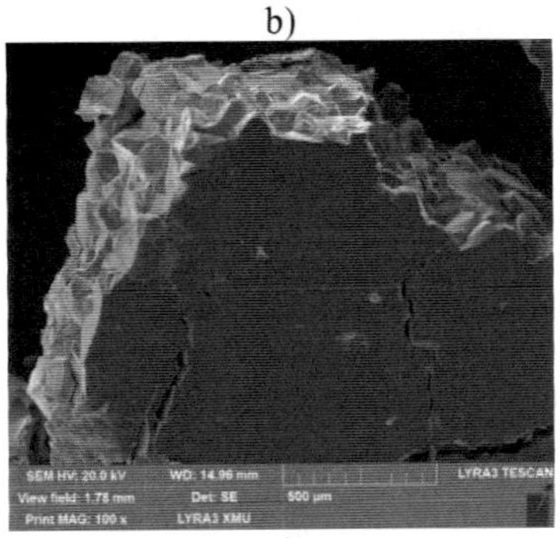

e)

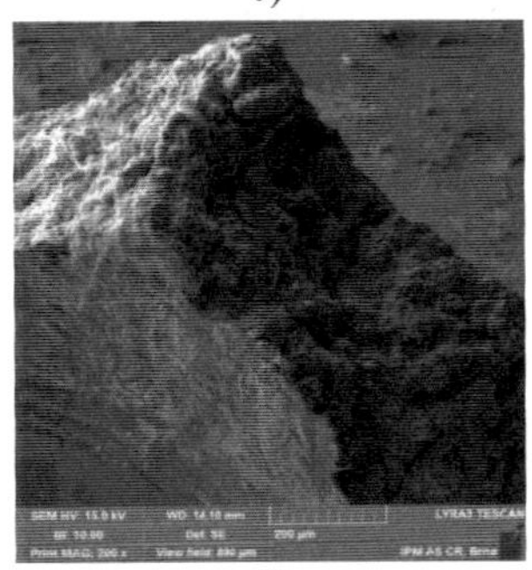

f)

Fig. 5. SPC specimens after the tests at 600 °C: a) Ti-0, 125 N, b) Ti-7, 300 N, c) Ti-4-Nb-4, 400 N, d) Ti-0, 100N sombrero like specimen from interrupted test, e) Ti-7 fracture, f) Ti-4-Nb-4 fracture.

4 Conclusions

The small punch testing of three Fe-Al based alloys with and without Ti and Nb additions enabled to determine and compare their strength and creep behavior using specimens prepared from very small button ingots. The results show that a significant improvement in the mechanical properties can be reached by adding a small amount of alloying elements. A superstructure like nanocrystalline morphology was found in Ti-7 alloy, however the limiting factor for the mechanical properties is the strength of the grain boundaries and brittleness even at the experimental temperature of 600 °C. This seems to be improved for Ti-4-Nb-4 alloy despite the superstructure was not clearly visible in TEM.

Acknowledgements

The authors thank Dr. D. Janičkovič (Institute of Physics, SAS Bratislava, SK) for cast ingots preparation. Financial support from the Czech Science Foundation (project No. 17-22139S) and the use of infrastructure within the project CEITEC 2020 (LQ1601) with financial support from the Ministry of Education, Youth and Sports of the Czech Republic under the National Sustainability Programme II is gratefully acknowledged.

References

[1] Moszner F, Peng J, Suutala J, Jasnau U, Damani M and Palm M 2019 *Metals* 9 847
[2] Jiraskova Y, Pizurova N, Titov A, Janickovic D and Friak 2018 *J. Magn. Magn. Mat.* 468 91–99
[3] Jiraskova Y, Pizurova N, Friak M, and Janickovic D Structural and magnetic study of Fe-Al20-(Ti, Nb) alloys. *Acta Phys. Polonica A* (submitted)
[4] Dymáček P, Dobeš F, Jirásková Y, Pizúrová N and Friák M 2019 *Theor. Appl. Fract. Mech.* 99 18–26
[5] Dobeš F and Milička K Small punch testing in creep conditions 2001 *J. Test. Eval.* 29 31–35
[6] European Committee for Standardization: Small Punch Test Method for Metallic Materials, CWA 15627. Part A: A Code of Practice for Small Punch Creep Testing and Part B: A Code of Practice for Small Punch Testing for Tensile and Fracture Behaviour, Documents of CEN WS21; Brussels, Belgium, 2007.
[7] García T E, Rodríguez C, Belzunce F J and Suárez C 2014 *J. Alloys Compd.* 582 708–717.
[8] Purmenský J and Matocha K Zkoušení malých vzorků ve fyzikální metalurgii, Metal 2001 10th International Metallurgical and Materials Conference 15.-17. 5. 2001, Ostrava, Czech Republic pp.1-13
[9] Holmström S, Li Y, Dymacek P, Vacchieri E, Jeffs S P, Lancaster R J, Omacht D, Kubon Z., Anelli E., Rantala J, Tonti A, Komazaki S, Naveena, Bruchhausen M, Hurst R.C, Hähner P, Richardson M and Andres D 2018 *Mater. Sci. Engng. A* 731 161–172
[10] Dymáček P, Li Y, Dobeš F and Stevens P 2019 *Mater. High Temp.* 36:2 157-164
[11] Dobeš F, Dymáček P and Friák M 2018 *Kovove Mater.* 56 205–212
[12] Dobeš F, Dymáček P and Friák M 2018 *IOP Conf. Ser.: Mater. Sci. Eng.* 461 012017

Intermetallics formation during hot dip galvanizing of high carbon steel

GOGOLA Peter, GABALCOVÁ Zuzana, SUCHÁNEK Henrich, KUSÝ Martin

Institute of Materials Science, Faculty of Material Science and Technology, Slovak University of Technology, Ulica Jána Bottu č. 2781/25, 917 24 Trnava, Slovakia, EU

peter.gogola@stuba.sk

Abstract. From the moment of immersion, the reaction between molten zinc and solid steel starts forming intermetallic phases. This diffusion-controlled process is largely responsible for the final phase composition of zinc coatings. Several literature sources describe this phenomenon for interstitial free steels, but high-carbon steels are rarely being used as substrates. Therefore, in this work high-carbon steel substrates were used. Multiple samples were created by hot-dipping at various immersion temperatures ranging from 450 to 490 °C and times from 24 to 60 s to investigate mainly the morphology of the obtained intermetallic phase layers. Investigation was carried out mainly by SEM on 20 condition, where several hundred sites were investigated in total to achieve statistically relevant information. It was found that while increasing the immersion time influences mainly the thickness of individual intermetallic phase layers, the temperature influenced mainly their morphology. It was also observed that these results are significantly different compared to ones found in literature for interstitial free steels.

Keywords: zinc, iron, intermetallic phases, high-carbon steel

1. Introduction

For specific applications high-carbon / high-strength steels are required [1, 2]. Hot-dip Zn coatings are applied for sufficient corrosion protection [3, 4]. Pure Zn is used to enable the formation of intermetallic layers as these are also adding to the corrosion protection and mechanical properties of the coating [3-6]. Many applications of hot-dip Zn coatings utilize alloying with Al. However, Zn coatings with as low as 0.2 wt.% Al are not forming a Fe-Zn based, but rather a very thin Fe-Al based intermetallic layer [3, 7].

Various hot-dip coating procedures may require different coating temperatures. These may be caused for example by heat capacity requirements on the hot-dip bath, or time constrains given by the production planning. This may cause that for various cases, the hot-dip temperature can vary between 450 up to 490 °C.

Due to these considerations it was decided to observe the formation of intermetallic layers at different hot-dip temperatures.

Similar research considering different temperatures and dipping times are usually found in the literature. However, these are mainly focused on low-carbon steels and give only a limited insight into the problematic of high carbon steels [3, 4, 8-11].

Based on the results summarized by Smirnov [3], both Si and C present in the steel substrate have the tendency to change both thickness and morphology of the Fe-Zn intermetallic layers.

2. Experimental procedure

A small hot-dip bath with about 5 kg of pure Zn melt was prepared in a vertical muffle furnace. In total 20 conditions were planned and executed. Hot-dipping was performed at various immersion temperatures ranging from 450 to 490 °C and immersion times were ranging from 24 to 60 s. Temperature of the melt was constantly monitored by a K-type thermocouple inserted directly into the melt.

Substrates of high-carbon steel (main alloying elements in wt.%: 0.92 C + 0.21 Si + 0.3 Mn + 0.2 Cr) with a thickness of 4 mm were cleaned by pickling in 20 vol.% aqueous solution of HCl acid for one minute followed by mechanical wiping and dipping in an aqueous solution of 150 g "double-salt" flux ($ZnCl_2.2NH_4Cl$) in 1000 ml of demineralized water for about 30 s. Samples were removed from the flux bath and left to dry on air.

As soon as the substrate was dry, it was dipped into the Zn melt. The sample was held in the Zn bath for the required amount of time while it was agitated to allow constantly fresh Zn melt to be in contact with its surface.

Samples were cooled in water immediately after withdrawal from the Zn-bath.

The procedure was repeated until two individual samples for each of the twenty conditions were available.

Microstructural analysis was performed by scanning electron microscopy (SEM) on cross sections of hot-dip samples. Samples were prepared using standard metallographic techniques including grinding on SiC papers, polishing on diamond paste and etching in a reagent of 0.5 vol.% Nital. Final grinding steps and complete polishing procedure had to be done using alcohol-based coolants and diamond pastes. Quantitative metallographic analysis including thickness measurements and determining the volume fraction of coexisting phases was performed by computerised image analysis using ImageJ 1.52 [12] and Gimp 2.10 software.

Only the total thickness of the intermetallic layers was measured. Thickness of the η layer was not investigated. Thickness of this layer is not controlled by diffusion (as in case of Γ, δ and ζ phases), but rather by dynamic forces acting on the coated steel substrate during raising from the Zn bath. This was out of scope for the current research.

Due to the rather uneven nature of the individual intermetallic layers observed, it was decided to follow up their area fraction instead of thickness. Three main microstructure features were distinguished as shown in figure 1.

Grains and areas corelated to individual phases or their morphological variations were manually traced out and filled in with corresponding colours as seen in figure 1. This enabled to calculate their corresponding area fraction and hence volume fraction. Such measurements were carried out on approx. 600 images in total as several sites were investigated on each sample.

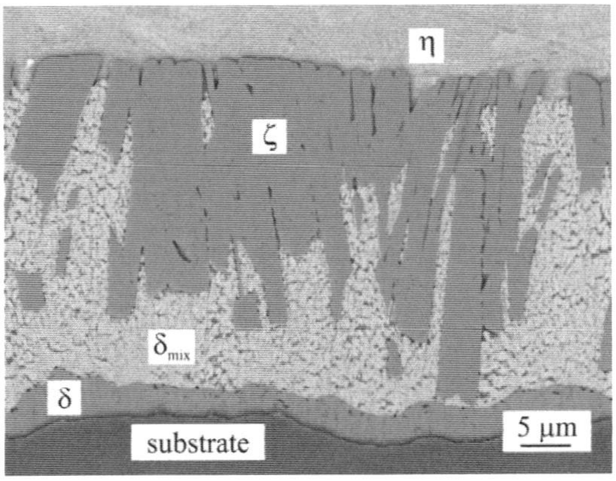

Figure 1. Example of an image processed for area measurement; gray: η – η phase layer, blue: ζ – particles and areas formed by ζ phase, green: δ_{mix} – areas formed by δ phase, but with a significantly different morphology, red: δ – compact δ layer.

Corelating these data with the overall intermetallic layer thickness enabled to calculate a geometrically idealized thickness for the individual microstructure features neglecting their uneven nature. These values were calculated simply by dividing the total thickness of the intermetallic layers into sections corresponding to the volume fraction of each microstructure feature. This way it was possible to visualise the influence of both immersion time and temperature much more clearly (as will be shown later in figure 8 and figure 10).

3. Results

A typical microstructure observed during our experiments is shown in figure 2. Main point of interest for this article was to observe the construction of the Fe-Zn intermetallic layers formed on a high-carbon steel substrate during the hot-dip process. EDX measurements were performed to confirm the composition of the individual phases observed (figure 2,
table 1). Following phases were distinguished based on inputs from literature: η (0.03 wt.% Fe), ζ (5.7-6.3 wt.% Fe) and δ (7.0-11.0 wt.% Fe) while the Γ phase (20.0-27.0 wt.% Fe) was not identified during our measurements. Basic properties of the investigated phases are listed in table 2 [6]. Based on this information, increasing the δ phase content should be preferred from a mechanical point of view.

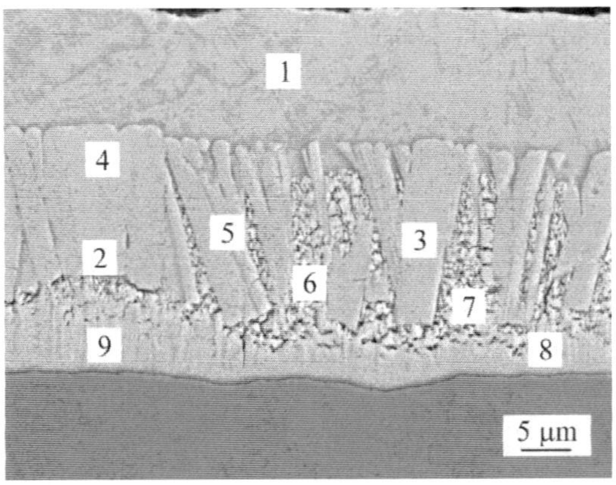

Figure 2. SEM image from the sample produced at 470 °C and 60 s; including EDX measurement points marked in the image.

Table 1. Chemical composition of the individual layers as measured by SEM/EDX.

Spectrum	Fe [wt.%]	Zn [wt.%]	Identified phase
1	0.42	99.58	η
2	6.92	93.08	ζ
3	6.65	93.35	ζ
4	6.17	93.83	ζ
5	6.55	93.45	ζ
6	7.93	92.07	δ
7	8.35	91.65	δ
8	10.59	89.41	δ
9	11.39	88.61	δ

Table 2. Basic properties of phases observed

	Phase	Fe content [wt.%]	Melting point [°C]	Crystal Structure	micro HV	Characteristic mechanical property
η	**Zinc**	<0.03	419	hexagonal	70-72	soft, ductile
ζ	**$FeZn_{13}$**	5.7-6.3	530	monoclinic	175-185	hard, brittle
δ	**$FeZn_7$**	7.0-11.0	530-670	hexagonal	240-300	ductile
Γ	**Fe_3Zn_{10}**	20.0-27.0	670-780	cubic	---	hard, brittle

3.1. Hot-dip galvanizing at 450°C

For all conditions prepared at 450 °C (figure 3) the compact δ layer and a rather constant ζ layer is visible. The δ layer is only locally interrupted as seen in figure 3b and d. ζ layer seems to have particles "floating" into the η layer (molten Zn during the experiment) as seen in figure 3b - d. Areas labelled δ_{mix} are only rarely observed. About 3 vol.% were measured at 60 s (figure 3d).

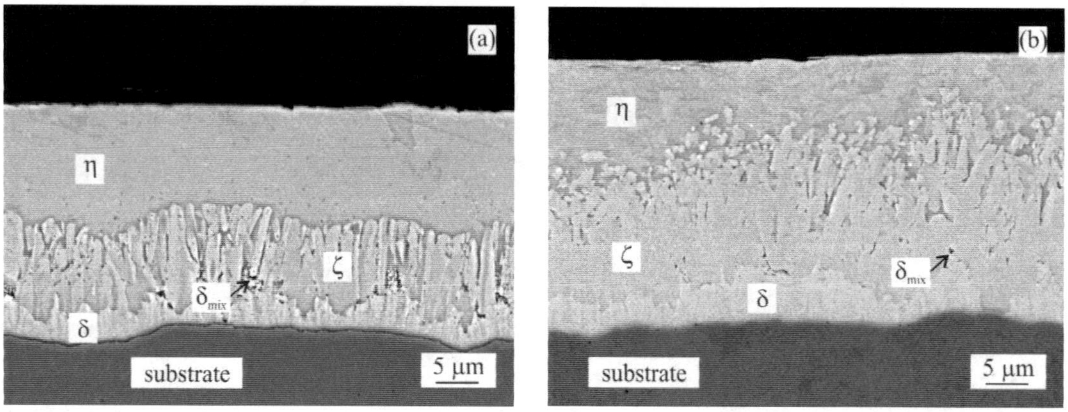

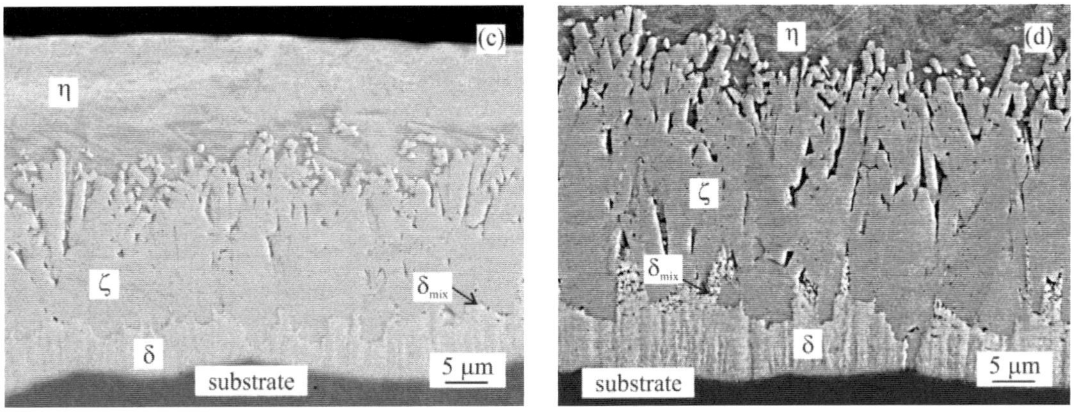

Figure 3. Cross section of samples produced at 450 °C: a) 24 s; b) 30 s; c) 40 s; d) 60 s.

3.2. Hot-dip galvanizing at 460°C

The conditions prepared at 460 °C (figure 4) are very similar to the ones prepared at 450 °C (figure 3). There was a slight increase in the volume fraction of δ_{mix} areas (as shown later in figure 9). Also, the ζ layer seems to be less compact. More individual particles are observed "floating" in the area of η phase (figure 4).

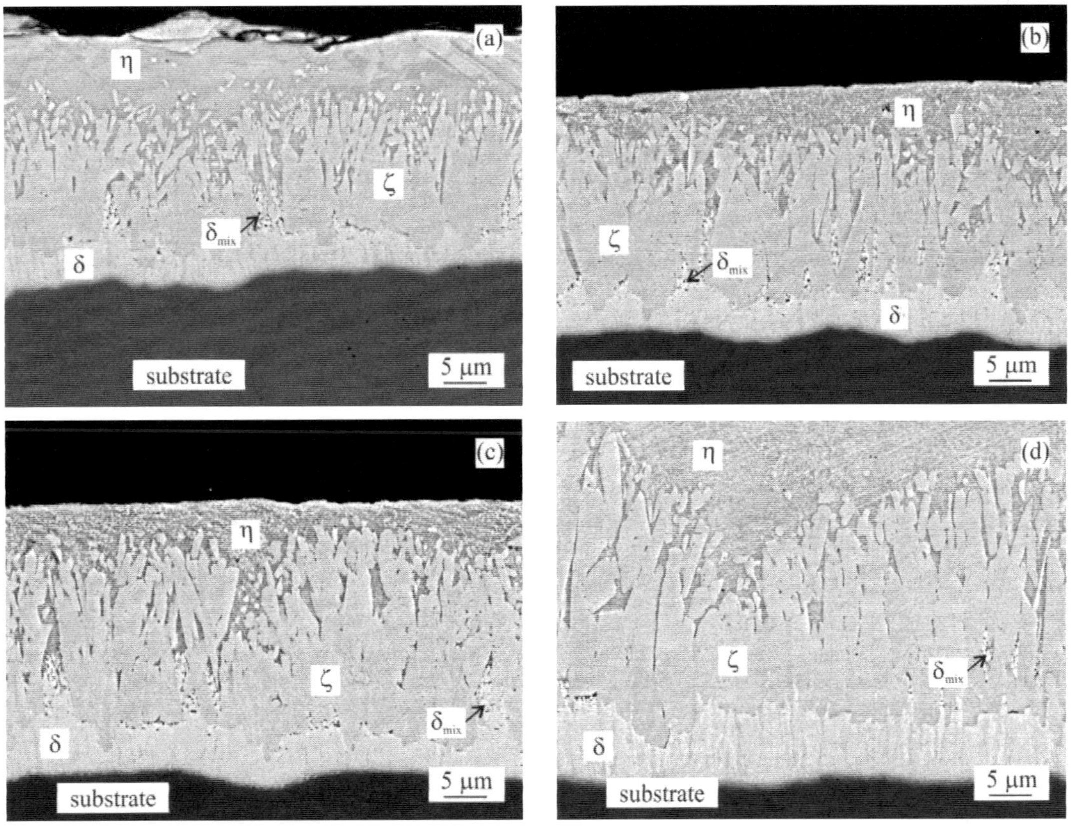

Figure 4. Cross section of samples produced at 460 °C: a) 24 s; b) 30 s; c) 40 s; d) 60 s.

IOP Conf. Series: Materials Science and Engineering **723** (2020) 012007 doi:10.1088/1757-899X/723/1/012007

3.3. Hot-dip galvanizing at 470°C

At 470 °C (figure 5) the fraction of δ_{mix} grew significantly. It is now clearly growing in between individual particles of ζ. On the other hand, no more "floating" ζ particles are observed compared to lower temperatures. The compact δ layer seems to be unchanged compared to previous conditions.

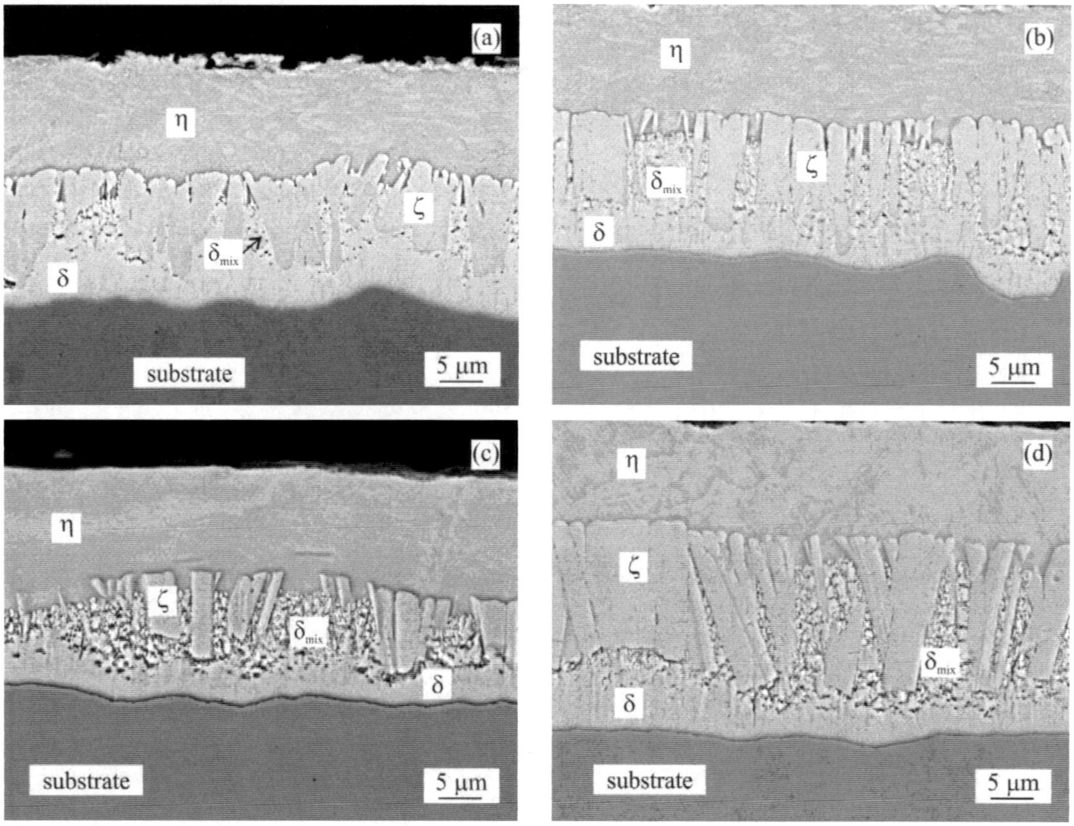

Figure 5. Cross section of samples produced at 470 °C: a) 24 s; b) 30 s; c) 40 s; d) 60 s.

3.4. Hot-dip galvanizing at 480°C

Conditions prepared at 480 °C (figure 6) are very similar to the ones prepared at 470 °C (figure 5). Volume fraction of δ_{mix} grew again mainly at the expense of ζ particles. Volume fraction of ζ particles in now only less than 30 % for the conditions prepared at 24, 30 and 40 s (figure 6a - c).

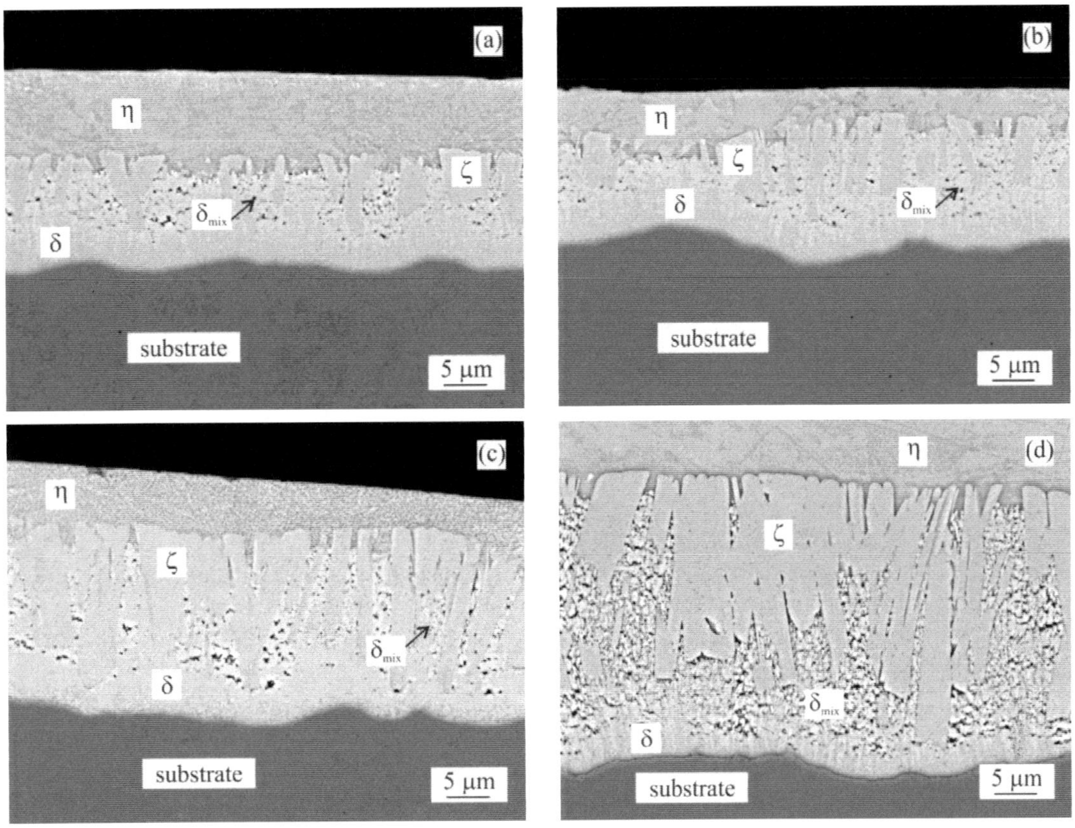

Figure 6. Cross section of samples produced at 480 °C: a) 24 s; b) 30 s; c) 40 s; d) 60 s.

3.5. Hot-dip galvanizing at 490°C

At 490 °C (figure 7) ζ particles were not observed at all for immersion times from 24 to 40 s (figure 7a - c). Most probably due to this fact also areas of δ_{mix} could not be distinguished. Only the compact δ layer was observed. At 60 s (figure 7d) all previously identified microstructure features were observed again.

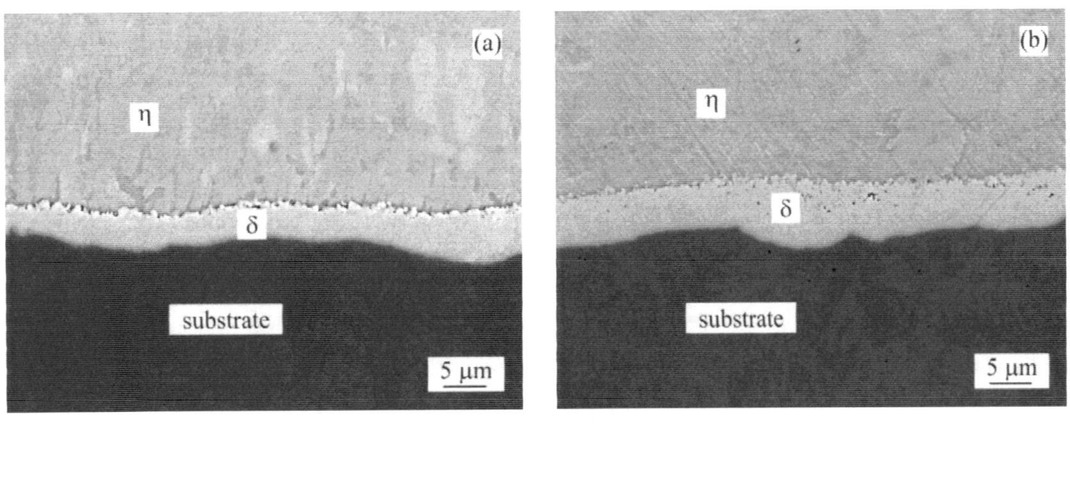

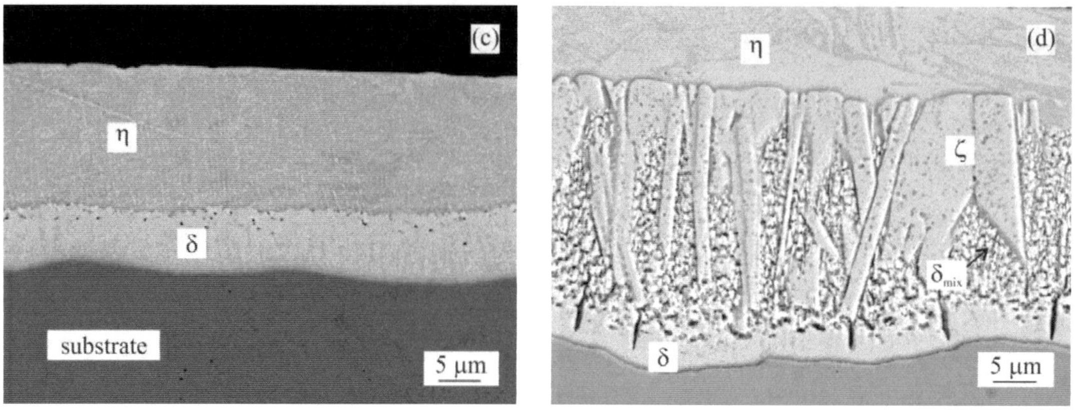

Figure 7. Cross section of samples produced at 490 °C: a) 24 s; b) 30 s; c) 40 s; d) 60 s.

3.6. *Summary of measured data*

Measured data can be summarized rather clearly as shown below in figure 8 and figure 9. In figure 8 it can be observed that for each temperature, geometrically idealized thickness of both δ and ζ phases is increasing with increasing immersion time. While as mentioned earlier, all areas formed by the δ phase were divided into areas of δ and δ_{mix} (for example figure 7d).

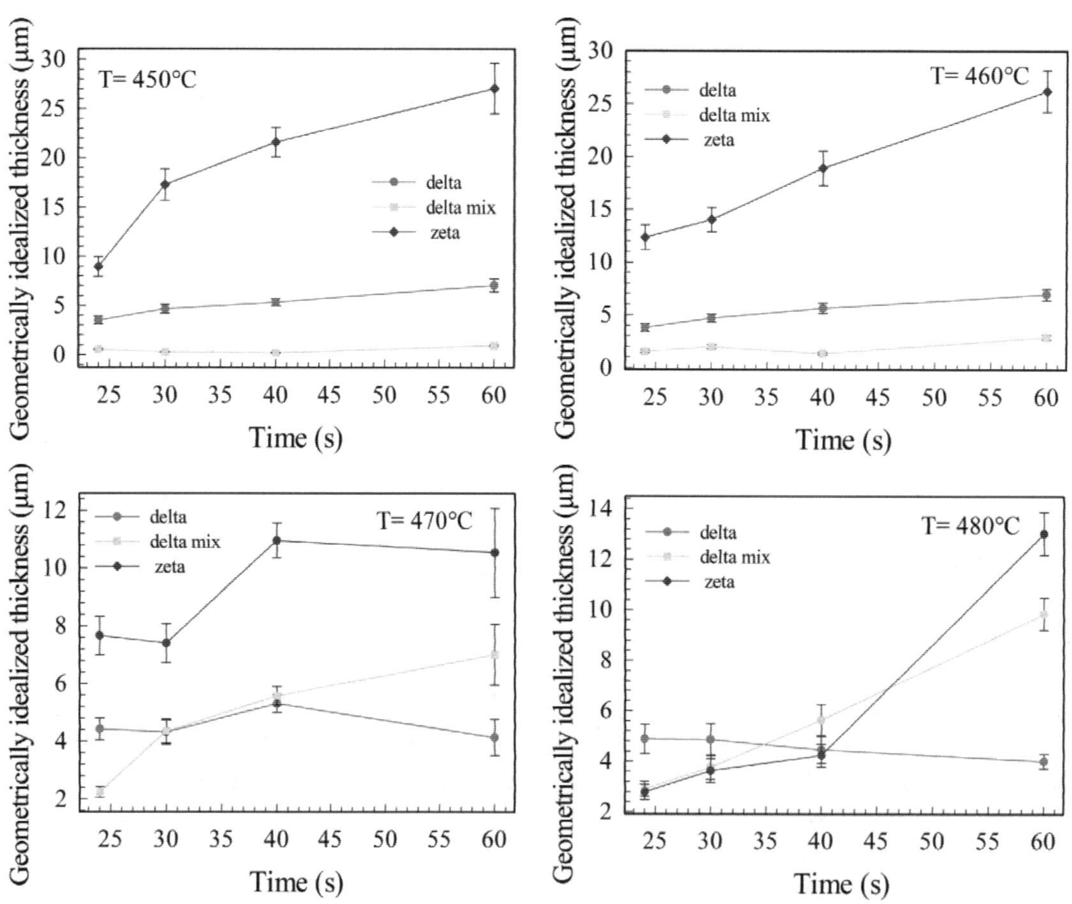

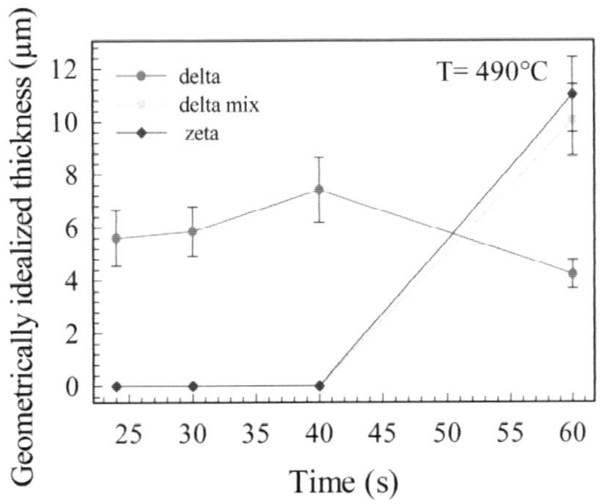

Figure 8. Geometrically idealized thickness calculated based on volume fraction at different temperatures: a) 450 °C; b) 460 °C; c) 470 °C; d) 480 °C; e) 490 °C.

The immersion time has a very significant influence on the overall intermetallic layer thickness. Therefore, only volume fraction of the individual microstructure features was used to assess the influence of temperature (figure 9). For the immersion times of 24, 30 and 40 s (figure 9a - c), increasing temperature promoted mainly an increase in the volume fraction of δ and δ_{mix} areas. Both increased their volume fraction at the expense of ζ phase areas. For all these conditions (figure 9a - c) the complete disappearance of ζ phase areas was observed at 490 °C. On the contrary, at the immersion time of 60 s (figure 9d) volume fraction of the compact δ layer is slightly decreasing with the increasing immersion temperature. Also, the volume fraction of ζ phase is decreasing significantly less. Even at 490 °C, over 40 vol.% of the microstructure is still formed by the ζ phase.

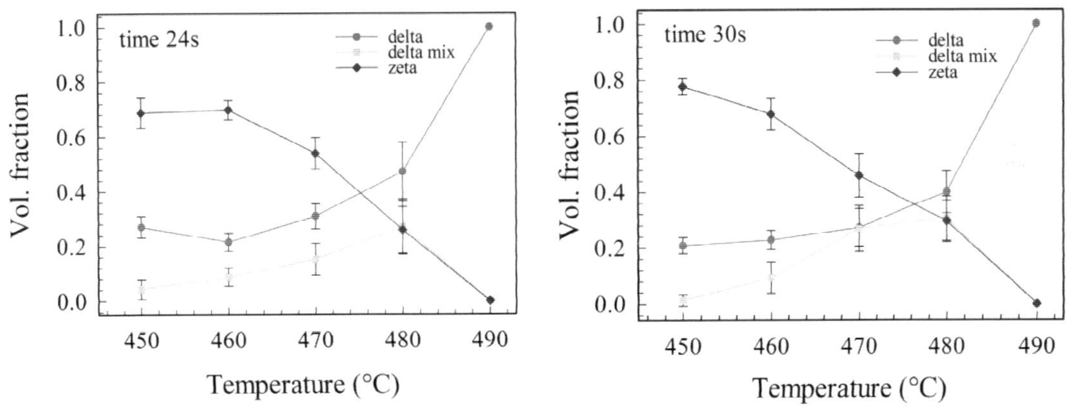

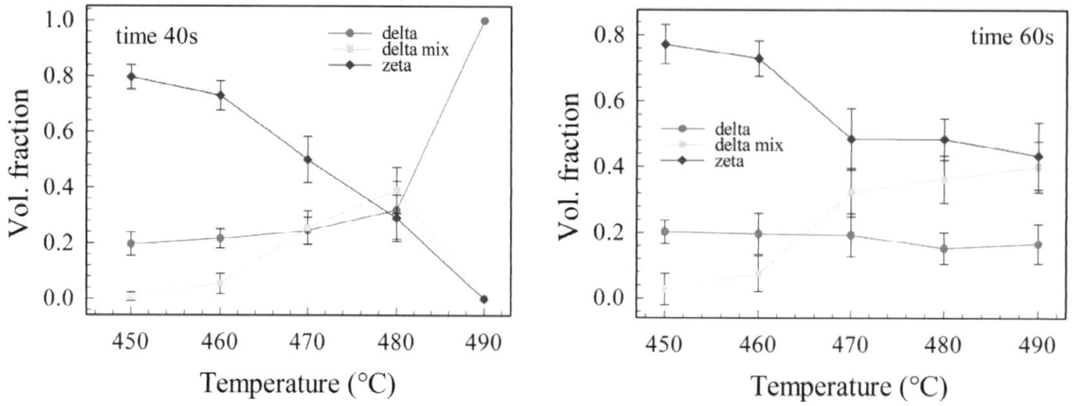

Figure 9. Volume fraction of individual microstructure features: a) 24 s; b) 30 s; c) 40 s; d) 60 s.

4. Discussion

Based on SEM / EDX measurements, it was found that the individual phases can be determined most reliably using BSE images. Their morphologies are clearly recognizable.

There is still lack of information about intermetallic layer's formation on hot-dip galvanised high-carbon steels. Literature sources deal mostly with low-carbon, or interstitial free steels as substrate [8, 9, 13-15]. Both thickness measurements and microstructures are different compared to our findings. Bakhtiari et.al. [9] identified maximum intermetallic layer thickness to be approx. 37 μm at 480 °C and 60 s immersion. These results were very similar to our measurements of approx. 36 μm, but ours was found already at 460 °C and 60 s. On the other hand, Bicao et.al. [8] found a significantly higher maximum thickness of approx. 60 μm, but again at 480 °C and 60 s similar to Bakhtiari [9].

We suspect, that our findings are influenced by higher Si (0.21 wt.%) as well as increased C (0.92 wt.%) content in our substrate material. Both of these elements are reported to influence the morphology as well as overall thickness of the alloy layer [3, 4].

Results of Kuklik [4] suggest that for interstitial free steels the maximum intermetallic layer thickness should be achieved between 480 to 490 °C. For a steel with over 0.12 wt. % Si this should be observed at 440 °C. These findings would correlate with the differences between our research and the results of both Bicao and Bakhtiari [8, 9].

Our overall morphology of the intermetallic layers was also different compared to those observed on low-carbon steel, or pure iron [4, 9, 8, 13]. Main difference was observed in the presence of the δ_{mix} areas. Similar microstructures were only observed in the work of Smirnov [3]. Also, rather good overlap was observed with the images taken by Kopyciński et.al. [10, 11] on hot-dip coated cast iron parts. Morphology of the intermetallic phases is comparable to our findings observed at 450 °C. However, overall intermetallic layer thickness is reported to be 42% lower on average. This is most probably caused by the rather different shape of their samples. Even the increased reactivity of their cast iron substrate (over 3 wt.% of C and over 2 wt.% of Si) could not counter act the need to heat the larger bulk.

In the work of Bakhtiari et. al. [9] also significant portion of Γ layer was identified. Thickness of this layer was ranging from 2 to 10 μm at 450 to 480 °C respectively. Similar results were reported by Šumec et. al. [13] with approx. 1.2 μm of Γ layer measured at 450 °C and 60 s. It was not possible to identify any Γ layer in our samples. This was probably caused by the increased carbon content of our steel. As reported by Smirnov [3], carbon can delay the formation of the Γ layer up to 7 min at 450 °C in high-carbon steels with 0.9 wt.% of C. As our immersion time was maximum 60 s, Γ layer probably did not form yet.

In samples prepared at 450 and 460 °C particles of the ζ layer seem to be "floating" away into the melt, hence the overall thickness is hard to measure. This also shows that the overall thickness of the intermetallic layer might be even more, but some parts of the ζ layer got detached and were washed away by the molten Zn. The same phenomena was observed also by other authors [4, 10, 11, 14]. These "floating" particles are not visible for temperatures above 460°C.

It can be observed that for each temperature, overall intermetallic layer thickness is increasing with increasing immersion time. On the contrary, overall intermetallic layer thickness is decreasing with increasing temperature as seen in figure 10.

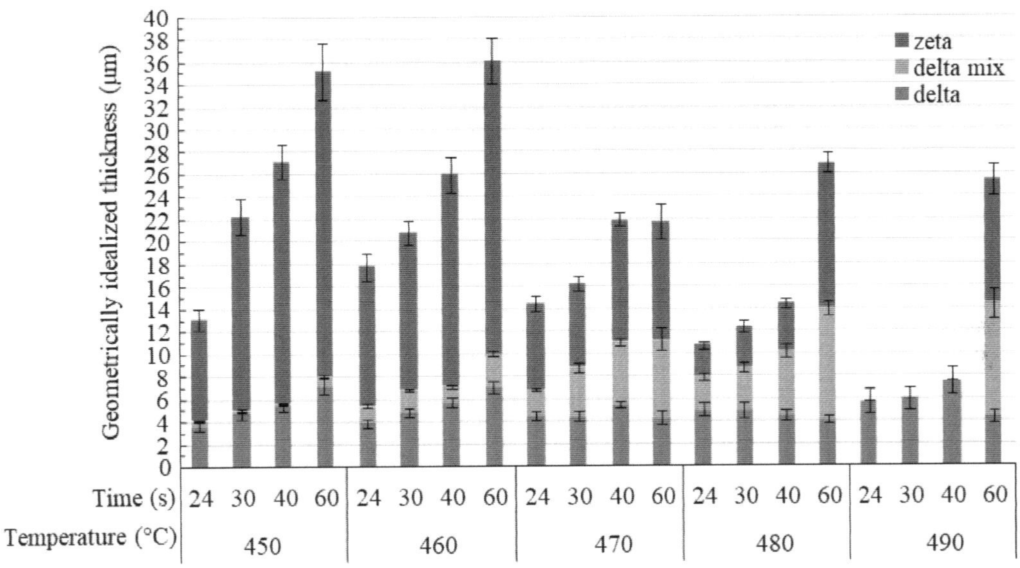

Figure 10. Geometrically idealized, total intermetallic layer thickness for all conditions investigated.

5. Conclusions

- Increasing dipping time from 24 to 60 s increases the intermetallic layer thickness, while raising the Zn-bath temperature from 450 up to 490 °C decreases the volume fraction of ζ phase, while volume fraction of δ phase is increasing.
- The δ phase is reported to be more ductile [3, 6], hence higher volume fraction of δ phase should be beneficiary for overall coating quality. Therefore, hot-dip galvanising at 480 – 490 °C for 24 – 40 s seem to be optimal parameters for industrial use.

Acknowledgements

This work was financially supported by the Project of VEGA Grant Agency of the Ministry of Education, Science, Research and Sport of the Slovak Republic and Slovak Academy of Sciences, No. 1/0490/18: "The effect of microstructure and phase composition on corrosion resistance of hot dip alloys". This publication is the result of the project implementation: CE for development and application of advanced diagnostic methods in processing of metallic and non-metallic materials - APRODIMET, ITMS:26220120048, supported by the Research & Development Operational Programme funded by the ERDF.

References

[1] Jie L, Hongbo L and Zhihua C 2017 *J. Constr. Steel Res.* **136** 110–27
[2] Singh R P, Mousumi M and Verma M K 2016 *Eng. Fail. Anal.* **70** 290–304
[3] Smirnov A V 1956 *Zinkování v Roztaveném Zinku* (Praha: Státní nakladatelství technické literatury) p 285
[4] Kuklik V and Kudlacek J 2014 *Žárové Zinkování* (Ostrava: Asociace českých a slovenských zinkoven) p 208
[5] Shibli S M A, Meena B N and Remya R 2015 *Surf. Coat. Tech.* **262** 210–5
[6] Dallin G W 2015 *Galvanizing 2015. Galvinfo Center – A Program of the International Zinc Association* **November 2015** p 8
[7] Maeda S 1996 *Prog. Org. Coat.* **28** 227–38
[8] Bicao P, Jianhua W, Xuping S, Zhi L and Fucheng Y 2008 *Surf. Coat. Tech.* **202** 1785–8
[9] Bakhtiari A, Toroghinejad M R and Ashrafizadeh F 2014 Metall. Mater. Eng. **20** 41–52
[10] Kopyciński D, Guzik E and Szczęsny A 2016 *Arch. Foundry Eng.* **16** 25–8
[11] Kopyciński D, Guzik E, Szczęsny A and Siekaniec D 2015 *Arch. Foundry Eng.* **15** 43–6
[12] Schneider C A, Rasband W S and Eliceiri K W 2012 *Nat. Methods* **9** 671–5
[13] Šumec M, Babinec M, Gogola P, Gondek J, Kusý M, Suarez L and Čička R 2018 *Metal 2018* (Ostrava: Tanger Ltd.) p 1270–4
[14] Jordan C E and Marder A R 1997 J. Mater. Sci. **32** 5593–602
[15] Jianhua W, Hao T, Bicao P, Xinming W, Fucheng Y and Xuping S 2009 *Mater. Charact.* **60** 1276–9

IOP Conf. Series: Materials Science and Engineering **723** (2020) 012008 doi:10.1088/1757-899X/723/1/012008

Heat treatment strategies for hot-rolled medium-Mn sheet steels

A. Grajcar, M. Morawiec, A. Skowronek, J. Dykas

Department of Engineering Materials and Biomaterials, Silesian University of Technology, ul. Konarskiego 18a, 44-100 Gliwice, Poland

Corresponding author: E-mail address: adam.grajcar@polsl.pl

Abstract. Heat treatment strategies for advanced high-strength medium-Mn sheet steels are addressed. A review of the available thermal cycles for cold-rolled and hot-rolled sheets are presented. The benefits from the intercritical annealing following hot-rolling are emphasized. The examples of intercritical annealing for the hot-rolled sheets of initial martensitic microstructure are provided. The heat treatment was performed in a temperature range of 680-720°C for various times from 1 to 5 hours. The material used in the experiment was a medium-Mn steel containing 0.16% C, 4.7% Mn, 1.6% Al, 0.2% Mo, 0.2% Si. The scope of the research included dilatometer tests, hardness tests and microstructural characterization. The effects of temperature and time on a stabilization of retained austenite are assessed. Intercritical annealing performed at 700 and 680°C allowed forming some stable retained austenite in the microstructure. Moreover, some coiling simulations for hot-rolled strips are described.

1. State of the art

Medium-Mn sheet steels are a newest answer of the scientists for growing demands of the automotive industry regarding high-strength, formable, cost effective steel sheets [1]. They offer not only a very satisfactory combination of strength and ductility but meet other important technological requirements concerning formability (stamping, bending, etc.) zinc galvanizing, weldability, etc. [2]. Their typical microstructure consists of ferrite and retained austenite. However, the formation of ferrite takes place under conditions of ultrafine-grained microstructures, which makes the polygonal ferrite very similar in its microstructural features to bainitic ferrite. This means a dislocation density, internal stresses, etc. The crucial structural constituent is retained austenite in the amount of 20-40%. Due to its metastability this phase upon cold straining (during forming) transforms into strain-induced martensite providing the TRansformation Induced Plasticity (TRIP effect) phenomenon. This effect is very well known for first generation AHSS (Advanced High Strength Steels) [3]. Final mechanical properties depend significantly on deformation temperature, strain rate and chemical composition [4]. The latter parameter controls to a great extent the mechanical stability of retained austenite and resulting mechanical properties. In practice, C and Mn contents in retained austenite and its grain size affect a kinetics of strain-induced martensitic transformation. Due to weldability reasons a carbon content is limited below 0.2% (more often to 0.1%) [5]. This makes difficult to obtain retained austenite at room temperature because of high martensite start temperature (M_s) of steel. Typical manganese contents from 5 to 12% allow to reduce the M_s [4]. However, the redistribution of C and Mn is beneficial during heat treatment schedules to effectively maintain the austenitic phase after final cooling to room temperature. Thus, different heat treatment strategies are needed for cold-rolled and hot-rolled steel sheets.

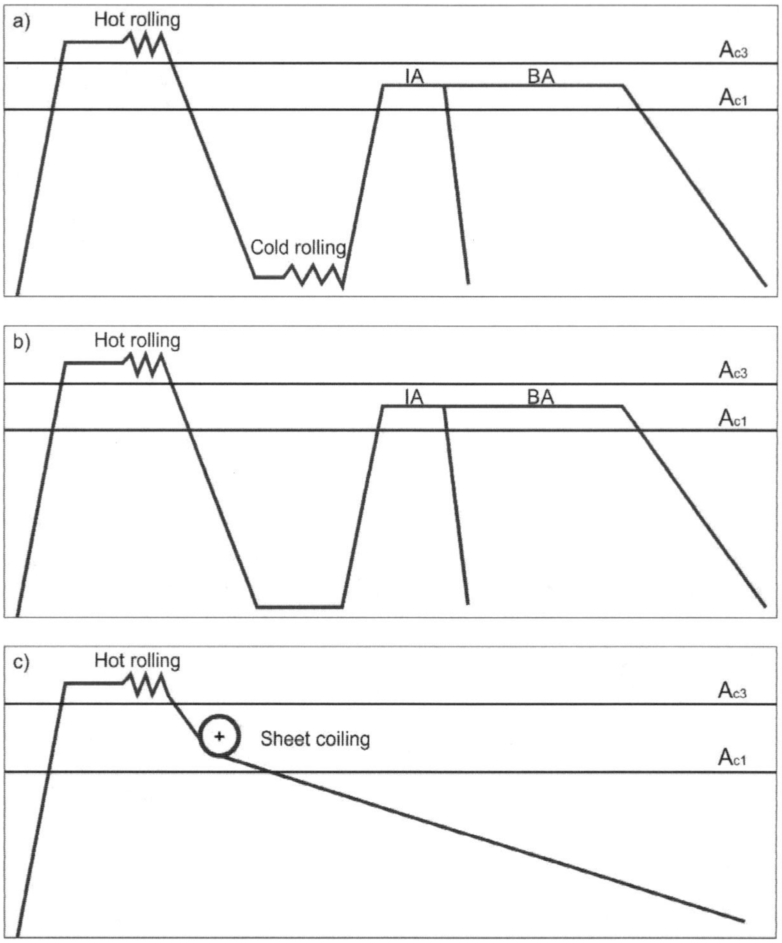

Figure 1. Heat treatment strategies for medium-Mn sheet steels; a) intercritical annealing (IA) or batch annealing (BA) following cold rolling, b) IA or BA following hot rolling, c) air cooling following hot rolling and sheet coiling.

1.1. Annealing of cold-rolled sheets

Typical heat treatment following cold-rolled sheets is shown in figure 1a. The conventional step requires hot rolling and cooling to room temperature. Medium-Mn steels have a martensitic structure after this step due to their high hardenability being a result of high Mn contents. The martensite is formed even at very slow cooling rates. Such a microstructure is very beneficial before an annealing step. This intercritical annealing (between A_{c1} and A_{c3}) is crucial in forming a desired microstructure. The low-C martensite is recovered and next recrystallized very fast forming polygonal ferrite. The ferrite is very fine-grained because it nucleates at martensite laths. That is why the medium-Mn steels show typical layered microstructures. Moreover, a remaining part of the martensite is subjected to an austenite reverted transformation (ART) forming again layered thin austenitic grains along the prior martensite laths. The fast formation of the ferrite decreases the M_s of the austenite due to C diffusion from the ferrite into austenite. In modern continuous annealing lines this short intercritical annealing (IA) step must be enough to decrease the M_s temperature below room temperature (RT). In batch annealing (BA) schedules the time available for diffusion is much longer. Hence, the C and Mn diffusion can take place during the annealing step, which enables to decrease the M_s more effectively. In other words, more cost effective lean medium-Mn steels (4-6% Mn) can be produced in this route.

1.2. Annealing of hot-rolled sheets

Typical heat treatment following hot-rolled sheets is shown in figure 1b. In this strategy a main difference is a dislocation density in the martensite before an annealing step. A lack of cold rolling is a reason of the smaller density of structural defects. Therefore, a presumable kinetics of ferrite formation and ART should be slower. Moreover, the formed ferrite-austenite mixture is more layered compared to cold-rolled sheets because of a lack of ferrite recrystallization step (or its limited extent). This affects a final morphology of structural products and time needed for efficient C and Mn redistribution. A lack of systematic studies is this field for hot-rolled sheet steels is a motivation of the research part of this study (next chapters).

1.3. Sheet coiling and air cooling

From an economical point of view it would be beneficial to produce ferrite / carbide-free bainite / retained austenite mixtures directly after hot rolling in just one technological step. Such a cheap heat treatment involving hot band coiling and subsequent air cooling of the coil is shown in figure 1c. This is a typical production strategy for hot-rolled strips used in the industry. Since the time of cooling is very long there is an opportunity to form some ferrite and carbide-free bainite fractions. If this happens the C and Mn redistributions to the remaining austenite are possible. Some laboratory trials are in progress. Further intense research is needed.

2. Material and experimental procedure

The chemical composition of the analyzed medium-Mn steel is as follows: 0.16% C, 4.7% Mn, 1.6% Al, 0.2% Mo, 0.2% Si. The steel was hot-forged (to a thickness of 20 mm) and next hot rolled and air cooled with a finishing rolling temperature of 850°C and a final sample thickness of ca. 5 mm [4]. The initial microstructure is purely martensitic (hardness ca. 525 HV1) due to the strong hardenability effect of manganese. The intercritical annealing was applied for hot rolled and machined samples. The heat treatment was performed by the means of dilatometry, which at the same time gives the information on a phase transformation kinetics upon heating and subsequent cooling. The samples in a shape of 5 mm diameter and 10 mm length were heated up to the desired temperatures in a range from 680 to 720°C (A_{c1} is 680°C and A_{c3} is 936°C) at a heating rate of 3°C/s, held for duration of 1, 2 and 5 hours to stimulate carbon and manganese diffusion. Finally, the samples were cooled to room temperature at the rate of 60°C/s. This heat treatment approach is presented in figure 2.

The whole process was carried out in vacuum. After the heat treatment the samples were cut in the middle part of the rod for the preparation for light and scanning electron microscopy. The samples were grinded with the use of 220, 500, 800 and 1200 grinding papers and polished with the use of 3 and next 1 μm diamond solution. Next, the samples were etched using 5% Nital. The Vickers hardness was measured with the load of 9.81 N.

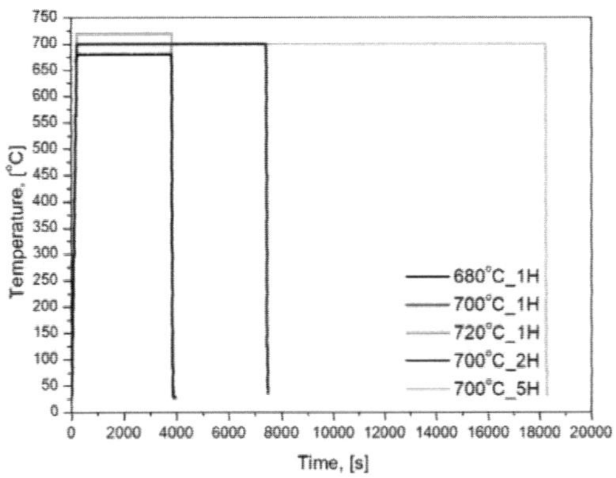

Figure 2. Heat treatment performed in the dilatometer for the hot-rolled medium-Mn steel.

3. Results and discussion

3.1. Dilatometric analysis

The first study was a dilatometric analysis of the samples subjected to different heat treatments. These results are presented in figure 3. According to the results it can be seen that temperature of 720°C does not ensure a full austenite stability. This leads to some martensite formation upon cooling. However, if the annealing temperature drops to 700 and 680°C the stability of austenite is enough to retain it at room temperature. This effect corresponded to the amount of austenite formed during the intercritical annealing.

When the amount of austenite increases during isothermal holding the amount of polygonal ferrite decreases at the same time. Because the ferrite dissolves a very small content of carbon the excess goes to austenite increasing its stability. If the amount of ferrite is small then the amount of available carbon is small too. This situation prevents the thermal stabilisation of austenite to room temperature. The results presented in figure 3 proves this theory. When the temperature of annealing is lowered the amount of ferrite increases together with excess carbon leading to a higher concentration of carbon in the austenite [6]. The stability of austenite and related carbon content influence martensitic transformation start and finish temperatures. The higher the amount of carbon in the austenite, the lower is the martensite start temperature. The idea is to increase the carbon content in the austenite to decrease the M_s below RT. In this situation it is possible to obtain a fraction of retained austenite in a final microstructure. The time increase in this case does not have a negative effect on a thermal stability of retained austenite. Even after 5 hours, the martensite start temperature is below room temperature. In this case not only carbon but also manganese favours the stability of austenite. Longer holding time gives more time for manganese atoms to diffuse into the austenite increasing its stability (manganese is an austenite stabilizer and it strongly reduces the M_s) [7-9].

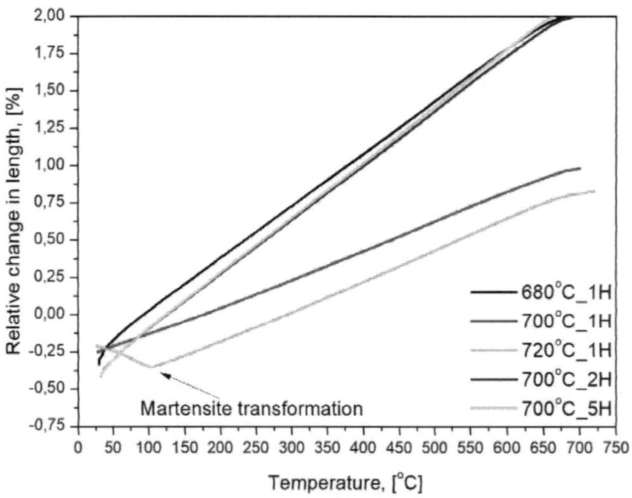

Figure 3. Dilatometric curves presenting the relative change in length during cooling the samples to room temperature following annealing at different intercritical temperatures and times.

3.2. Microstructure investigation

After the dilatometry study the light and electron scanning microscope investigations were performed. The microstructural results are presented in figure 4. The microstructure registered after the annealing at 720°C is composed of fresh martensite created during cooling, ferrite laths and a small fraction of retained austenite. The austenite is present because according to the dilatometry analysis the martensite finish temperature is below RT. Hence, the martensitic transformation is not finished. The presence of retained austenite was confirmed using XRD results (the results will be presented elsewhere).

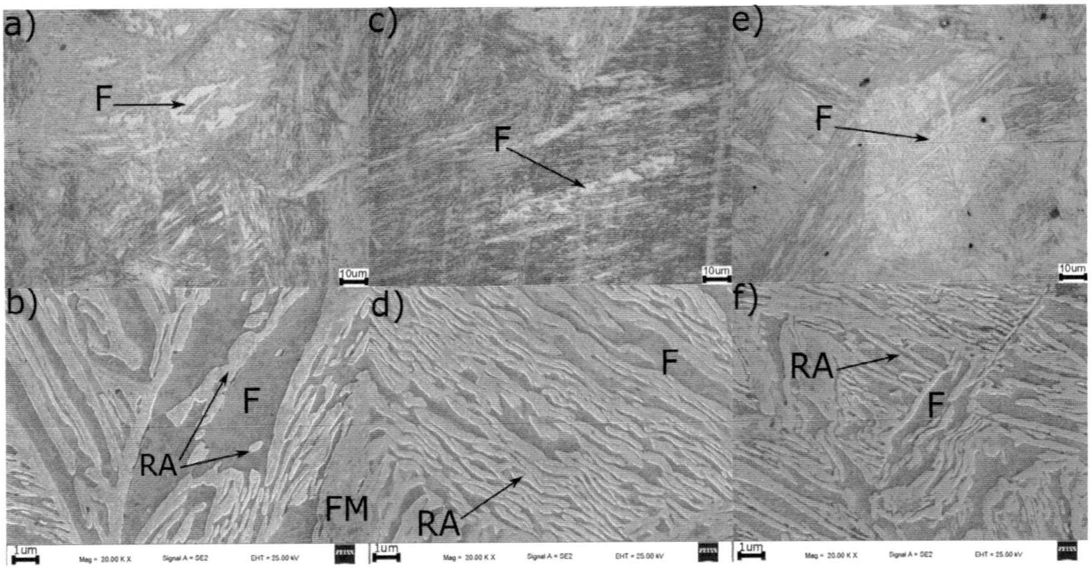

Figure 4. Microstructure of 5Mn steel subjected to intercritical annealing (for 1 hour) between 720°C and 680°C (magnification 1kx and 20kx); a-b) 720, c-d) 700 and e-f) 680°C; F-ferrite, FM-fresh martensite, RA-retained austenite.

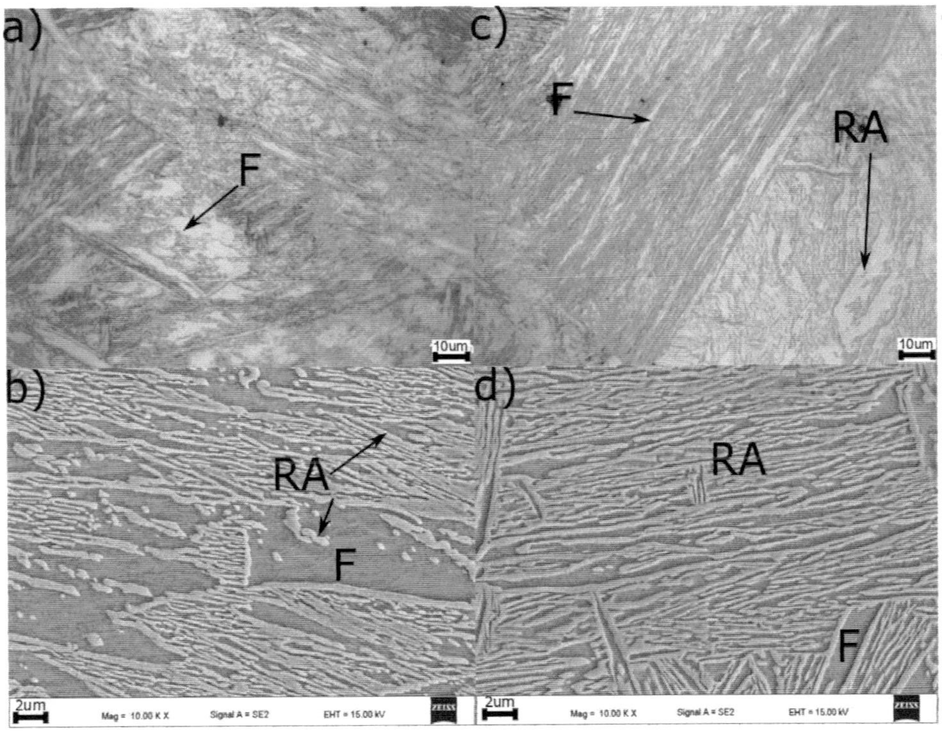

Figure 5. Microstructure of 5Mn steel subjected to intercritical annealing at 700°C (magnification 1kx and 10kx) for: a-b) 2 hours and c-d) 5 hours; F-ferrite, RA-retained austenite.

At 700 and 680°C temperatures the microstructure is composed of ferritic and austenitic laths without martensite. The lath-like morphology is present because the phases inherit the initial martensite morphology. During heating the ART (austenite reverse transformation) occurs leading to lath-shaped ferrite and austenite [6, 10]. The samples annealed at 680°C are characterized by a wavy microstructure (figure 4f). The nature of such a behaviour requires further research. Moreover, the micrographs show an increase in the ferrite phase fraction together with lowering the intercritical annealing temperature from 720 to 680°C (the difference is not high). The increase of holding time affects an amount and morphology of the ferrite laths (figure 5). The polygonal ferrite formed in the intercritical region grows together with prolonging holding time from 1 to 5 hours due to longer diffusion processes. The preliminary retained austenite fractions measured using XRD at 700 and 680°C are between 20 and 30% (further research is in progress).

However, even 5 hours annealing is not enough for the microstructures to undergo a morphological transformation from a lath-like to globular morphology. Similar results were obtained by Yunbo et al. [6]. This kind of lath morphology positively influences the thermal stability of retained austenite [11,12].

3.3. Hardness results

The last step of the analysis was hardness measurement. These results are presented in figure 6. The results are in accordance with dilatometry and microstructure investigations. The lowest hardness was obtained at 680°C, which has the highest ferrite fraction. When the temperature of intercritical annealing is increased the amount of ferrite decreases resulting in an increase of the steel hardness. The hardness is increased not only because of lower ferrite fraction but by the presence of fresh martensite formed after annealing at 720°C.

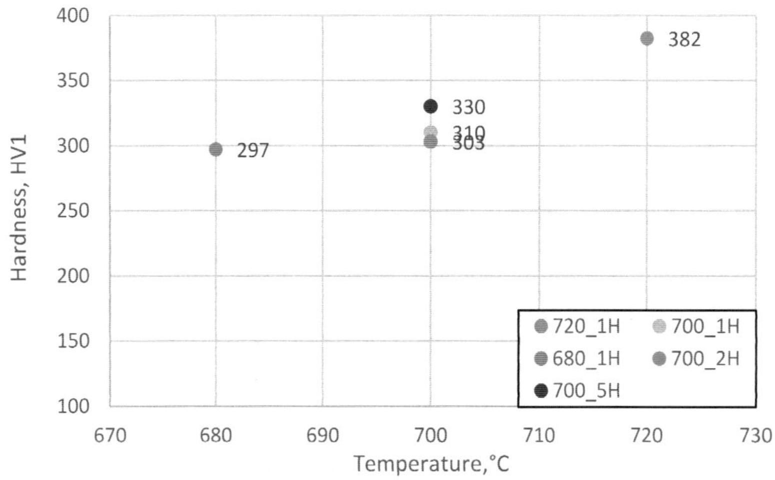

Figure 6. Hardness results of 5Mn steel intercritically annealed at different temperatures and time.

4. Conclusions

The strategies for cold-rolled and hot-rolled medium-Mn sheet steels were reviewed. The differences in processing routes and microstructural features were indicated. The research part of the study dedicated to the intercritical annealing of hot-rolled 5Mn steel revealed that the temperature and time strongly affect microstructure-properties characteristics. At 720°C the microstructure was composed of ferritic and fresh martensite laths. At 700 and 680°C the austenite was stable to room temperature. It means that 700°C annealing is enough to increase the carbon content of the austenite and subsequent decrease of the martensite start temperature below room temperature. The lath-like morphology is the result of austenite reverse transformation from the initial martensite structure. The 5 hour annealing is not enough for the microstructure to change from a lath-type to more granular morphology. It is enough to increase the ferrite lath size a little bit. The decrease of annealing temperature from 720 to 680°C leads to some higher amount of ferrite phase and a resulting hardness decrease.

Acknowledgements
The financial support of the National Science Center, Poland, is gratefully acknowledged, grant no. 2017/27/B/ST8/02864.

References

[1] Schneider R, Steineder K, Krizan D and Sommitsch Ch 2018 Effect of heat treatment on the microstructure and mechanical properties of medium-Mn-steels *Mater. Sci. Technol.* vol **34** p 1 – 9
[2] Lisiecki A, Burdzik R and Siwiec G 2015 Disk laser welding of car body zinc coated steel sheets *Arch. Metall. Mater.* vol **60** p 2913 – 22
[3] Kvackaj T 2000 The review of steel and steel products flat-rolled production *Metalurgija* vol **39** p 185 – 9
[4] Grajcar A, Skrzypczyk P and Woźniak D 2014 Thermomechanically rolled medium-Mn steels containing retained austenite *Arch. Metall. Mater.* vol **59** p 1691 – 7
[5] Gorka J 2018 Assessment of the weldability of T-welded joints in 10mm thick TMPC teel using laser beam *Materials* vol **11** doi: 10.3390/ma11071192
[6] Kim J, Lee G, Choi I and Lee S 2001 Effects of heat treatment and alloying elements on the microstructure and mechanical properties of 0.15 wt pct C transformation-induced plasticity-aided cold-rolled steel sheets *Metall. Mater. Trans. A* vol **32**A p 505 – 14
[7] Kucerova L, Jirkova H and Masek B 2017 The effect of alloying elements on microstructure of 0,2%C TRIP steel *Mater. Sci. Forum* vol **891** p 209 – 13
[8] Lasek S and Mazancowa E 2013 Influence of thermal treatment on structure and corrosion properties of high manganese triplex steel *Metalurgija* vol **52** p 441 – 4
[9] Shu Y, Liu X, Liang T and Zhao Y 2018 The effects of the initial microstructure on microstructural evolution, mechanical properties and reversed austenite stability of intercritically annealed Fe-6.1Mn-1.5Si-0.12C steel *Mater. Sci. and Eng. A* vol **712** p 324 – 32
[10] Xiaochuan X, Chen B, Wang H and Wang L 2013 The effect of morphology on the stability of retained austenite in quenched and partitioned steel *Scr. Mater.* vol **68** p 321 – 4
[11] Grajcar A, Kwaśny W and Zalecki W 2015 Microstructure-property relationships in TRIP aided medium-C bainitic steel with lamellar retained austenite *Mater. Sci. Technol.* vol **31** p 781 – 94

IOP Conf. Series: Materials Science and Engineering **723** (2020) 012009 doi:10.1088/1757-899X/723/1/012009

Optimization of test specimen dimensions for thermal power station exposure device

E Chvostová[1], J Horváth[2], P Konopík[1], S Rzepa[1] and D Melzer[1]

[1] COMTES FHT a.s., Průmyslová 955, 334 41 Dobřany, Czech Republic

[2] UJP PRAHA a.s., Nad Kamínkou 1345, 156 10 Praha – Zbraslav, Czech Republic

E-mail: eva.chvostova@comtesfht.cz

Abstract. Degradation of metallic materials can decrease mechanical properties of metal constructions and components in thermal power plants during their operation and consequently lead to loss of safety and reliability. It is difficult to remove material from the pressure system of the equipment being operated. Therefore, the effort is to design, produce and subsequently install exposure channels, which will allow to remove and evaluate samples of exposed material without intervention into the pressure system. The exposure device for pressures and steam temperature in supercritical blocks, which would contain sufficient test material to produce conventional test specimens, is unrealistic in size, energy and economics. The way is to optimize the dimensions of the exposure device according to the miniaturized test specimens, which, in addition to the tensile test and the structural evaluation, can also carry out other required tests, such as bending impact tests with transition temperature, fracture toughness and fatigue tests. This paper deals with optimization of test specimens for the internal dimensions of the exposure device and shows the applicability of small sample methodologies for selected materials.

1 Introduction

The use of relatively new steels and alloys in thermal power plants, caused by the pressure to achieve the highest possible efficiency, also entails significant technical risks in the future. The gradual deterioration of the mechanical properties caused by the combined stresses, in particular the temperature exposure, can significantly limit the residual life of the pressure systems. The knowledge of the progress of material degradation and the linkage between the structural state and the mechanical and creep parameters can be obtained by evaluating the exposed samples in time periods The residual lifetime of these machines and metal constructions, mainly turbines, boilers and steam pipelines, can be examined using standard mechanical tests as well as semi-destructive techniques using mini-specimens. Testing of operating constructions demands minimizing the volume of extracted material. For this reason, sub-sized specimen techniques have several advantages over standard ones, e.g., when the volume of tested material is insufficient or for the development of new alloys to ensure the lowest cost possible. Furthermore, testing of small samples allows for local properties to be evaluated (weld joints) and mechanical properties of in-service components investigated [1]. The data obtained using the newest measurement equipment and mini-specimen techniques can be used for calibration of fracture locus and in FEM simulations [2, 3].

The aim of the paper was to examine the effect of specimen dimensions on selected mechanical properties of materials applied in the power industry. Fundamental mechanical properties are tested using various miniature specimen testing methods, such as tensile test, fatigue test, impact test and fracture toughness test. The results for standard size specimens were compared to specimens of reduced size. Based on this research, the possibilities and limitations of sub-size specimen methods and the application of these methods to testing components in the power industry are discussed. However, despite the potential small sample technique limitations, exposure device without specimens miniaturisation could not be realized at all.

2 Mechanical testing of sub-sized specimens

In this paper, mechanical properties of materials were investigated using mechanical tests on standard and small specimens. Applicability and reliability of sub-sized specimen techniques are discussed on the basis of the following mechanical tests: tensile test, impact test with transition temperature evaluation, fatigue test and fracture toughness evaluation.

2.1 Mini-Tensile Test (M-TT)

Tensile test is a basic and the most common method for mechanical properties evaluation. Using sub-sized specimen techniques there is a possibility to evaluate all standard mechanical characteristics including anisotropy. Furthermore, MTT can be advantageously used for strain rate sensitivity investigation of conventionally or additively manufactured materials [4-6].

In order to determine the tensile properties of small amounts of material, the geometry of sub-sized specimen was proposed [1]. For this type of specimen, it was necessary to design grips to enable gripping of the specimen in a universal testing machine. The deformation was measured using contactless measuring system ARAMIS, which is based on the Digital Image Correlation (DIC) method. A stochastic pattern is applied to the specimens for optical monitoring of displacement, which can be recorded by one (2D) or more (3D) cameras. During the test, the DIC software is tracking the pattern changes in each image in comparison with the reference image. Thanks to this method, it is possible to measure deformation and displacement of the whole specimen as well as locally in selected regions.

In this study, the tensile tests were performed on standard and sub-sized specimens. Different kinds of alloyed and non-alloyed ferrous and non-ferrous materials were tested. The results of the tensile tests are presented in **Figure 1**. The resulting curves show that the values of the yield strength, ultimate tensile strength and elongation for standard specimens are in agreement with the results of the small specimens in a wide range of stresses.

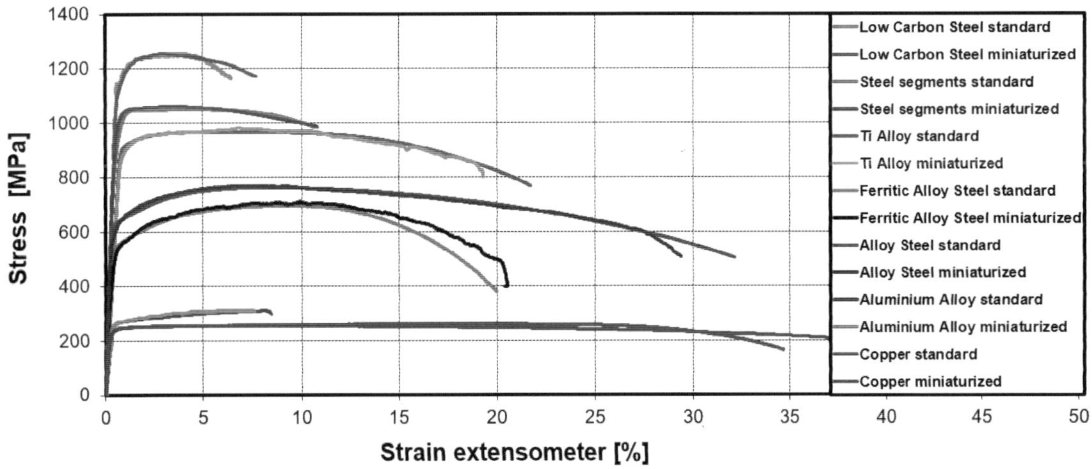

Figure 1. Engineering stress-strain curves of tested materials using standard and MTT test.

2.2 Miniature fatigue test

Fatigue fracture can occur in many types of metal structures, e.g., turbines or reactor vessels [1, 7]. Therefore, it is necessary to determine the maximum stress values, at which no damage will appear. The fatigue limit can be affected by impurities of the material as well as surface and environmental effects [8, 9].

Applicability of sub-sized specimens in fatigue testing was verified using XCrNiMoAlTi12-11-2 steel, which is commonly found in the power industry. Standard specimens were tested according to ASTM E466 and assessed according to ASTM E739 standard. All tests were performed using servo-hydraulic machine MTS BIONIX with a capacity of 25 kN at room temperature. The tests were conducted in tension-compression mode with controlled force and load ratio R=-1. The frequency of 50 Hz was applied. The fatigue test was conducted until failure or reaching 10^7 cycles, if failure did not appear. Based on the obtained results, the Wöhler curve was constructed, where different stress amplitudes are plotted with a number of cycles to failure on the semi-logarithmic scale. The test results are shown in **Figure 2**. The results for both batches of specimens are in agreement. The fatigue limit received as a result of small specimen testing equals 710 MPa and for standard specimens it is 720 MPa. Sub-sized specimen results show a higher scatter than standard specimens. The scatter exhibited by both batches of specimens can be caused by the surface condition and the contaminations of steel.

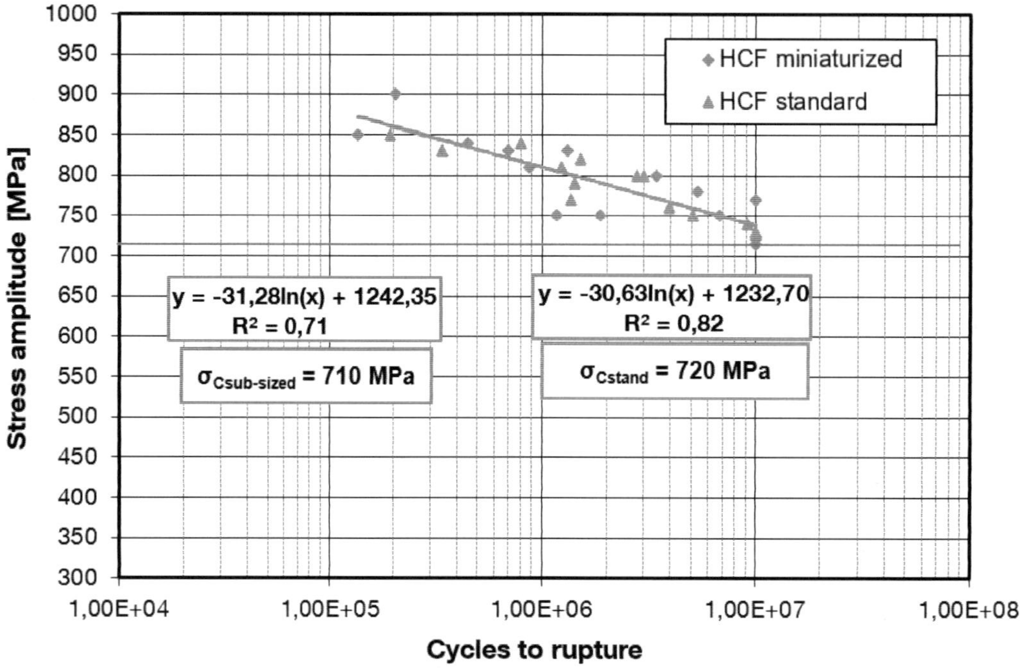

Figure 2. Fatigue test results of standard and small specimens of XCrNiMoAlTi12-11-2 steel.

2.3 Impact test with FATT evaluation using mini-Charpy specimens

The impact tests were performed to determine ductile-to-brittle transition temperature. Many materials exhibit a transition temperature, which can be a reason of their limited application due to an increased risk of brittle fracture [10, 11].

Evaluation of DBTT was determined on the basis of impact tests using small (3x4x27 mm) and standard (10x10x55 mm) Charpy specimens. The transition temperature was calculated applying the equation (1), where C value is constant:

$$FATT_{standard} = FATT_{sub\text{-}sized} + C \tag{1}$$

It is necessary to calculate the value of coefficient C using mini-Charpy specimens, which expresses a shift of FATT related to the specimen dimensions. A weld joint consisting of two heat-resistant chromium-molybdenum steels and weld metal, commonly used in power engineering, was chosen as an experimental material. The tests were performed in 5 regions of the heterogeneous weld according to the EN ISO 14556:2016 standard. The value of C was established at C=60±2°C. The tests were performed using an instrumented Charpy pendulum with impact energy of 300 J and 15 J. The results of the impact tests with calculated C values are summarized in **Table 1**.

Table 1. Results of FATT transition temperature evaluation using standard Charpy and mini-Charpy specimens.

Material	FATT [°C]		Coefficient C
	Specimen dimensions [mm]		
	10x10	**3x4**	**°C**
Basic Material 1	13.1	-48.4	**61.5**
Heat Affected Zone 1	-11.7	-69.9	**58.2**
Weld Material	52.1	-8.7	**60.8**
Heat Affected Zone 2	21.1	-39.4	**60.5**
Basic Material 2	30.6	-30.0	**60.6**

2.4 Fracture toughness evaluation using mini-Charpy specimens

Fracture mechanics is an important field of material investigation nowadays. It is helpful when common testing methods are not able to fully describe material properties. For metal components, such as rotors, pressure vessels and building constructions, fracture toughness testing is one of the main tools for determining the applicability of the material [12-17].

Standard specimens for a 3-point bending test and mini-Charpy specimens were manufactured from common steel 34CrNiMo6. Before testing, a precracking procedure on each specimen was performed with crack length a/W=0.5 and final value K=16 $MPa\cdot m^{1/2}$. Due to the loss of the constraint effect, side-grooves were applied. Tests were performed and J-R curves were evaluated according to ASTM E1820 standard using a multiple specimen test method. For each test different ductile crack growth was achieved. The crack length after the test was marked by the heat tinting process. Subsequently, specimens were cooled down in liquid nitrogen and cracked in the temperature region of brittle fracture. The crack length was measured using digital image processing. Based on the known crack length, the geometry of the specimen and applied force, the J-R curves were obtained and fracture toughness evaluated. The curves for both geometries are presented in **Figure 3**. Due to small dimensions of mini-Charpy specimens, the criteria established by the ASTM standard were not met. The evaluation of fracture toughness cannot be valid, therefore the provisional value $J_C=J_Q$ was used. Despite this fact, the constructed curve for small specimens is in agreement with the curve for standard specimens for a 3-point bending test and estimated J_Q values differ by 1.7%.

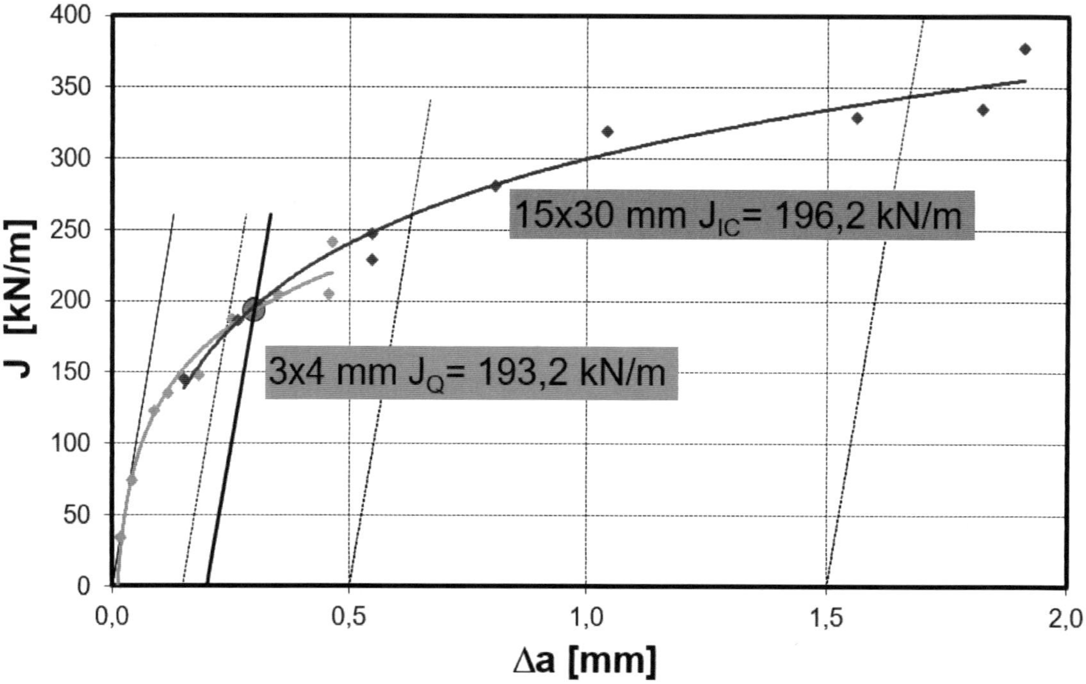

Figure 3. Results of fracture toughness evaluation using standard 3PB specimens and mini-Charpy specimens.

3 Summary and conclusions

The relatively small size of the exposure device makes performing standard mechanical tests of in-service components impossible. It leads to need of developing miniature sample techniques. The aim of this study was to compare the results of selected mechanical tests using standard and small specimens to show the applicability of sub-sized specimen techniques in mechanical testing.

The stress-strain curves obtained for small specimens are fully comparable with the results of standard tests. All tensile test results show very good compliance with yield strength and ultimate strength and elongation for all tested materials.

Mini-Charpy specimens were used to determine fracture toughness by a multiple specimen test method. Although the crack extension Δa in the J-R curve did not reach the value of 0.5 mm, the resulting fracture toughness J_Q was in good agreement with the fracture toughness J_{IC} determined using standard 3PB specimens.

To determine the transition temperature using mini-Charpy specimens, calculation of the coefficient C was necessary. The obtained transition temperatures for different regions of the weld joint were shifted by 60±2°C.

Fatigue limits obtained for standard and small specimens are almost identical. The performance of the fatigue test for both batches of specimens yielded reliable results without using any correlation.

All presented applications show that it is possible to obtain reliable and repeatable results using sub-sized specimen techniques. Therefore, the exposure device can be designed based on miniature specimen size.

Acknowledgments

This paper was created by project Evaluation of degraded steels for the construction of turbines and superheaters of power boilers (for the need to predict residual service life time and local repairs) No.: FV40166, financed by the Ministry of Industry and Trade.

References

[1] Dzugan J, Konopik P, Rund M and Prochazka R 2015 Determination of local tensile and fatigue properties with the use of sub-sized specimens *PVP-2015* **1A** ed Xu SX, Hojo K, Cipolla RC (New York: Amer Soc Mechanical Engineers)

[2] Kumar K, Pooleery A, Madhusoodanan K, Singh RN, Chakravartty JK, Dutta, BK and Sinha RK 2014 Use of Miniature Tensile Specimen for Measurement of Mechanical Properties *ICONS-2014* (Kalpakkam) **86** ed Jayakumar T, Sandhya R, Rao BPC, Bhaduri AK (Amsterdam, Elsevier Science BV) pp. 899-909

[3] Dzugan J, Spaniel M, Prantl A, Konopik P, Ruzicka J and Kuzelka J 2018 Identification of ductile damage parameters for pressure vessel steel *NUCL ENG DES* **328**, pp. 372-80

[4] Konopik P, Rund M and Dzugan J 2015 Investigation of the mechanical behaviour of zirconium alloy at different strain rates using sub-size tensile specimens *M2D2015* ed Gomes JFS, Meguid SA (Porto: INEGI-FEUP) pp.277-84

[5] Byun TS, Kim JH and Chi SH 1998 Effect of specimen thickness on the tensile deformation properties of SA508 C1.3 reactor pressure vessel steel *ASTM STP* **1329** ed Corwin WR, Rosinski ST, VanWalle E (West Conshohocken/PA USA: American Society Testing And Materials) pp. 575-87

[6] Gussev MN, Busby JT, Field KG, Sokolov MA and Gray SE 2015 Role of Scale Factor During Tensile Testing of Small Specimens *ASTM STP* **1576** ed Sokolov MA, Lucon E (West Conshohocken/PA USA: ASTM International) pp. 31-49

[7] Connolley T, Mchugh PE and Bruzzi M 2005 A review of deformation and fatigue of metals at small size scales *Fatigue Fract Eng M* **28** pp. 1119-52

[8] Schijve J 2009 *Fatigue of structures and materials* Springer (Amsterdam) pp. 36-47

[9] Wan HY, Chen GF, Li CP, Qi XB and Zhang GP 2019 Data-driven evaluation of fatigue performance of additive manufactured parts using miniature specimens *J MATER SCI TECHNOL* **35** pp. 1137-46

[10] Rzepa S, Bucki T, Konopik P, Dzugan J, Rund M and Prochazka R 2017 Influence of specimen dimensions on ductile-to-brittle transition temperature in Charpy impact test *IOP Conf. Ser.: Mater. Sci. Eng.* **179** (Bristol: IOP PUBLISHING LTD)

[11] Kayano H, Kurishita H, Kimura A, Narui M, Yamazaki M and Suzuki Y 1991 Charpy impact testing using miniature specimens and its application to the study of irradiation behavior of low-activation ferritic steels *J. Nucl. Mater.* **179** pp. 425-8

[12] Valo M, Planman T and Wallin K 1998 The applicability of small and ultra-small fracture toughness specimens for material characterization *ASTM STP* **1329** ed Corwin WR, Rosinski ST, VanWalle E pp. 196-213

[13] Lucon E 2001 Material damage evaluation of primary power plant components using sub-size specimens *Adv. Eng. Mater.* **3** pp. 291-302

[14] Sokolov MA and Nanstad RK 1996 On impact testing of subsize Charpy V-notch type specimens *ASTM STP* **1270** ed Gelles DS, Nansta RK, Kumar AS, Little AE pp. 384-414

[15] Scibetta M, Lucon E and Van Walle E 2002 Optimum use of broken Charpy specimens from surveillance programs for the application of the master curve approach *Int. J. Fract.* **116** pp. 231-44

[16] Li X, Liu Y and Tao X 1987 Experimental study of the effect of specimen size on the JR resistance curve of ductile materials *INT J PRES VES PIP* **30** pp. 293-8

[17] Bucki T, Rzepa S, Melzer D, Konopik P, Dzugan J and Rund M 2018 The effect of strain rate on master curve of the martensitic chromium steel *METAL 2018* (Ostrava: TANGER LTD) pp. 641-5

The effect of NPs addition on the photocatalytic and antibacterial effectivity of composite TiO_2/SiO_2 paint

M Jakubickova[1] M Petrzilkova[1] B Amartuvshin[1] L Kejzlarova[1] and P Kejzlar[1]

[1] Technical University of Liberec, Studentska 1402/2, 461 17, Liberec, Czech republic

michaela.jakubickova@tul.cz

Abstract. TiO_2 in the anatase allotropic modification is well known for its photocatalytic activity. When the anatase is irradiated by UVA light, it forms an electron-hole pair which can participate in redox reaction. Thus, anatase-based layers are suitable for decomposition of organic compounds. Surface self-cleaning ability as well as antimicrobial efficacy are therefore key features for TiO_2 based layers. The present paper deals with a study of ZnO, CuO and Ag NPs addition on the antibacterial and photocatalytic effectivity of TiO_2/SiO_2 paint.

1 Introduction

Building facades, especially at high air humidity and poorly sunlit areas, are under strong attack of microorganisms like bacteria, algae and fungi. These microorganisms cause not only structural disintegration of the facades and subsequent functional malfunction, but often also disrupt the aesthetic appearance of the buildings.

There are few possibilities how to treat such affected areas. The biofilm can be removed physically (e.g. high-pressure water cleaning or mechanical brushing) or chemically (e.g. chlorine-based products). Unfortunately, these methods are effective only in the short term. In addition, often repeated mechanical cleaning destroys the façade itself and the use of chemical products is environmentally harmful. Effective solution offers the use of photocatalytic paints. If the light with appropriate wavelength hits the surface of photocatalytic active substance, electron-hole pairs are created. These pairs further generate free radicals (e.g. hydroxyl radicals OH) which can undergo secondary reactions. These include the gradual oxidative degradation of organic substances including microorganisms. Finally, even the most complex organic molecules are transformed into harmless simple inorganic compound, such as carbon dioxide, water and corresponding mineral acids. [1]

The photocatalytic paint under the name Balclean® used in the present experiment was developed in 2016 on the Technical University of Liberec in cooperation with Czech Academy of Science, BARVY A LAKY TELURIA, s.r.o. and PRAGOTHERM, servis fasad s.r.o. [2]

Key properties of paints used for cleaning of facades include high photocatalytic effectivity and stability under environmental conditions. Effectivity of such paints can also benefit from antimicrobial activity when the coating is not exposed to UV radiation.

The nanoparticles of some metals represent inhibitors of effects for many microorganisms. In particular, ZnO, CuO and Ag nanoparticles are known for their cytotoxicity to eukaryotic and prokaryotic cells. Due to their size, nanoparticles can penetrate through the cell membranes into the cytosol and then target multiple parts of the cell where their action induces complex processes. [3]

In bacterial strains such as E.Coli and S.Aureus, silver nanoparticles attach directly to the cell wall, where increases breathability and subsequently disrupts metabolism. [4,5] CuO nanoparticles are known for their antimicrobial effects (toxic to fungi, viruses and bacteria) and are mainly used as antimicrobial agents. They are effective against both susceptible microorganisms and microorganisms exhibiting resistance to antibiotics used for treatment. They have been shown to have the ability to kill more than 99 % of gram-positive or gram-negative bacteria when exposed for 4 hours when administered at the appropriate dose. CuO nanoparticles show greater cytotoxicity to gram-negative E. coli than gram-positive S. aureus. [6] Another metal oxide which disrupts the cell membrane is ZnO. When the ZnO particles have small size, the specific surface area of the particles increases, promoting their penetration into cells through cell membranes. The penetrated ZnO particles exhibit high solubility in both extracellular and intracellular fluids. Consequently, the metabolism of the cell is impaired, which is attenuated towards the external environment. [7]
The aim of this research is to determine the effect of selected nanoparticles on antibacterial and photocatalytic activity of the Balclean® paint.

2 Experimental

Three types of laboratory prepared nanoparticles were added into the Balclean paint in amount of 1 %. Namely, Ag, ZnO and CuO were selected as suitable materials. For the purposes of Ag nanoparticles was chosen method of chemical reduction of silver nitrate (AgNO3), which is found predominantly on the findings from the work [8]. Sensitive trisodium dihydrates (C6H5Na3O7, 2H2O) as well as formaldehyde (CH2O with stabilizing effect) were used as reducing agents, which can be eliminated from the final product by washing and centrifugation.
Reactions receive equations:

$$12AgNO_3 + 4C_6H_5O_7Na_3 + 6H_2O \rightarrow 12Ag^o + 4C_6H_5O_7H_3 + 12NaNO_3 + 3O_2$$

$$2\,AgNO_3 + H_2O + HCHO \rightarrow 2\,Ag^o + 2\,HNO_3 + HCOOH$$

The chemical precipitation method was chosen for the preparation of ZnO nanoparticles. The exact procedure used in the preparation of ZnO nanoparticles is reported in the paper [9].
The reaction proceeds according to the equation:

$$ZnSO4*7H_2O + 2NaOH \rightarrow Zn(OH)_2 + Na_2SO_4 + 7H_2O$$

$$Zn(OH)_2 + 2H_2O \rightarrow [Zn(OH)_4]^{2-} + 2H^+ \rightarrow ZnO + 3H_2O$$

For the preparation of CuO nanoparticles the chemical precipitation method was also chosen.
The reaction proceeds according to the equation:

$$Cu(NO_3)_2*3H2O + 2NaOH \rightarrow Cu(OH)_2 + 2NaNO_3 + 3H_2O$$

$$Cu(OH)_2 \rightarrow CuO + H_2O$$

The results and structure of the used particles are shown in Figures 1-3. A Bandeline ultrasonic homogenizator was used to homogenize the particles in solution. Using this apparatus, the agglomerates of the particles in solution were separated into individual nanoparticles and at the same time the resulting solution was homogenized. The prepared suspension was subsequently transferred to an ultrasonic homogenizer where it was further homogenized and dispersed at a frequency of 20 kHz for 5 minutes.

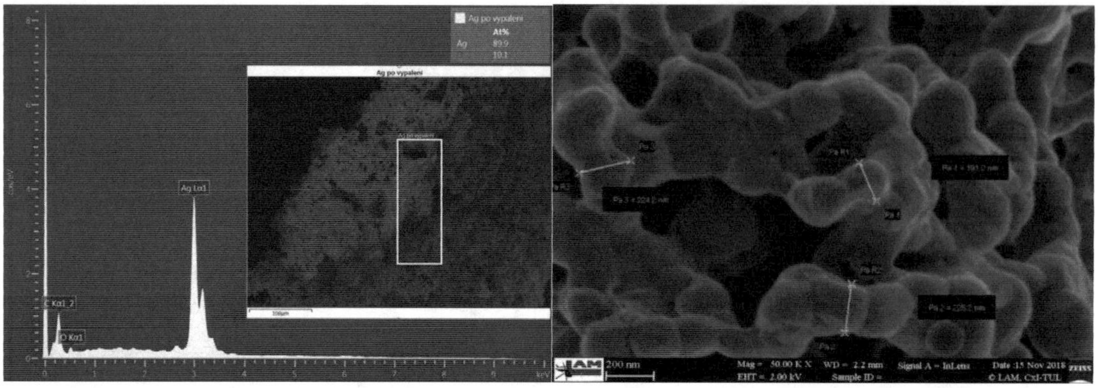

Figure 1. EDS analysis and SEM image showing the structure of Ag nanoparticles.

Figure 2. EDS analysis and SEM image showing the structure of ZnO nanoparticles.

Figure 3. EDS analysis and SEM image showing the structure of CuO nanoparticles.

A matt glass with dimensions of 50 x 50 x 1 mm was selected as a suitable substrate. 100 μl from prepared nanoparticle suspensions was pipetted onto the surface of the glass plate cleaned by ethanol. The solution was spread over the substrate using a brush.

The antibacterial activity was measured according to ISO 27447/2009. The antibacterial properties of the samples were measured using a Gram-negative bacterial culture of Eschericha coli CM 3954.

Using McFarland dilutions, the initial cell concentration was 3×108 cells/ml. Appropriate inoculated samples were placed in Petri dishes. 150 µl of prepared inoculum was pipetted onto the samples and spread evenly over the surface of the sample. Petri dishes were covered with a transparent film. Half of the Petri dishes with inoculated test samples were incubated in the dark while the other half were incubated under the UVA lamp (intensity 0.25 mW / cm2). Samples for direct rinsing were processed without incubation for 4 hours, after which the inoculum was applied to the glass plate, immediately rinsed with 10 ml of saline and diluted in decimal series. After the exposure period, samples were transferred to shaking tubes. The samples were then rinsed with 10 ml saline and mixed thoroughly on a Vortex shaker. By decadic series, samples of bacteria concentrations were diluted. This procedure is shown in figure 4. Subsequently, 1 ml was pipetted into Petri dishes. Then, these Petri dishes were embedded with Nutrient agar and incubated in a thermostat at 37 °C for 48 hours. After 48 hours, Petri dishes were removed from the thermostat and viable bacteria count (CFU) was performed. In addition, an agar inoculum was inoculated to control whether the bacterial culture was vital.

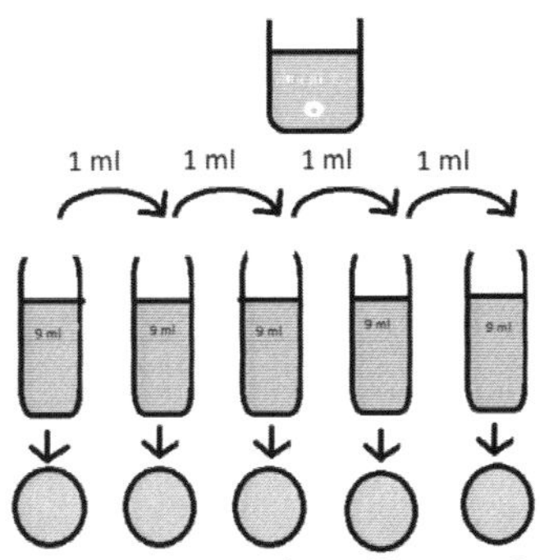

Figure 4. Schematic procedure of decimal dilution.

Methylene blue degradation was measured according to ISO 10678; 2010. Prior to testing, samples were exposed to UV radiation. The irradiation lasted 23 hours at an intensity of 1 mW / cm^2. Samples were placed in dishes with 80 ml of 20 µmol/dm^3 methylene blue (MB) solution and left in the dark for 16 hours. After 16 hours, 3 ml of MB solution were removed from each dish. Consequently, the concentration was measured with a UV / VIS spectrophotometer at a wavelength of 665 nm till the MB concentration dropped below 10 µmol/dm^3.
Subsequently, in all dishes, the old MB solution was replaced with 80 ml of a new 10 µmol/dm^3 MB solution and the dishes were covered with a transparent plastic lid.
The plates (glass plates with photocatalytically active layer and 10 µmol/dm^3 MB solution) were sequentially placed at a distance of 15 cm from the sample surface under an UV lamp (intensity 2 mW/cm^2), while the reference sample was kept in the dark. Subsequently, from the beginning of exposure to the first sample, after 20 minutes, 3 ml of MB solution were taken from each plate together with the reference sample and their concentration was measured by UV/VIS spectrophotometer at a wavelength of 665 nm. The exposure time of the sample was 3 h under UV.

Results and discussion

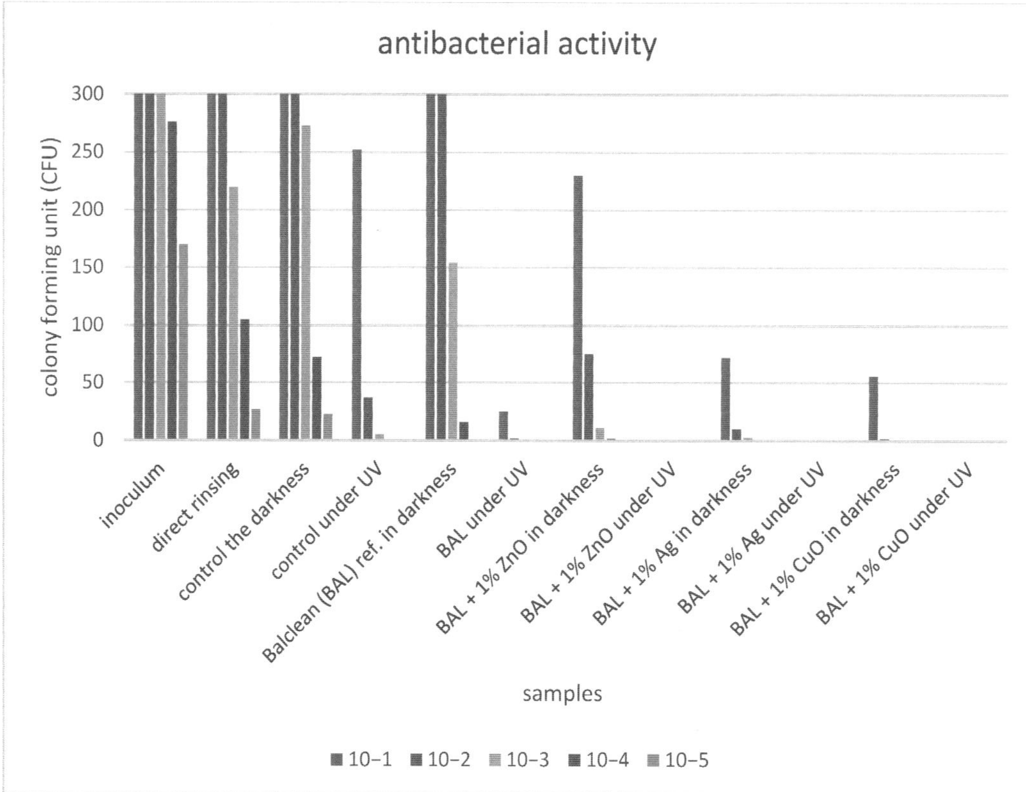

Figure 5. antibacterial activity of samples.

The results of antibacterial activity tests are summarized in figure 5. The samples left in the dark show a significant decrease in CFU compared to the reference sample (RS), suggesting the bactericidal properties of the added nanoparticles.

The result from the inoculum confirmed that the bacterial strain E. coli is viable or vital when cultured on Nutrient agar. In the direct rinse sample (glass plate without photocatalytic coating), we observe a high amount of bacterial colonies surviving about 300 to a dilution of 10^{-3}, which is comparable to the control glass sample (without photocatalytic coating) incubated in the dark. The same results showed a reference coat named Balclean incubated in the dark, showing that there was no photocatalytic effect. However, a significant decrease in the number of bacterial colonies was observed in the control incubated under UV lamp at all dilutions compared to the control incubated in the dark. This suggests that a UV lamp at 0.25 mW/cm^2 produced a germicidal effect on the bacteria and thereby killed, requiring a reduction in radiation intensity.

In a sample coated with a Balclean reference paint incubated under a UV lamp, we observe a noticeable bacterial death compared to its reference sample incubated in the dark, due to the photocatalytic properties of the Balclean paint. For other samples (Balclean with 1% Ag, Balclean with 1% ZnO and Balclean with 1% CuO) complete bacterial mortality was recorded in case of incubation under UV lamp. These samples also killed bacteria even when incubated in the dark. This can be explained by the toxic effects of Ag, CuO and ZnO nanoparticles, since these nanoparticles all have a germicidal effect on E. coli.

IOP Conf. Series: Materials Science and Engineering **723** (2020) 012010 doi:10.1088/1757-899X/723/1/012010

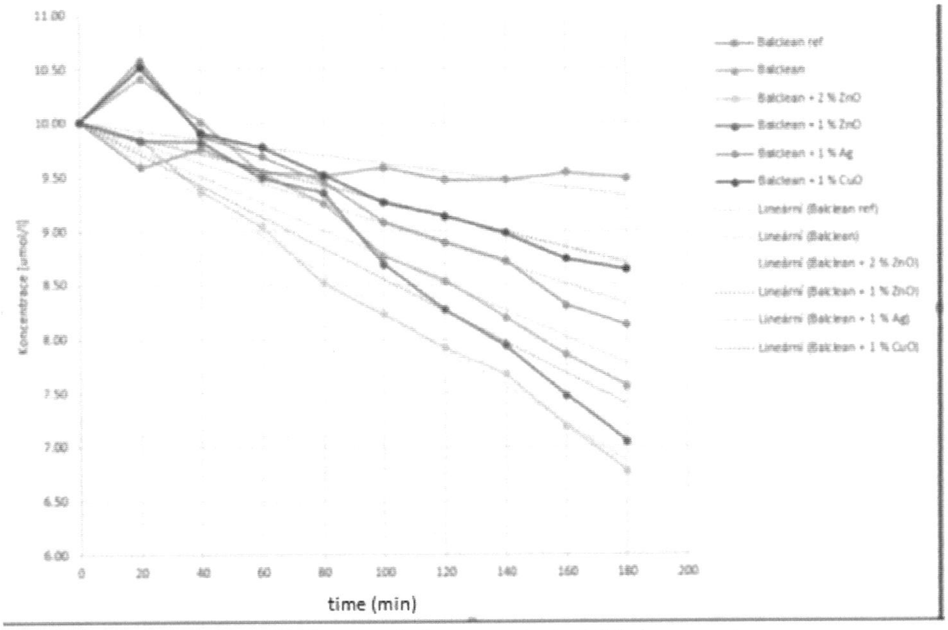

Figure 6. Graph of photocatalytic MB degradation.

MB degradation results are summarized in figure 6. In some cases there was observed slight increase in MB concentration after 20 minutes which was caused by MB desorbtion. The results clearly demonstrate high photocatalytic activity of BAL and BAL with ZnO sample. With the addition of CuO and Ag we observed a decrease of photocatalytic activity, which may be caused by an increase in the proportion of photocatalytically inactive particles (shading effect). In addition, the silver particles were too large (see figure 1).

3 Conclusion

The application of photocatalytically active coatings seems to be a very effective alternative for the reduction of undesirable ingrowth of facades, plasters and other surfaces by microorganisms. Nanoparticles of Ag were synthetized used chemical reduction method. For the preparation of ZnO and CuO nanoparticles, we chose the precipitation method, which successfully synthesized oxidic nanoparticles with the corresponding ratio of element Zn and Cu to oxygen, which has been confirmed by EDS analysis. A significant increase in the antimicrobial activity was achieved. With regard to the elimination of the number of CFUs in the samples left in the dark, it can be concluded that there was a significant germicidal effect of the added particles. A significant decrease in the incidence of bacterial colonies was also observed in samples exposed to UV radiation, which can be attributed to a combination of photocatalytic effect of the layer and germicidal effect of the added particles.

The negative effect of Ag and CuO addition on photocatalytic activity was confirmed also by methylene blue decomposition tests. Here, an increase in photocatalytic efficiency was observed in the case of ZnO addition, which can be explained by an increase in the total photocatalyst proportion in the resulting SiO_2 / TiO_2 / ZnO composite system. The increase in photocatalytic activity in the composite system is probably due to the fact that the coupling of the two catalysts results in a heterotransition between TiO_2 and ZnO, resulting in an increase in the recombination time and hence the photocatalytic efficiency of the system. The results of methylene blue digestion tests also confirmed the negative effect of CuO and Ag addition.

Acknowledgement
This research paper was performed with financial support of Faculty of Mechatronics, Imformatics and Interdisciplinary studies at the Technical University of Liberec.

Theoretical data for presenration of the results of the search for funds from the state budget using the Ministry of Industry and Trade of the Czech Republic within the TRIO program in project FV20234 „Innovative photocatalytic screeds and concrete additives“.

References
[1] SÁZAVSKÁ, Tereza, Jan ŠUBRT, Michaela JAKUBIČKOVÁ a František PETERKA. Photocatalytic Coatings - Promising Way to Improve a Quality of Urban Building Surfaces. Chemistry-Didactics-Ecology-Metrology [online]. 2015, 20(1-2), 113-122 [cit. 2019-08-09]. DOI: 10.1515/cdem-2015-0012. ISSN 2084-4506.
[2] BARVY A LAKY TELURIA, s.r.o., et al. Nátěrová hmota s fotokatalytickou funkcí (užitný vzor) Skrchov CZ, udělen 2017 – 02 – 15 CZ30342
[3] MCSHAN, Danielle, Paresh C. RAY a Hongtao YU. Molecular toxicity mechanism of nanosilver. Journal of Food and Drug Analysis [online]. 2014, 22(1), Nanomaterials - Toxicology and Medical Applications, 116–127. ISSN 1021-9498.
[4] MCSHAN, Danielle, Paresh C. RAY a Hongtao YU. Molecular toxicity mechanism of nanosilver. Journal of Food and Drug Analysis [online]. 2014, 22(1), Nanomaterials - Toxicology and Medical Applications, 116–127. ISSN 1021-9498.
[5] KHAN, S. Sudheer, Amitava MUKHERJEE a N. CHANDRASEKARAN. Studies on interaction of colloidal silver nanoparticles (SNPs) with five different bacterial species. *Colloids and Surfaces B: Biointerfaces*[online]. 2011, **87**(1), 129-138 [cit. 2019-08-11]. DOI: 10.1016/j.colsurfb.2011.05.012. ISSN 09277765.
[6] GRIGORE, Madalina Elena, Elena Ramona BISCU, Alina Maria HOLBAN, Monica Cartelle GESTAL a Alexandru Mihai GRUMEZESCU. Methods of Synthesis, 87 Properties and Biomedical Applications of CuO Nanoparticles. Pharmaceuticals [online]. 2016, 9(4) [vid. 2019-01-14]. ISSN 1424-8247.
[7] Hidden toxicity of zinc oxide nanoparticles | Atlas of Science [online]. [vid. 2019-04-21]. Available from: https://atlasofscience.org/hidden-toxicity-of-zinc-oxide-nanoparticles/
[8] YIN, Hengbo, Tetsushi YAMAMOTO, Yuji WADA a Shozo YANAGIDA. Largescale and size-controlled synthesis of silver nanoparticles under microwave irradiation. Materials Chemistry and Physics [online]. 2004, 83(1), 66–70. ISSN 0254-0584. Dostupné z: doi:10.1016/j.matchemphys.2003.09.006
[9] SHARMA, Deepali, Sapna SHARMA, B. S. KAITH, Jaspreet RAJPUT a Mohinder KAUR. Synthesis of ZnO nanoparticles using surfactant free in-air and microwave method. Applied Surface Science [online]. 2011, 257(22), 9661–9672. ISSN 01694332. Dostupné z: doi:10.1016/j.apsusc.2011.06.094

Influence of higher partitioning temperatures on mechanical properties of heat treated high-strength steel alloyed with 1.3 % chromium

Tomáš Janda[1], Karolína Burdová, Michal Peković, Adam Stehlík

University of West Bohemia, Regional Technological Institute, Univerzitní 22, 301 00 Pilsen, Czech Republic

[1]jandat@rti.zcu.cz

Abstract. This paper deals with the innovative Quenching and partitioning (QP) heat treatment of low-alloyed chromium steel, and is especially focused on the higher temperatures of partitioning as well as longer holding times and its influence on mechanical properties of the material. As part of the experiment various heat treatment, metallographic analysis, hardness measurement, X-ray diffraction phase analysis (to determine retained austenite content), tensile test and Charpy impact test were performed. In QP treated specimens the best combination of tensile strength and elongation was observed after quenching in 200 °C salt bath and partitioning at 250 °C for 30 minutes. This specimen showed the tensile strength above 1900 MPa, elongation of 14 % and also good impact toughness (34 J). Equally good values of impact toughness (36 J) were also observed in the specimens partitioned at 300 °C or 320 °C for holding times up to 30 minutes. On the other side longer partitioning times at the temperature of 320 °C affected the elongation and impact toughness negatively.

1 Introduction

Current automotive structures and their safety requirements place great demands on the material and its mechanical properties. In the high strength steels considerable ductility and good fatigue properties together with high ultimate tensile strength (UTS) are often required. In order to achieve such a combination of mechanical properties the application of the innovative methods of processing is necessary. A convenient technology of heat treatment (HT) may be represented by quenching and partitioning (QP), nanobainitisation or some other derived processes utilizing the beneficial properties of film morphology retained austenite (RA) in the microstructure [1, 2]. These methods depend on precise parameters of processing. In the QP processing it is a cooling rate, temperature of austenitization and quenching media during the quenching step as well as temperature and holding time of the immediately following partitioning. After the quenching a martensitic microstructure with increased RA content is obtained. During the partitioning step diffusion processes (carbon migrating from supersaturated martensite to RA) results in stabilization of RA in the microstructure [3-5]. Due to film morphology RA in martensitic or bainitic microstructure composite-like microstructure with increased ductility and toughness is obtained [5-7].

Some experiments with the 42SiCr steel were carried out using QP process on thermomechanical simulator [8, 9] and also in the real process. Different quenching temperatures and quenching media were tested in order to change cooling rate and also different temperatures and holding times of partitioning were applied [9-11]. The martensitic/bainitic microstructures with thin film RA were obtained. A goal of the study was to establish relations between the mechanical properties and microstructure development with respect to the method of HT. As described in this paper, previously tested [9, 11] relatively low temperatures and short holding times was completed with higher temperatures and longer holding times of partitioning in order to clarify the development of microstructure and ductility during such processing.

2 Experiment

As the experimental material high-strength steel alloyed with 1.3 % chromium and 2.03 % silicon was chosen (table 1). Increased silicon content should suppress carbide precipitation and helps to stabilize austenite during partitioning [10]. Chromium improves hardenability and together with carbon ensures high strength of the steel. The steel was further alloyed by some amount of manganese which should delay the carbide precipitation and pearlite transformation [9]. The suitability of the chemical composition for QP process was verified elsewhere [8, 12]. The phase transformation temperatures were calculated in JMatPro software.

Table 1. Chemical composition and transformation temperatures of the experimental steel.

C [wt%]	Si [wt%]	Mn [wt%]	P [wt%]	S [wt%]	Cr [wt%]	Mo [wt%]	Ni [wt%]	Al [wt%]	Nb [wt%]	Ms [°C]	Mf [°C]
0.43	2.03	0.59	0.009	0.004	1.33	0.03	0.07	0.008	0.03	298	178

The experimental material was supplied as the 10 mm thick rolled strip from which the specimens of approximately 110×80×10 mm were cut using waterjet. The initial microstructure was martensitic/bainitic with small amounts of pearlite and free ferrite (figure 1). Its hardness was 391 HV10. Prior to the experimental HT, the specimens were homogenized at 1100 °C for 4 hours in a furnace with an argon atmosphere. The resulting microstructure was pearlitic with a small amount of granular ferrite at prior austenite grain boundaries (figure 1) and the hardness was 280 HV10.

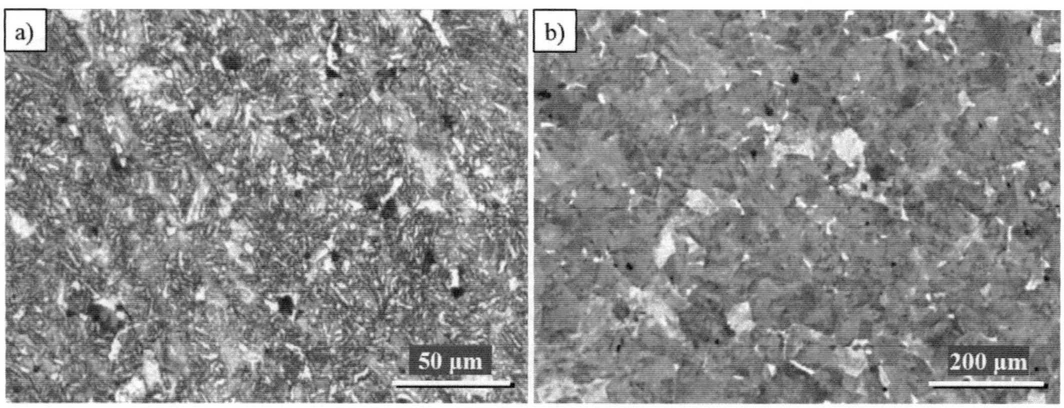

Figure 1. Microstructure of a) initial stock, b) homogenized steel.

Experimental program was divided into two stages. At the first stage, six QP process sequences were designed in order to describe the influence of partitioning temperature on evolution of the microstructure and their mechanical properties. The specimens were processed in a furnace without protective atmosphere. Austenitization at 950 °C for 20 minutes and quenching to the salt bath of 200 °C were applied. Partitioning at various temperatures (PT - 250, 300 and 320 °C) and holding times (15, 30 minutes) followed (figure 2). These parameters were chosen on the basis of results of previous experiments [8, 11]. The second stage of the experiment was focused on the long holding times (up to 72 hours) applied at the partitioning temperature of 320 °C (figure 2) in order to study diffusion processes and their influence on the mechanical properties of the steel.

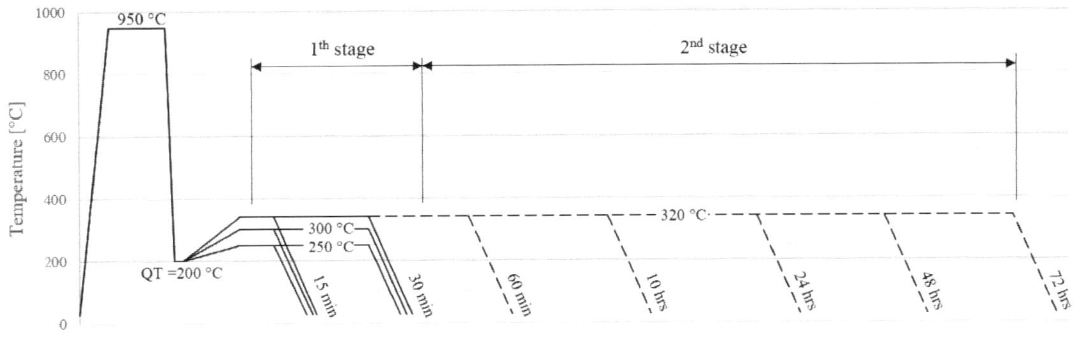

Figure 2. HT sequences of first and second stage of the experiment.

After HT microstructure was examined using light optical and scanning electron microscopy and its hardness was measured. X-ray diffraction phase analysis was carried out using AXS Bruker D8 Discover diffractometer equipped with HI-STAR detector and a cobalt cathode ($\lambda_{K\alpha}$ = 0.1790307 nm). Measurements were performed in the center of metallographic sample in a range of diffraction angles 2ϑ from 25 to 105°. To determine mechanical properties static tensile test on the round bar specimens 5.0 mm in diameter and a gauge length of 25 mm and Charpy V-notch test on specimens 55x10x10 mm was carried out. The tests were performed on three samples. The mean values as well as standard deviations were calculated from the results.

3 Results and discussion

Metallographic analysis revealed martensitic microstructures with small amounts of bainite same in all the specimens (figure 3). No specimen showed signs of extensive precipitation of carbides.

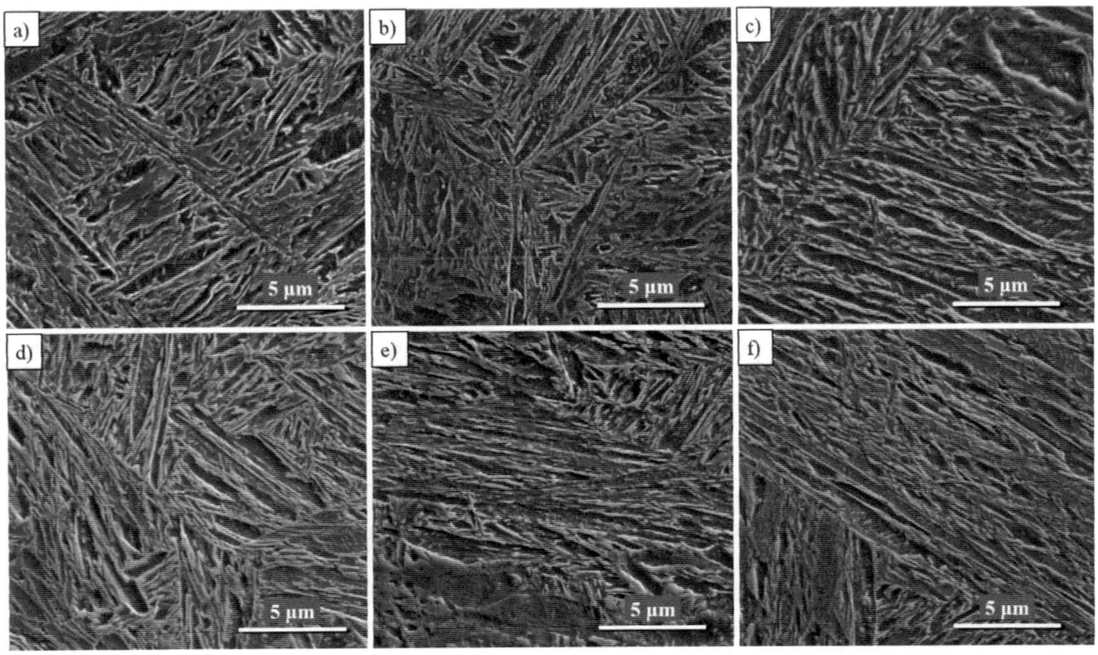

Figure 3. Martensitic/bainitic microstructures after partitioning at: a) 250 °C for 15 min, b) 300 °C for 15 min, c) 320 °C for 15 min, d) 320 °C for 60 min, e) 320 °C for 24 h, f) 320 °C for 72 h.

Any coarse RA austenite particles were not observed on SEM micrographs. Therefore, with respect to high content of RA (determined by X-ray diffraction analyzes), the film RA on grain boundaries can be supposed. The thickness of these films may be around tens of nanometers.

3.1 Effect of partitioning temperature (short holding times)

Using the X-ray diffraction analyses the RA content of selected specimens were measured. The specimens partitioned at the temperatures of 250 °C and 300 °C showed lower RA content - 11% and 10.5 %, respectively (figure 4). An increase of the RA content was observed at a temperature of 320 °C (12.5% for 15 minute hold; 13% for 30 minute dwell). With respect to not very large measurement differences it can be assumed that the portion of RA only slightly increased with increasing PT.

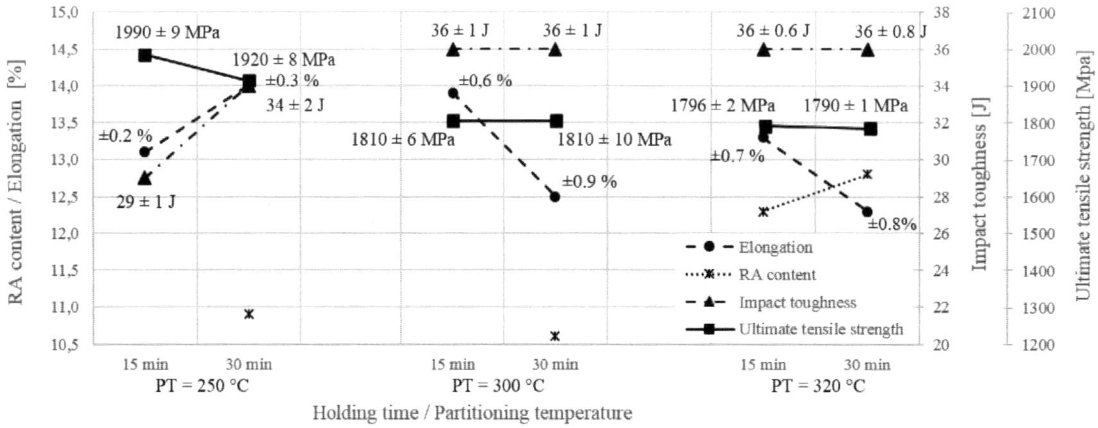

Figure 4. The UTS, elongation, impact toughness and RA content vs. PT and holding time.

A similar tendency was also observed in the results of Charpy impact tests. The specimens partitioned at 250 °C showed lowest impact toughness and at the same time an increasing tendency along with the increasing in holding times - 29 J and 34 J for 15 min and 30 minute dwell, respectively. Partitioning at the temperatures of 300 °C and 320 °C provided higher impact toughness: 36 J equally for both temperatures and holding times (figure 4). This increase can be attributed both, to a higher portion of RA and to a higher stage of martensite decomposition (higher intensity of diffusion processes). This assuming can be further supported by decrease of the UTS from 1990 MPa at the PT 250°C below 1800 MPa at the PT of 300 °C and 320 °C (figure 4). A rather inconsistent trend of elongation is due to small differences in measured values and relatively large standard deviations not reliable.

3.2 Effect of long holding times (partitioning temperature 320 °C)

A specific course of RA content (and also impact toughness and elongation) values was observed in these specimens. At first the values decreased with increasing time of partitioning but at very long dwells they slightly raised again (figure 5). The course of impact toughness corresponded quite well with the RA content evolution. The elongation measurement results were somewhat worse (for its relatively small range of measured values and quite a large standard deviations). The maximum values were observed at the holding times of 15 and 30 minutes: RA content above 12%, impact toughness 36 J and elongation above 12 %. During longer dwells on PT all these characteristics decreased quite rapidly. The RA content was reduced down to 11% beyond the holding time of 10 h. Such a long dwell at 320 °C probably caused exceeding of carbide nucleation energy limit which had been artificially increased by special alloying. It would enable diffusion of carbon from saturated stabilized RA to nearby carbides and hence the decomposition of some portion of RA.

The values of impact toughness and elongation also decreased down to 28 J and 12 % after 24 and 10 hours, respectively. However, even longer partitioning times caused again an increase up to 31 J and above 13%, respectively. The primary decline of ductility correlate quite well with the course of RA content and shows clearly the relation of film morphology RA content and the ductility of the steel. The loose of ductility can be also caused by rod-like morphology of carbides that could precipitated on martensite lathes boundaries. With even longer hold on the PT coarsening of the carbides and transformation into their globular form as well as further recovering and starting recrystallization could occur. This would explain the slow increase of elongation and impact toughness at the holds longer than 24 hours. The UTS was only slightly affected by the change of holding time. Gradual decrease from the 1800 MPa (for 15 minutes hold on PT) to 1740 MPa (for 72 hours hold) was observed.

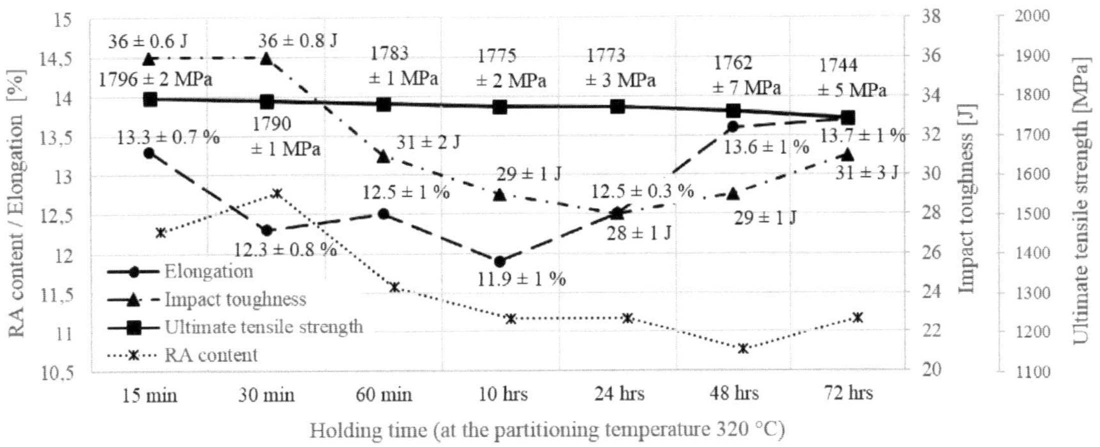

Figure 5. The UTS, elongation, impact toughness and RA content vs. hodling time at PT 320 °C.

4 Conclusion

Described experiment dealt with the QP process of high-strength steel alloyed with 1.3 % chromium and 2.03 % silicon. It especially focused on the effect of higher partitioning temperatures (PT) and long holds and their impact to the microstructure and mechanical properties. The main task was to investigate whether the microstructures with even higher retained austenite (RA) content and so higher ductility and toughness could be obtained. Higher PT (300 and 320 °C) examined in the first stage of experiment resulted in only little contribution to the impact toughness - from 34 J to 36 J (30 min hold) but at the same time it caused a decrease in ultimate tensile strength (UTS) from 1990 to values around 1800 MPa. At this stage prolonged hold (especially at lower PT) appeared to be more beneficial. The long holds (at PT of 320 °C) in the second stage of experiment influenced the elongation and impact toughness rather negatively. A decrease in elongation, impact toughness and also the RA content at holding times longer than 30 min together with only gradually declining UTS indicated slow start of precipitation and following partial decomposition of RA. Holds longer than 48 hours resulted in slight increase in elongation and impact toughness again, and further slow decrease of UTS. The recovery and recrystallization processes obviously played the main role here. To confirm these theory further analyzes employing e.g. transition electron microscopy and detail study of the microstructure will follow in the next step.

Acknowledgements
This contribution has been prepared under project LO1502 'Development of the Regional Technological Institute', under the auspices of the National Sustainability Programme I of the Ministry of Education of the Czech Republic aimed at supporting research, experimental development and innovation.

References

[1] Edmonds D V et al. 2006 Quenching and partitioning martensite - A novel steel heat treatment *Materials Science and Engineering: A* Vols. **438–440** pp. 25–34 DOI: 10.1016/j.msea.2006.02.133

[2] Bhadeshia H K D H 2005 High Performance Bainitic Steels *Materials Science Forum* Vols. **500-501** pp. 63-74

[3] Li H Y and Jin X J 2010 Microstructural Evolution of Medium Carbon Steels during the Quenching and Partitioning Process *Materials Science Forum* Vols. **654-656** pp. 86-89

[4] Xie Z J et al. 2014 Stabilization of retained austenite by the two-step intercritical heat treatment and its effect on the toughness of a low alloyed steel *Materials and design* Vol. **59** pp. 193-198

[5] XU Y et al. 2014 Microstructure evolution and mechanical properties of a hot-rolled directly quenched and partitioned steel containing proeutectoid ferrite *Materials Science & Engineering: A* Vol. **607** pp. 460–475

[6] Jirková H et al. 2014 Influence of metastable retained austenite on macro and micromechanical properties of steel processed by the Q-P proces. *Journal of Alloys and Compounds* Vol. **615** pp. S163–S168 DOI: 10.1016/j.jallcom. 2013.12.028

[7] Wang Y 2010 Study of a novel Ultra-High Strength Steel with adequate ductility and toughness by Quenching-Partitioning-Tempering Process *Materials Science Forum* Vol **654** pp. 37-40

[8] Kučerová L, Jirková H and Káňa J 2016 The suitability of 42SiCr steel for quenching and partitioning process *Manufacturing Technology* Vol. **16** No. 5 pp. 984 – 989 ISSN: 1213-2489

[9] Jirková H, Kučerová L and Mašek B 2015 The effect of chromium on microstructure development during Q-P process *Materials Today: Proceedings* Vol. **2** supl. 3 pp. 627-630 ISSN: 2214-7853

[10] Kučerová L et al. 2009 Optimization of Q-P Process Parameters with Regard to Final Microstructures and Properties *Annals of DAAAM for 2009 & Proceedings of the 20th international DAAAM symposium* pp. 1035-1036 ISSN: 1726-9679

[11] Janda T, Jirková H, Jeníček Š, Kučerová L, 2019 Influence of cooling rate on microstructure and mechanical properties of 42SiCr steel after Q&P process *Manufacturing technology (In print)*

[12] Kučerová, L, Jirková H, Mašek B 2014 The Effect of Alloying on Mechanical Properties of Advanced High Strength Steels *Archives of Metallurgy and Materials* Vol **59** pp. 1189-1192 ISSN: 1733-3490

IOP Conf. Series: Materials Science and Engineering **723** (2020) 012012 doi:10.1088/1757-899X/723/1/012012

Combination of press-hardening and isothermal holding in the treatment of high-strength steels

Hana Jirková*, Kateřina Opatová, Štěpán Jeníček, Ludmila Kučerová

University of West Bohemia, RTI - Regional Technological Institute, 301 00 Pilsen, Czech Republic

*hstankov@rti.zcu.cz

Abstract. Today, high-strength parts for the automotive industry are mainly produced by the press-hardening technology. In order to reduce costs, savings in production and optimal materials are sought. One option involves the use of high strength steels of various chemical compositions. Two high strength steels were studied in the present experiment. The first one was CMnSi steel, a typical low-alloy TRIP steel with a carbon content of 0.2 %; alloyed with manganese and silicon. The second one was 42SiCr steel, a member of the group of martensitic steels, with a carbon content of 0.42 %. In addition to manganese and silicon, this steel is also alloyed with chromium. The CMnSi steel has not proved to be very sensitive to changes in process parameters, achieving an ultimate tensile strength of more than 950 MPa and an elongation of over 10%. 42SiCr steel was found to respond to both heating temperature and deformation, showing different mechanical properties. With the right combination of process parameters, an ultimate tensile strength of over 2100 MPa was achieved.

1. Introduction

Press-hardening, a thermomechanical treatment method which combines conventional hot forming and cold pressing, has been evolving rapidly in recent years [1]. Using this technique, high-quality automotive parts can be produced and treated to strengths over 1500 MPa [1-3]. These include parts such as front bumper, tail bumper, A-pillar, B-pillar, C-pillar, roof frame, floor frame, door panel, and door anti-beam [2-4]. At the same time, new materials are sought and developed with a view to reduced weight, less emissions and improved passenger safety. The first materials used for press hardening operations were boron steels similar to the 22MnB5 grade [1, 5]. Grades such as USIBOR1500, Docol Boron 02 and B1500HS followed suit. Several companies are in the process of developing new materials with compositions based on Mn-Mo-B, Mn-Cr, Mn-Cr-B, Mn-W-Ti-B and other concepts [1]. Steels of this kind are known as PH steels (press hardening steels).

Other grades which can be used for this process include advanced high-strength steels in which retained austenite is produced purposefully in order to improve their properties. These are known as TRIP and QP steels [6-9]. When subjected to deformation, these steels exhibit the TRIP effect (transformation induced plasticity). TRIP effect occurs in the course of cold deformation, which includes impact and crash scenarios. In such an event, retained austenite transforms to strain-induced martensite [10, 11]. As a result, these steels can absorb large amounts of crash energy, which makes them good candidates for safety components for cars. To achieve the desired mechanical properties to these steels, special heat treatment routes are needed which involve isothermal holding. Using appropriate treatment parameters, strengths up to 1100 MPa and elongations of more than 30% can be achieved in TRIP steels. In QP steels, these values can reach 2000 MPa and 10–14%, respectively. However, to be able to obtain these properties in these steels, one needs to test and understand their behaviour during press hardening and, where relevant, during isothermal holding.

2. Experimental Programme

Two high-strength steels have been selected for the present experiments (table 1). One of them was a TRIP steel which contained 0.2% C, manganese and silicon. These alloy additions are typical of this group of steels. The other was a martensitic steel from the class of materials intended for Q&P processing. It contained more than 0.4% C, manganese, silicon, and chromium at a level of 1.33%, the purpose of which was to provide solid solution strengthening and better hardenability. The semi-product for the experiments was 1.8 mm sheets with untreated surfaces.

The initial structure of TRIP steel contained ferrite and pearlite and had a hardness of 190 HV10. Its ultimate strength and elongation were 627 MPa and 26%, respectively. The QP steel in initial state had a pearlitic microstructure with a small amount of proeutectoid ferrite. Its hardness was 222 HV10, ultimate strength was 668 MPa and the elongation level was 25%.

Their phase transformation temperatures were determined using a Bähr dilatometer; on specimens which were 5 mm in diameter and 10 mm in length. The A_{c3} temperature was 909°C in the TRIP steel and 844°C in the QP steel. The M_s was determined at a cooling rate of 30°C/s. In the TRIP steel, it was 387°C. In the QP steel, it was 305°C.

Table 1. Chemical compositions of experimental steels (wt. %).

	C	Mn	Si	Al	Nb	P	S	Ni
TRIP steel	0.21	1.4	1.8	0.006	0.002	0.007	0.005	0.07
QP steel	0.42	0.68	1.96	0.008	0.07	0.01	0.01	0.07

2.1. Sequences for simulating press hardening operations

Testing the use of new materials directly in production can prove difficult both economically and technologically. One of alternatives to such attempts involves testing new material processing techniques in the laboratory and under conditions closely approaching the real-world process. Based on this reasoning, this study was carried out using a thermomechanical simulator, for which process sequences can be constructed from data measured in real production.

For the present study, time-temperature profiles, handling times and forming tool temperatures were measured in a real-life press hardening process. Based on this data, physical simulation sequences as shown in figure 1 were developed. These sequences included soaking for 180 seconds. Two soaking temperatures were proposed and used: 860°C and 940°C. The temperature 860°C was below the A_{c3} temperatures of both materials whereas the temperature of 940°C was in a single-phase region. Soaking was followed by cooling which simulated free cooling of a sheet stock while being transferred from the furnace to a tool. The duration of this cooling step was chosen to be 7 s. The sheet forming step was simulated by tensile deformation of the specimen at $\varepsilon = 0{,}15$. In order to explore the effects of strain on microstructural evolution, the deformation was applied at three different temperatures: 680°C, 750°C and 800°C. Deformation was followed by rapid cooling to room temperature at a rate equal to the cooling rate in a closed tool maintained at room temperature. With the goal of improving elongation values, specimens of both steels were treated using additional sequences in which an isothermal hold was integrated in the final cooling phase. In the TRIP steel sequences, which involved deformation at various temperatures, quenching ended at 425°C, followed by holding for 600 seconds. This temperature was above the M_s. Generally, holding at temperatures in this range leads to formation of bainitic structures and to higher amounts of retained austenite. This particular temperature was chosen on the basis of earlier experiments [12, 13].

The same step was added to the sequences for the QP steel. The isothermal temperature was 250°C, which was between the M_s and M_f. The resulting sequence corresponds to Q-P processing, which also stabilises retained austenite and improves toughness in martensitic steels [14, 15].

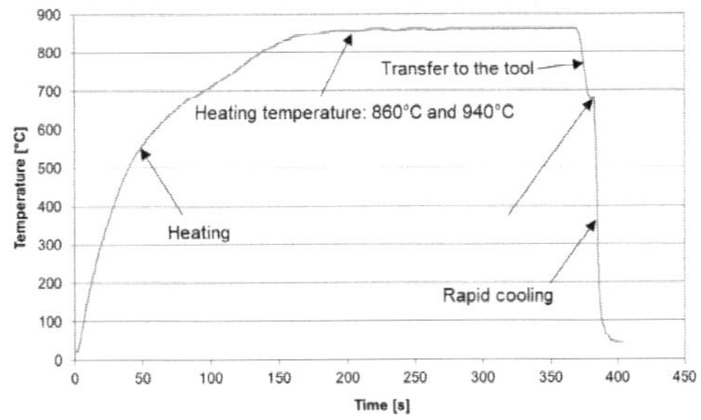

Figure 1. Schematic illustration of a press-hardening sequence.

Microstructures of the specimens were examined in metallographic sections made in the rolling direction and prepared by standard techniques. They were documented using an optical microscope (OM) and a scanning electron microscope (SEM). Microstructural features were revealed by etching with Nital. The amount of retained austenite was measured by means of XRD in the automatic powder diffractometer AXS Bruker D8 Discover with a position-sensitive area HI-STAR detector and a cobalt X-ray source ($\lambda K\alpha$ = 0.1790307 nm). Mechanical properties were evaluated using measurement of HV10 hardness and tensile testing.

3. Results and Discussion

3.1. Press hardening simulations without isothermal holding

In the TRIP steel, press hardening with both soaking temperatures without an isothermal step produced a martensitic-ferritic microstructure with a small amount of retained austenite and a hardness between 282 and 290 HV10 (table 2, figure 2a-c). Deformation at 680°C failed to promote formation of pearlite (figure 2a). No bainite was found, although its presence is quite common in TRIP steels. The amount of retained austenite was very small, 3-4%. It was a result of the absence of isothermal holding in the bainitic transformation region. This step promotes bainite formation and stabilises retained austenite. The resulting large amounts of martensite led to high strengths, from 860 to 946 MPa, but the elongation levels were low: 9-13% (table 2).

Table 2. Parameters of press hardening sequences without isothermal holding; mechanical properties of specimens.

Heating temp. [°C]	Def. temp. [°C]	TRIP steel					QP steel				
		$R_{p0.2}$ [MPa]	UTS [MPa]	A_{20} [%]	HV10 [-]	RA [%]	$R_{p0.2}$ [MPa]	UTS [MPa]	A_{20} [%]	HV10 [-]	RA [%]
860	680	480	933	12.2	282	4	647	1222	5.2	320	5
	750	456	885	13.1	282	4	616	1323	3.7	410	6
	800	464	942	8.9	284	3	665	1487	3.6	451	6
940	680	480	861	11.1	274	3	1454	2150	1.5	673	5
	750	471	909	10.8	281	4	1441	2068	1.5	661	3
	800	464	946	10.8	290	4	1413	2103	1.8	663	4

Soaking and deformation temperatures had a notable impact on the second steel (table 2). The sequence which involved the lower soaking temperature, 860°C (which was below the A_{c3}), and the

lowest deformation temperature, 680°C, produced not only martensite and ferrite but also pearlite (figure 2d). Hardness was 320 HV10 and ultimate strength and elongation were 1222 MPa and 5%, respectively. Sequences with higher deformation temperatures, 750°C and 800°C eliminated pearlite from the microstructure. The microstructure then consisted of martensite, ferrite and a small amount of retained austenite (figure 2e). The presence of a larger volume fraction of martensite led to higher hardness of more than 400 HV10. Ultimate strength was higher as well: 1323 MPa and 1487 MPa for deformation temperatures of 750°C and 800°C. Elongation levels were around 6% (table 2). The volume fraction of retained austenite was 5-6%.
Sequences with a soaking temperature of 940°C, which lies in the austenite region, led to higher mechanical properties. In all cases, martensite dominated in the resulting microstructure, being complemented by very small amounts of ferrite and retained austenite (table 2, figure 2f). Since most ferrite had been eliminated the ultimate strength was much higher. The ultimate strength exceeded 2000 MPa and hardness was higher than 660 HV10. The microstructure consisted of fresh martensite, and therefore elongation was very low, less than 2%.

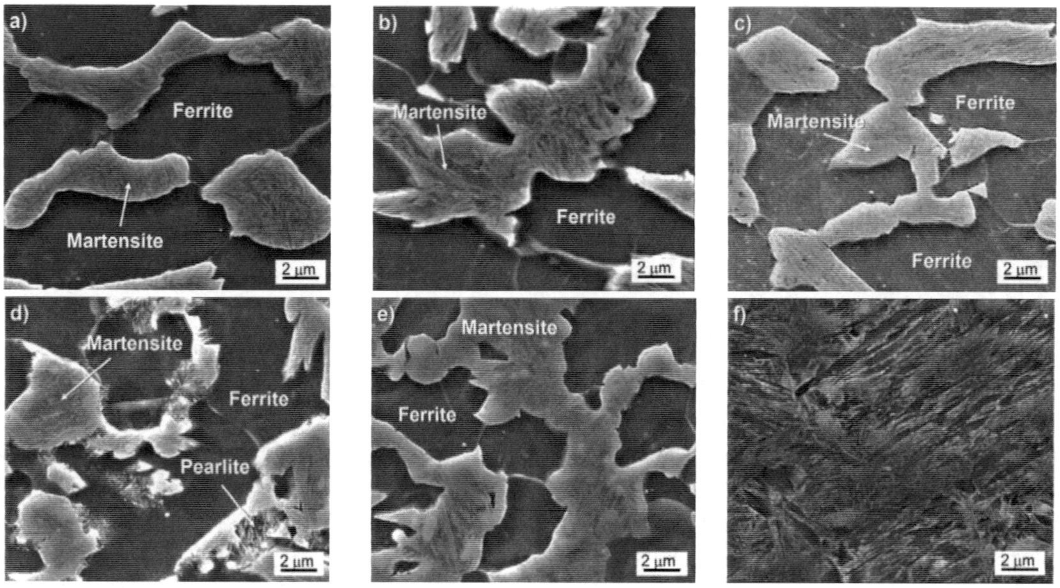

Figure 2. Microstructures produced by press-hardening sequences without isothermal holding: a) TRIP steel: 860°C – 680°C, b) TRIP steel: 860°C – 800°C, c) TRIP steel: 940°C – 800°C, d) QP steel: 860°C – 680°C, e) QP steel: 860°C – 800°C, f) QP steel: 940°C – 800°C.

3.2. Effects of an isothermal hold inserted in the cooling between the forming temperature and RT
With a view to improving elongation levels in both experimental materials, physical simulation sequence were developed and tested which included an isothermal step applied in the course of cooling to room temperature (table 3, figure 3a,b). For TRIP steel, this involved sequences with soaking temperatures of 860°C as well as 940°C. In sequences involving soaking at 860°C, two deformation temperatures were tested: 680°C and 800°C. The specimens were held at 425°C for 600 seconds. In all cases, this produced a small amount of bainite (figure 3a) and to a larger amount of retained austenite, approx. 7%. Though minor, this change in the microstructural composition resulted in higher elongation, by approximately 5%, i.e. to values of 17 and 19% (table 3). The ultimate strength was approx. 100 MPa lower than in the relevant reference case. The isothermal step was also reflected in the nature of fracture surfaces. In the earlier cases, there were brittle fracture with cleavages. Now the fracture was ductile and showed dimples (figure 3c, d).

For QP steel, the isothermal hold was only incorporated in the sequence with a soaking temperature of 940°C (table 3). The purpose of the isothermal step was to facilitate partitioning of carbon between super-saturated martensite and retained austenite. The temperature and time were 250°C and 600 seconds. Tempered martensite and about 9% retained austenite (figure 3b) were the result of this sequence. This hold at the partitioning temperature led to a notably higher elongation, 8%, and lower ultimate strength, 1885 MPa (table 3). In this case, too, isothermal holding influenced the nature of fracture surfaces. The mixed fracture changed to ductile fracture with dimples without signs of brittle failure (figure 3 e,f).

Table 3. Effects of isothermal step during cooling to RT on mechanical properties of TRIP steel.

Steel	Heating temp. [°C]	Def. temp. [°C]	Hold [°C/s]	$R_{p0.2}$ [MPa]	UTS [MPa]	A_{20} [%]	HV10 [-]	RA [%]
TRIP steel	860	680		408	800	18.6	254	6
		800	425/600	383	830	17.2	250	7
	940	800		362	750	19	244	6
QP steel	940	800	250/600	1085	1885	8.3	598	9

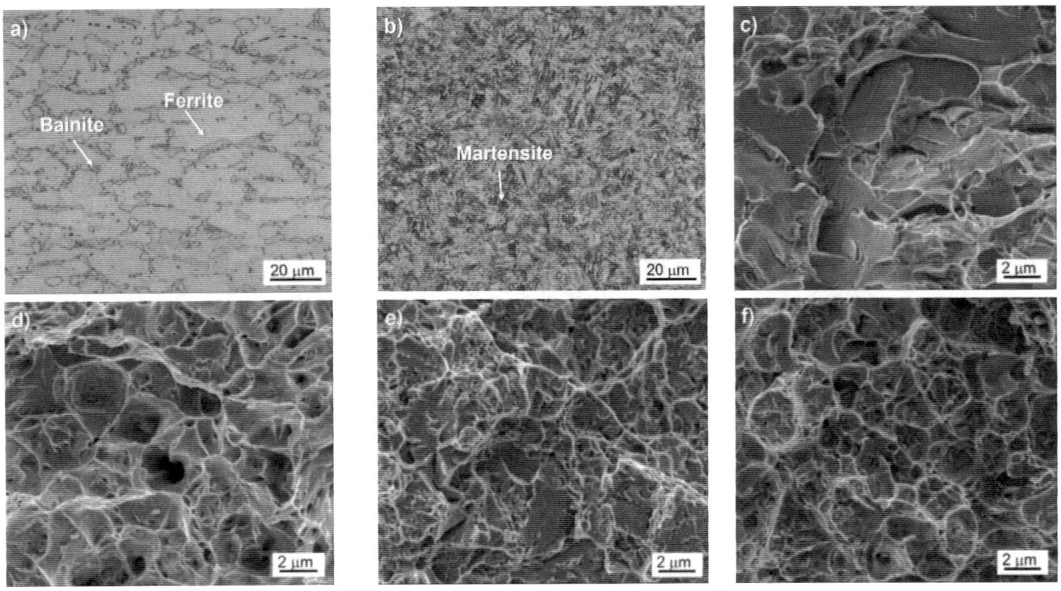

Figure 3. Microstructures produced by press-hardening sequences with isothermal holding: a) TRIP steel: 940°C – 800°C, b) QP steel: 940°C – 800°C; Fracture surfaces - c) TRIP steel: 940°C – 750°C (no isothermal hold), d) TRIP steel: 940°C – 800°C (with isothermal hold), e) QP steel: 940°C – 800°C (no isothermal hold), f) QP steel: 940°C – 800°C (with isothermal hold).

4. Conclusions

Sequences which simulated press hardening were carried out on two high-strength steels with different levels of carbon and alloying elements. Various soaking temperatures and deformation temperatures were used. In TRIP steel, which contained 0.2%, microstructure was not affected by changes in soaking and deformation temperatures to any appreciable extent. In all cases, the resulting microstructures were of ferritic-martensitic type and contained a small amount of retained austenite. Ultimate strengths were between 860 and 940 MPa, with elongations of more than 12%. Where isothermal holding was applied in the bainitic transformation region, bainite formed slightly more readily and a higher amount of retained austenite was stabilised. As a result, elongation increased to 19%.

In QP steel, changes in both soaking and deformation temperatures had a considerable impact on mechanical properties. To obtain the highest possible mechanical properties, soaking had to be performed at 940°C, which eliminated ferrite. Then, ultimate tensile strengths above 2000 MPa and elongation levels of 2% were achieved. Elongation was then improved by incorporating an isothermal step, resulting essentially in a Q&P process. Consequently elongation rose to 8% while the ultimate strength was 1885 MPa.

Acknowledgements

The present contribution has been prepared under project LO1502 "Development of the Regional Technological Institute" under the auspices of the National Sustainability Programme I of the Ministry of Education of the Czech Republic aimed to support research, experimental development and innovation.

References

[1] Hu P, Ying L and He B 2017 Hot Stamping Advanced Manufacturing Technology of Lightweight Car Body (Singapore: Springer) chapter 3 pp 45-94

[2] Karbasian H and Tekkaya A E 2010 A review on hot stamping *J. of Mat. Proc. Tech.* **10** 2103

[3] Mori K, Bariani P F, Behrens B A, Brosius A, Bruschi S, Maeno T, Merklein M and Yanagimoto J 2017 Hot stamping of ultra-high strength steel parts *CIRP Annals – Manuf. Tech.* **66** 755

[4] Hu P, Ying L and He B 2017 Hot Stamping Advanced Manufacturing Technology of Lightweight Car Body (Singapore: Springer) chapter 2 pp 19-44

[5] Sparrer Y, Rüskamp A, Tenié A, Lian J, Münstermann S and Bleck W 2019 Proc. of 7th International conference: Hot sheet metal forming of high-performance steel (Lulea) (Auerbach: Verlag Wissenschaftliche Scripten) p 15

[6] Jirková H, Opatová K, Káňa J, Bublíková D and Bystrianský M 2018 Integration of Press-Hardening Technology into Processing of Advanced High Strength Steels *Mat. Sci. Forum* **941** 317

[7] Grajcar A and Krztoń H 2011 Effect of isothermal holding temperature on retained austenite fraction in medium-carbon Nb/Ti-microalloyed TRIP steel *J. of Achiev. in Mat. and Manuf. Eng.* **49/2** 391

[8] Seo E J, Cho L and De Cooman B C 2014 Application of quenching and partitioning (Q&P) processing to press hardening steel *Metall. and Mat. Tran. A* **45** 4022

[9] Jirková H, Opatová K, Jeníček Š, Vrtáček J, Kučerová L and Kurka P 2019 Use of Multi-phase TRIP Steel for Press-hardening Technology *Acta Metall. Slov.* **25** 101

[10] Gu X, Xu Y, Peng F, Misra R D K and Wang Y 2019 Role of martensite/austenite constituents in novel ultra-high strength TRIP-assisted steels subjected to non-isothermal annealing *Mat. Sci. and Eng. A* **754** 318

[11] Kong H, Chao Q, Rolfe B and Beladi H 2019 One-step quenching and partitioning treatment of a tailor welded blank of boron and TRIP steels for automotive applications *Materials & Design* **174** 107799

[12] Mašek B, Staňková H, Nový Z, Meyer L W and Kracík A 2009 The Influence of Thermomechanical Treatment of TRIP Steel on its Final Microstructure *J. of Mat. Eng. and Per.* **18** 385

[13] Jeníček Š, Vorel I, Káňa J, Opatová K, Rubešová K, Kotěšovec V and Mašek B 2017 IOP Conference Series-Materials Science and Engineering (Chemnitz) vol. 181 (Bristol: IOP Publishing Ltd) p 1

[14] Edmonds D, He K, Rizzo F C, De Cooman B C, Matlock D K and Speer J G 2006 Quenching and partitioning martensite-A novel steel heat treatment *Mat. Sci. and Eng. A.* **438–440** 25

[15] Bublíková D, Jeníček Š, Vorel I and Mašek B 2017 IOP Conference Series-Materials Science and Engineering (Pilsen) vol. 179 (Bristol: IOP Publishing Ltd) p 1

IOP Conf. Series: Materials Science and Engineering **723** (2020) 012013 doi:10.1088/1757-899X/723/1/012013

The effect of mechanical surface pre-treatment on the strength of the adhesive joint of high strength sheets

P Kejzlar [1] O Seibert [1] and T Pilvousek [2]

[1] Technical University of Liberec, Studentska 1402/2, 461 17, Liberec, Czech Republic

[2] Skoda Auto a.s., tř. Václava Klementa 869, 293 01 Mladá Boleslav, Czech Republic

E-mail: pavel.kejzlar@tul.cz

Abstract. Nowadays, in the car body parts production, the quenching is often replaced by adhesive joints. The adhesive joints have a lot of advantages – they allow to keep the protective layer, enable to join different material and prevent from the formation of heat affected zones. With respect to passenger's passive safety, plenty of car body parts are manufactured of high strength steel sheets. The body parts are hot-formed at a temperature of 900 °C to the desired shape followed by subsequent rapid quenching to achieve the required strength. The sheets are covered with an AlSi 60/60 layer which should protect them from high-temperature oxidation during the forming process. However, brittle intermetallic interlayers form during the high-temperature treatment due to the reaction of iron from the substrate with Al/Si from the protective cover, which negatively affects the final strength of adhesion joint. The paper deals with the comparison of different possible ways of removing the layer on the final strength.

1 Introduction

The key joining technologies used in the automotive are welding, riveting and adhesive joining. In the last years, we can notice an increase in the ratio of adhesive joints. The adhesive joints show plenty of benefits. The application of adhesive joints can decrease of car weight, doesn't affect negatively the material mechanical properties, and, opposite to the other methods, enables joining parts having different materials (e.g. metal to polymer). Kilograms of different adhesives can be found in today's cars starting from car bodies up to interiors fulfilling different functions like adhesives, sealings and/or shock absorbents. The current market offers a huge variety of adhesives differing in their composition, application procedure or their reinforcement. The selection of optimal adhesive for the specific application is very complicated and a lot of different parameters and requests have to be considered and took into account. Thus, all the new application of adhesive joints has to be thoroughly tested for their strength, durability, and feasibility. [1]

The high strength car body parts are typically made of micro-alloyed martensitic boron-manganese steels. [2] These parts are formed to their final shape by hot-stamping technology. The steel sheets are austenitized at temperatures around 900 °C and then simultaneously rapidly formed and quenched in a die. [3, 4]. Because of undesirable oxidation and decarburization during the thermomechanical treatment the sheets are often protected by Al-Si coating. [5, 6]

The huge application potential of adhesives in car bodywork production is where it is necessary to preserve anticorrosion protection of original parts. In fact, many protective layers can be damaged

IOP Conf. Series: Materials Science and Engineering **723** (2020) 012013 doi:10.1088/1757-899X/723/1/012013

by conventional welding and then can become the starting point for unwanted corrosion. The usage of proper adhesives offers an advantageous alternative, which enables to achieve similar or even higher strength. The application of adhesives allows to replace and supply function of parts for vibration damping and shock absorption. However, the application of adhesives is sometimes limited, because most of them require a clean surface with good integrity for achieving sufficient strength and durability. The failure of adhesive joints can appear in different areas through different mechanisms. The typical adhesive joint failure causes are insufficient adhesion (connected to adhesive/substrate wettability) and insufficient cohesion (low mechanical strength of adhesives). [1] The Al-Si layer protects the steel substrate from high-temperature oxidation, but at stamping temperatures occurs diffusion between Fe-substrate and Al-Si layer [7]. As well known, in the Fe-Al system appear some intermetallic phases (Figure 1) which show very undesirable properties at room temperature as considerable brittleness and low thermal and electric conductivity. [8, 9]

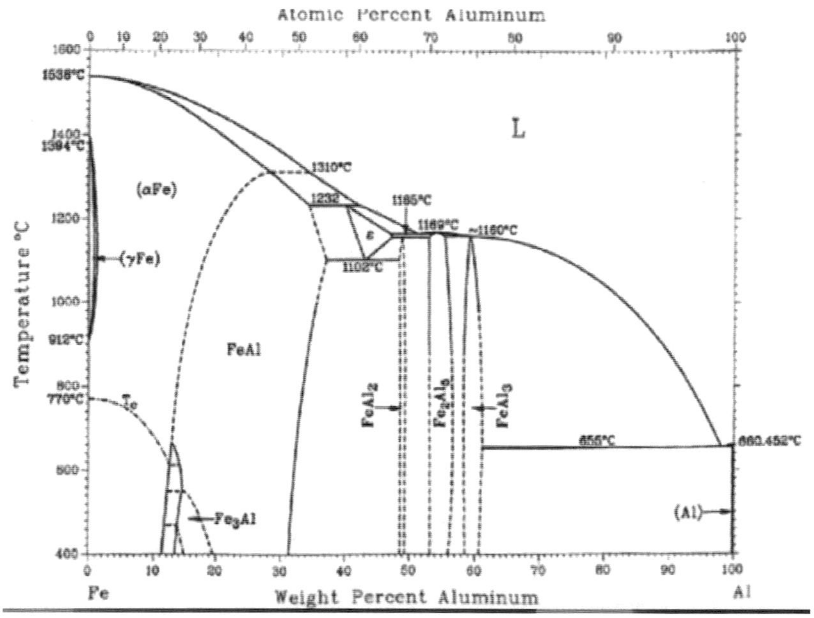

Figure 1. A binary Fe-Al diagram showing the presence of a plenty of different intermetallic phases corresponding to alloy's chemical composition [9].

The presence of cracks observed on metallographic cross-sections (Figure 2) made of these materials indicates insufficient cohesion of these layers leading to delamination at even low mechanical load. It is obvious, that the presence of these intermetallic multi-phase multilayers has a detrimental effect on the final mechanical strength of adhesive joints and thus this brittle multi-layer should be removed before the adhesive application to allow achieving desired mechanical strength.

The present paper deals with the evaluation of the effect of different ways of removing the intermetallic multilayer on the mechanical strength of adhesive joints.

2 Experimental

Rectangular samples of high-strength steel MBW 1500 with dimensions of 25×150 mm with a thickness of 1.2 mm were quenched from the temperature of 900 °C / 8 min; massive steel plates were used for rapid cooling to achieve conditions corresponding to the car body parts manufacturing process. The formation of the intermetallic Fe-Al-Si multiphase layer is evident from Figure 2.

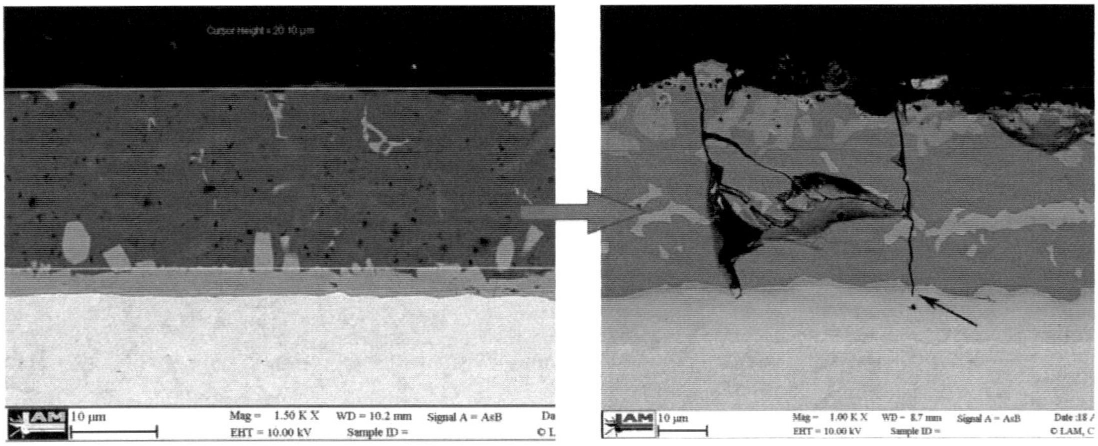

Figure 2. The Al-Si layer in the inicial state (A) and the formed Al-Si-Fe multiphase layer (B) formed during the heat treatment. The cracks observed after the quenching clearly indicate considerable brittleness and thus low cohesion of the formed layer.

Three types of intermetallic layer removal techniques were tested. 1 set of samples were untreated as a reference. For the second sample set, the intermetallic layer was removed using a rotary steel brush. The third set was treated by sandblasting using SiO_2 sand with a grain size of 40 µm followed by cleaning using high-pressure air. The surface of the last sample set was treated (Figure 3) using diode pumped fiber laser JK Fiber, with a wavelength in the range of 1070-1090 nm. The laser beam was applied to a square area of 1 cm^2. The laser intensity was set to 30 W and the focus point was set 4 mm above the surface. The gap between the individual laser tracks was 0.2 mm and the laser moved at 500 mm/min. The goal of the laser treatment was to re-melt and homogenize the material and remove the surface cracks.

Figure 3. The laser-treated area on the sample surface.

The samples were joined using Betamate 1040 adhesive. Copper wires with a diameter of 0.2 mm were used to set the required adhesive thickness. The samples were fixed using metal clamps (Figure 4) and subsequently cured at 180 °C for 2 hours in a laboratory furnace.

Figure 4. Sample ready for the curing.

The adhesive joint strength was evaluated by uniaxial shear test (Figure 5), where the maximum tension was calculated as $\sigma_{max} = \frac{F_{max}}{S}$, where F_{max} is the maximal mechanical load and S is the adhesive joint area.

Figure 5. A schema of adhesion joint loading during the measurement of mechanical strength.

Further, the image analysis was applied to calculate the area corresponding to the different failure mechanisms, where CF represents the fraction of area of cohesion failure; DF failure caused by layer delamination and AF represents adhesion failure.

3 Results and discussion

The morphology of the failure surface is shown in Figure 6. The used adhesive shows excellent adhesion and that why there isn't observable any area showing adhesive failure at adhesive / substrate interface. Instead, two other failure mechanism took part. We can easily distinguish two different colours manifesting cohesive failure of the interlayer (grey) and cohesive failure of the used adhesive (blue). It should be noticed, that the maximum strength of the adhesive joint was achieved if its strength is limited by the cohesive strength of the used adhesive or adherent.

In Figure 6A we can see that the failure is mostly caused by delamination of the brittle intermetallic layer (visible as grey area). This is in good coincidence with the results shown in Figure 2B (cracks present in the intermetallic layer formed during the heat treatment).

After the brushing (Figure 6B) we can notice a slight increase in the proportion of blue area, which marks cohesion failure of the used adhesive. Quite large grey area indicates the insufficient intensity of brushing (the intermetallic layer wasn't removed properly).

After the sandblasting (Figure 6C), the failure area is almost completely blue, what indicates that only adhesive cohesion failure took part and maximal strength of the joint was achieved.

In the case of laser treated sample (Figure 6D), we can observe central laser treated square in the blue colour, which means that the laser treatment eliminated layer delamination. Around the laser treated square is grey area indicating the layer delamination.

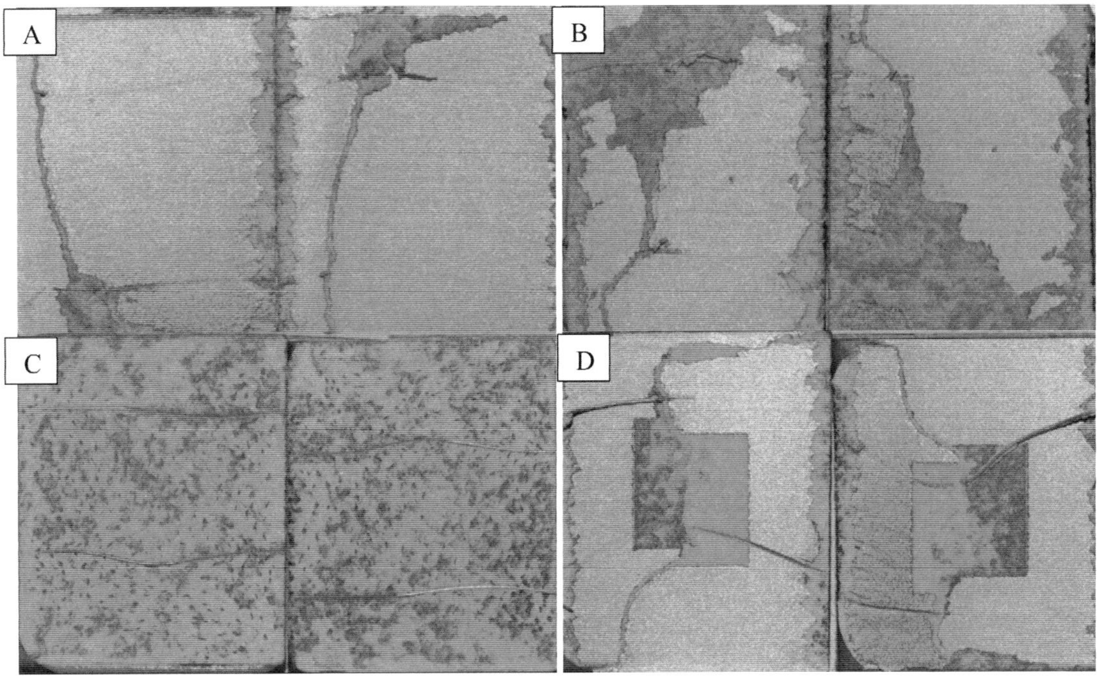

Figure 6. Fracture surface structures. A) Untreated sample; B) after brushing; C) sandblasted; D) laser treated.

Table 1. The values of measured σ_{max} and their relative change achieved by different methods and area fractions corresponding to different failuramechanisms.

	σ_{max} [MPa]	$\sigma_{max}/\ \sigma_{ref}$ [%]	Failure mechanism
Untreated (ref.)	32.2	-	CF (24 %) + DF (76 %)
Brushed	34.0	5.6	CF (27 %) + DF (69 %) + AF (4 %)
Sand blasted	36.7	14.0	CF (94 %) + DF (2 %) + AF (4 %)
Laser	31.6	-1.9	CF (20 %) + DF (76 %) + AF (4 %)

The results of the mechanical test are summarized in Table 1. If we take untreated (quenched) sample into account as a reference, we can say, that brushing (↑ σ_{max} 5.6 %) and sandblasting (↑ σ_{max} 14 %) led to the enhancement of mechanical strength. The lower strength of brushed specimens compared to the sandblasted confirmed the result from failure surface observation – the brushing led only to partial intermetallic layer removing (decrease of delamination failure area only by 7 %) and thus it can be estimated, that more intensive brushing is necessary to bring more significant strength improvement. The mechanical removing of the brittle layer, especially in the case of sandblasting, changed the mechanism of failure from the layer delamination (from 76 % to 2 %) to cohesion failure of the adhesive (from 24 % to 94 %). The laser treatment showed to have a detrimental effect on the total strength of the adhesive joint (↓ σ_{max} 1.9 %)). Considering the failure area morphology, we can estimate, that the strength of directly laser treated area is higher because of cohesion failure of the adhesive, however in the surrounding material the failure mechanism corresponds to the intermetallic layer delamination. In the case of laser treatment, the decrease in the total strength could be explained by the fact that the heat induced during the laser treatment process caused further disintegration of the surrounding intermetallic layer (outside the laser treated square). Based on this it can be concluded, that it is necessary to apply the laser treatment on the whole joined area to eliminate the negative effect of the laser-affected surrounding area.

Despite the best results achieved through the sandblasting (14 % increase in the strength), it is necessary to consider technical aspects connected to the individual tested methods. The sandblasting is connected with a need for separation of the layer removal step from the subsequent adhesive joining due to production of high dust levels and the risk of contamination or damage of other surfaces. In the case of brushing, there is a need to optimize the process with respect to the layer thickness and brush hardness. Another issue is again the production and swirling of particles / dusts generated by the brushing or grinding process.
For these reasons, the laser homogenization of the layer seems to be the most promising method allowing to solve problems connected with the formation of the intermetallic layer. As obvious from the image, local laser re-melting of the layer prevented the layer delamination and pure adhesive cohesion failure was observed in the laser treated area. During the laser treatment, there are no particles released into the environment. In addition, laser cleaning technology can be relatively easily incorporated into the current joining technology. Of course, before the implementation into the adhesive joining technology, further optimization of laser cleaning technological parameters is required.

Conclusion

The strength of adhesive joints is limited by the presence of intermetallic multiphase layer formed during the forming and subsequent heat treatment. Undesirable properties of this layer (especially brittleness) bring a need to remove or modify this layer before the adhesive joining. The effect of three different possibilities on failure mechanism and final adhesive joint strength was evaluated and discussed. Brushing led to only partial removing of the intermetallic layer and thus only slight increase in the strength was observed (+5.6 %). The intermetallic layer was successfully removed by sand blasting leading to the increase of the mechanical strength by 14 %. Despite the laser homogenization showed decrease in the final mechanical strength -1.9 %, it was assessed as the most perspective method. In the laser treated square, the failure mechanism was changed from the intermetallic layer delamination to cohesion failure of the used adhesive and thus allowed to fully exploit the potential of the used adhesive. However, the cohesion of the surrounding intermetallic layer was negatively affected by the introduced heat and thus laser treatment of the whole joined area is necessary. In addition, the possibility of an accurate setting of the operating parameters of the laser beams makes it a promising candidate for incorporation into the current joining technology.

Acknowledgement

This paper was supported by the project of Technical University of Liberec SGS "The study and evaluation of structures and properties of metallic and non-metallic materials"

References

[1] ABBOTT, Steven. Adhesion Science: Principles and Practice. DEStech Publications, 2015. ISBN 1605951781.
[2] GUI, Zhong Xiang, Yi Sheng ZHANG, Hong Qing LI a Ming Tu MA. Hot Stamping and Blank Designing for a Vehicle Bumper Using Ultra High Strength Steel (UHSS). Advanced Materials Research [online]. 2013, 690-693, 2240-2244 [cit. 2019-08-02]. DOI: 10.4028/www.scientific.net/AMR.690-693.2240. ISSN 1662-8985
[3] MERKLEIN, M. a J. LECHLER. Investigation of the thermo-mechanical properties of hot stamping steels. *Journal of Materials Processing Technology* [online]. 2006, **177**(1-3), 452-455 [cit. 2019-08-02]. DOI: 10.1016/j.jmatprotec.2006.03.233. ISSN 09240136.
[4] MERKLEIN, M., J. LECHLER a M. GEIGER. Characterisation of the Flow Properties of the

Quenchenable Ultra High Strength Steel 22MnB5. *CIRP Annals* [online]. 2006, **55**(1), 229-232 [cit. 2019-08-02]. DOI: 10.1016/S0007-8506(07)60404-1. ISSN 00078506.

[5] JEON, Yong-Phil, Hyung-Yoon SEO, Jong-Deok KIM a Chung-Gil KANG. Experimental analysis of coating layer behavior of Al–Si-coated boron steel in a hot bending process for IT applications. *The International Journal of Advanced Manufacturing Technology* [online]. 2013, **67**(5-8), 1693-1700 [cit. 2019-08-02]. DOI: 10.1007/s00170-012-4602-5. ISSN 0268-3768.

[6] SUEHIRO M, KUSUMIK K, MIYAKOSHI T, et al. Properties of aluminum-coated steels for hot stamping. Japan: Nippon Steel Technical Report, 2003. Available from: https://www.nipponsteel.com/en/tech/report/nsc/pdf/n8805.pdf

[7] GUI, ZhongXiang, WeiKang LIANG a YiSheng ZHANG. Enhancing ductility of the Al-Si coating on hot stamping steel by controlling the Fe-Al phase transformation during austenitization. *Science China Technological Sciences* [online]. 2014, **57**(9), 1785-1793 [cit. 2019-08-02]. DOI: 10.1007/s11431-014-5576-3. ISSN 1674-7321.

[8] MCKAMEY, C. G. Iron Aluminides. STOLOFF, N. S. a V. K. SIKKA, ed. *Physical Metallurgy and processing of Intermetallic Compounds*[online]. Boston, MA: Springer US, 1996, 1996, 351-391 [cit. 2019-09-05]. DOI: 10.1007/978-1-4613-1215-4_9. ISBN 978-1-4612-8515-1.

[9] Hugh BAKER, Editor, ASM Handbook, Volume 3: Alloy Phase Diagrams. ASM International, Materials Park, OH, 1992

IOP Conf. Series: Materials Science and Engineering **723** (2020) 012014 doi:10.1088/1757-899X/723/1/012014

The experimental investigation of behaviour of expanded polystyrene (EPS)

Omid Khalaj*, Seyed Mohammad Amin Ghotbi Siabil, Seyed Naser Moghaddas Tafreshi, Miloslav Kepka, Tomáš Kavalir, Michal Křížek, Štěpán Jeníček

*Regional Technological Institute, University of West Bohemia, Univerzitní 8, 301 00, Pilsen, Czech Republic

*Email: khalaj@rti.zcu.cz

Abstract. Low-strength substrates and anthropogenic soils are always an issue in civil engineering. Based on the soil layer types, several methods could be used to improve the basic/foundation layer however it would be difficult to make sure if the specified requirements are achieved. Nowadays, Expandable Polystyrene (EPS) as a lightweight material found as a substitution for traditional methods like soil replacement, soil mixing, using piles driving and other treatment techniques. This paper will demonstrate the static properties of EPS foams in a view point of construction material which will be a key for the future study of these materials. A series of compression tests were carried out on different types of EPS foam to study the effect of EPS geofoam density on the mechanical behaviour of these materials.

1. Introduction

Expanded Polystyrene (EPS) geofoam is a type of cellular polymeric material with a history of successful applications in civil engineering. It is used in different applications wherever it is exposed to various kinds of stresses [1]. In the 1960s, Norway started to use the EPS as a lightweight material to build the new reads on a low strength subgrades. Since then, other characteristics of EPS have made it ideal material to substitute other improvement method in different aspects of construction which need to reduce vertical and lateral stresses with acceptable compression strength. This meant a rapid increase in the use of EPS civil projects such as embankment construction [2-11], slope stability [12-14], retaining structures [15], bridge abutments [16-18], buried pipes and culvert [19-23].

Recently, increased work has been done on the compressible inclusion function of EPS geofoam. The high compressibility specification of EPS makes it an ideal material for reducing vertical and horizontal stress within its porous microstructure while there is almost no side deformation. Continuing investigation on this role of EPS has shown that the compressive strength of EPS geofoam is highly depend on material density, strain rate and the confining stress [24] which is still under the investigation of researchers. This study is intended to provide an understanding of the mechanical behavior of EPS geofoam subjected to monotonic compression loading with different strain rates. A series of uniaxial tests were carried out on cylindrical EPS geofoam specimens with different densities varying from 14.4 kg/m^3 to 28.8 kg/m^3. In addition, a laser sensor was used to measure the side deflection of the EPS samples to obtain a better understanding of the behavior of EPS geofoam.

2. Experimental procedure

The EPS blocks used in this study were produced by the company IZOPOL Dvořák, s.r.o. (a local manufacturer in the Czech Republic) with 4 different densities (Table 1) which were cut into cylinders using a water-jet cutting system.

Table 1. Material properties.

Properties	**EPS 70**	**EPS 100**	**EPS 150**	**EPS 200**
Density (kg/m^3)	13.5	18.0	23.0	28.0
Initial elastic modulus, E_i (MPa)	2.5	4	5	7.5
Compressive resistance@10% axial strain (KPa)	70	110	135	200

A universal hydraulic loading system equipped with high speed precise data acquisition was used to apply the uniaxial vertical loading. The applied load was measured by an inline load transducer with a capacity and accuracy of 100 kN and 0.02% which was double checked by an oil pressure sensor inside the hydraulic cylinder. The vertical movement of the hydraulic cylinder is controlled by an internal displacement transducer which can control the movement of the hydraulic piston with an accuracy of 0.01%. In order to monitor the side deformation of EPS samples, two laser scanners with measurement span and accuracy of 10 mm and 640 points/profile were installed on the side of the sample, free from the frame movement, to obtain the precise side measurement of the middle part of the samples. The whole system is controlled by the central data acquisition system with a sampling rate of 25 Hz - which covers the precise measurement for static tests. Figure 1 shows the real view of the testing system.

Figure 1. Testing equipment.

The first series of tests were carried out to check the performance of the loading system, loading frame, all the transducers, measuring and control system in addition of repeatability of the tests. The repeatability of the tests shows less than 2% difference, which confirms that the whole loading and measuring system works properly. The second group of tests were carried out to investigate the effect of strain rate on the behavior of the stress-strain curves of EPS. The loading was applied with different strain rates of 0.001, 0.01 and 0.1 s^{-1} up to the maximum available span/load capacity of the hydraulic loading system. The sample height and diameter was 30 cm (H/D = 1) with various densities of EPS so at the same time, the effect of EPS density on the mechanical behavior of EPS geofoams was also investigated.

3. Results and discussion

Three strain rates of 0.1, 0.01 and 0.001 s^{-1} were selected to evaluate the effect of loading speed. The samples' height and diameter were 300 mm and the EPS density was varied from 13.5 to 28.0 kg/m^3. Figure 2 shows the stress-strain curve for different strain rates for first 10% strain and up to maximum compression stress. According to figure, EPS geofoam exhibits larger compressive strength with increasing strain rate and the difference in the compressive strength between the selected rates increases with increasing strain amplitude. It can be understood that, as the strain rate increases, the increment of increase in the compressive strength of EPS samples gradually decreases. When the rate of applied pressure is slow, EPS bubbles have enough time to deform and eventually destruct under pressure. As the loading rate increases, the bubbles are forced to contract evenly under the confinement and a smaller number of them might become damaged (comparable to what happens to saturated soil during consolidation). Thus when pressure is applied in a gentle manner, more EPS bubbles are destroyed and therefore a lower compressive strength is observed. The described mechanism seems to be more valid for lower density EPS geofoam, as the bubbles are larger. The structure of denser EPS geofoam consists of less air and thus might be less sensitive to the loading rate, but this requires further investigation.

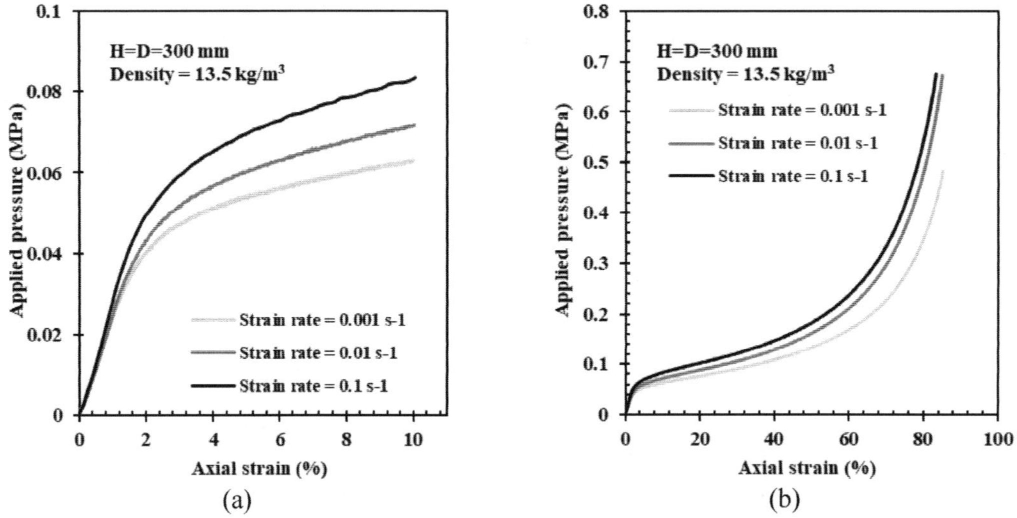

Figure 2. Stress-strain curve for different strain rates (a) up to 10% strain, (b) up to maximum compression

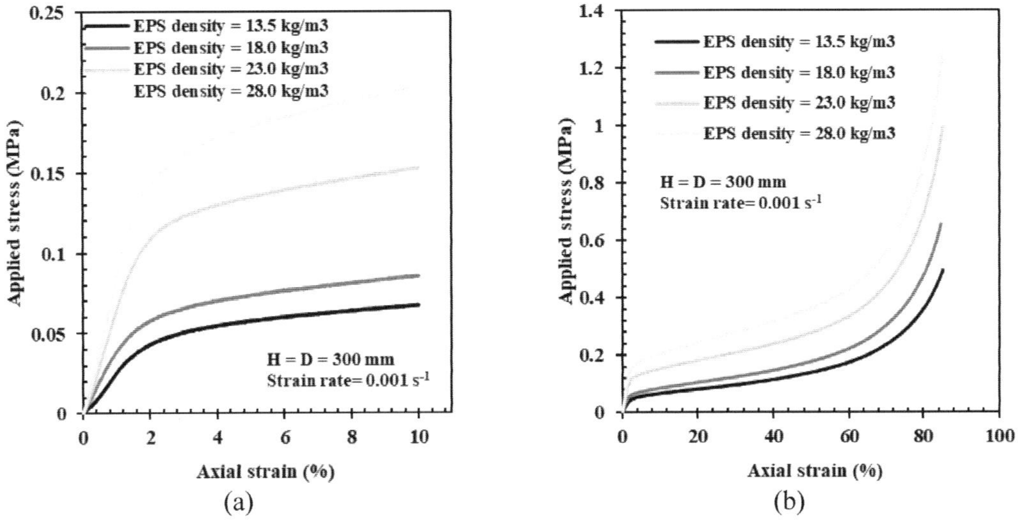

Figure 3. Stress-strain curve for different densities (a) up to 10% strain, (b) up to maximum compression

Figure 3 shows the overall stress-strain curves for various densities of EPS geofoam. As expected, the compressive strength of EPS geofoam increases with increasing EPS density. EPS with densities 13.5 and 18.0 kg/m^3 show nearly close compression resistance rather than the other densities in lower strain however at higher strain (over 50%), the difference in properties became clearer. The maximum compressive strength are approximately 0.5, 0.65, 0.95 and 1.3 MPa for EPS with densities 13.5, 18.0, 23.0 and 28.0 kg/m^3 respectively.

Figure 4 displays the variation of Young's modulus (Elastic modulus) for different densities of EPS geofoam with different strain rates. It's clear that the larger elastic modulus achieved at highest density EPS, while by increasing the strain rate, the elastic modulus increased. On the other hand, the influence of the strain rate on the Young's modulus of the EPS samples increases with increasing EPS density. The reason could be the elastic moduli which obtained from the elastic region of the plots that are not affected by the bubbles interaction phenomena on the overall plots (Fig. 2 and 3). When the EPS material is elastic (1% strain), the governing influential factor depends on the EPS material itself.

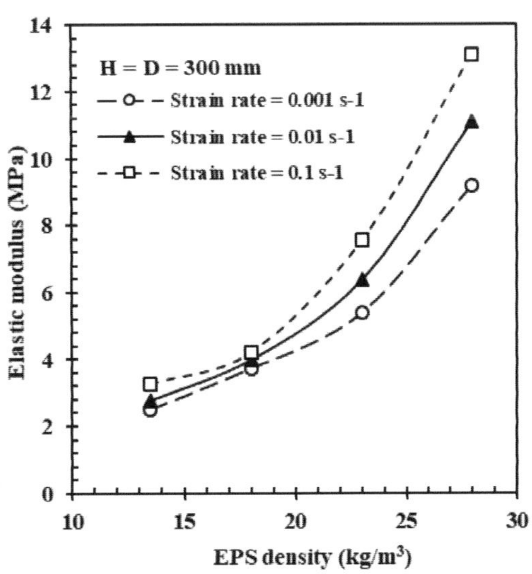

Figure 4. Variation of EPS geofoam elastic modulus with EPS density at various strain rates.

4. Conclusion

Different series of tests was performed on EPS geofoam samples to investigate the effect of sample density and strain rate on the stress-strain behaviour and elastic modulus of EPS geofoam. All the tests were performed on one type of EPS geofoam purchased from a regional producer in the Czech Republic, so there would be slight change if the other production from other producers will use. However, the general trend of behaviour was in good agreement with previous research. The results show that with increasing strain rate, both elastic modulus and compressive strength of the EPS sample increase. The elastic modulus is more sensitive to the strain rate for denser EPS, while the overall sample strength over the plastic strain region is more sensitive to the strain rate for lighter EPS, which can be attributed to the damage to air bubbles as the applied pressure increases. On the other hand, the elastic modulus of the EPS samples increases with increasing density of EPS geofoam, and can be related to it using a simple linear function.

Acknowledgements

The present contribution has been prepared under project LO1502 "Development of the Regional Technological Institute" under the auspices of the National Sustainability Programme I of the Ministry of Education of the Czech Republic aimed to support research, experimental development and innovation.

References

[1] N. Abu-Hejleh, J. G. Zornberg, V. Elias, and J. Watcharamonthein, "Design Assessment of Founders-Meadows GRS Abutment Structure," in *Proc., 82nd Annual TRB Meeting*, 2003.

[2] R. J. N. R. R. L. P. Aaboe, "13 years of experience with expanded polystyrene as a lightweight fill material in road embankments," vol. 61, pp. 21-27, 1987.

[3] M. Duskov, "Use of expanded polystyrene (EPS) in flexible pavements on poor subgrades," in *Proceedings of the international conference on geotechnical engineering for coastal development*, 1991, pp. 783-788.

[4] T. Frydenlund, "E.,"Expanded Polystyrene, A lighter Way Across Soft Ground" Norwegian Road Research Laboratory," Internal Report1991.

[5] S. Ghotbi Siabil, S. Moghaddas Tafreshi, A. Dawson, and M. J. G. I. Parvizi Omran, "Behavior of expanded polystyrene (EPS) blocks under cyclic pavement foundation loading," vol. 26, no. 1, pp. 1-25, 2018.

[6] D. Neguccy and M. Sun, "Reducing lateral pressure by geofoam (EPS) substitution," in *Proceedings of international symposium on EPS (Expanded Poly-Styrol) construction method (EPS Tokyo 96)*, 1996.

[7] G. Refsdal, "Plastic foam in road embankments: future trends for EPS use," Internal Report, Norwegian Road Research Laboratory, Oslo, Norway1985.

[8] S. Selvakumar, B. J. G. Soundara, and Geomembranes, "Swelling behaviour of expansive soils with recycled geofoam granules column inclusion," vol. 47, no. 1, pp. 1-11, 2019.

[9] S. M. Tafreshi, S. A. J. W. A. o. S. Ghotbi, Engineering, I. J. o. C. Technology, Environmental, Structural, Construction, and A. Engineering, "Efficiency of Geocell Reinforcement for Using in Expanded Polystyrene Embankments via Numerical Analysis," vol. 11, no. 9, pp. 1217-1221, 2017.

[10] S. M. Tafreshi and S. A. G. Siabil, "Behavior of Expanded Poly Styrene (EPS) with Experimental and Numerical Methods," in *5th International Conference on Geofoam Blocks in Construction Applications*, 2019, pp. 231-237: Springer.

[11] T. Van Dorp, "Building on EPS geofoam in the 'low-lands'. Experiences in Netherlands," in *Proceedings of international symposium on EPS (Expanded Poly-Styrol) construction method (EPS Tokyo 96)*, 1996.

[12] D. Negussey, *Slope stabilization with geofoam*. Geofoam Research Center, Syracuse University, 2002.
[13] M. Oliaei, H. J. E. J. o. E. Tohidifar, and C. Engineering, "Seismic stability of slopes reinforced with sleeved and unsleeved piles," pp. 1-29, 2018.
[14] S. Srirajan, "Recycled content and creep performance of EPS geofoam in slope stabilization," Syracuse University, 2001.
[15] A. F. Elragi, "Selected engineering properties and applications of EPS geofoam," ed: State University of New York. College of Environmental Science and Forestry ..., 2000.
[16] P. McDonald and P. Brown, "Ultra lightweight polystyrene for bridge approach fill," in *Proceedings of the 11th Southeast Asian geotechnical conference, Singapore*, 1993, pp. 664-668.
[17] H. Skuggedal and R. J. P. X. Aaboe, Florence, "Temporary overpass bridge founded on expanded polystyrene," vol. 2, 1991.
[18] D. Williams and R. Snowdon, "A47 Great Yarmouth Western Bypass: performance during the first three years," 0266-7045, 1990.
[19] S. Bartlett, "Performance of a geofoam embankment at 100 south, I-15 Reconstruction Project, Salt Lake City, UTAH."
[20] O. Khalaj, M. Azizian, S. M. Tafreshi, and B. Mašek, "Laboratory Investigation of Buried Pipes Using Geogrid and EPS Geofoam Block," in *IOP Conference Series: Earth and Environmental Science*, 2017, vol. 95, no. 2, p. 022002: IOP Publishing.
[21] O. Khalaj, S. Moghaddas Tafreshi, B. Mask, A. R. J. G. Dawson, and Engineering, "Improvement of pavement foundation response with multi-layers of geocell reinforcement: Cyclic plate load test," vol. 9, no. 3, pp. 373-395, 2015.
[22] O. Khalaj, S. M. A. G. Siabil, S. N. M. Tafreshi, and H. Jirkova, "Performance Evaluation of Pavements Constructed on EPS Geofoam Backfill Using Repeated Plate Load," in *IOP Conference Series: Earth and Environmental Science*, 2019, vol. 221, no. 1, p. 012007: IOP Publishing.
[23] Q. Ma, J.-J. Zheng, H.-l. J. E. J. o. E. Xiao, and C. Engineering, "Analysis of pressure on the roof of a culvert underneath a ditch with compressible material covered by a geosynthetic layer," pp. 1-17, 2019.
[24] H. Gao, J. Liu, and H. J. I. J. o. G. E. Liu, "Geotechnical properties of EPS composite soil," vol. 5, no. 1, pp. 69-77, 2011.

IOP Conf. Series: Materials Science and Engineering **723** (2020) 012015 doi:10.1088/1757-899X/723/1/012015

Optimization of workability technological testing for open-die forging

Jakub Kotous[1a], Václav Kubec[1b], Michal Duchek[1c], Tomáš Studecký[1d]

[1]COMTES FHT a.s., Dobřany 995, Dobřany, 334 41

E-mail: [a]jkotous@comtesfht.cz, [b]vkubec@comtesfht.cz, [c]mduchek@comtesfht.cz, [d]tstudecky@comtesfht.cz

Abstract. Developing an open-die forging process for a poor-workability alloy has always been a challenge. This paper presents a rapid method that meets this purpose. A simplified forging test was considered for assessing the limit states of material in open-die forging. It can be employed in industrial forge shops. A V-shaped testing die was designed using finite element (FE) simulation. The V-shape geometry is convenient for strain evaluation. Rectangular blocks will be tested across a wide range of working temperatures. Based on the test results, limit states can be identified. Then friction testing method was optimized. It is also simple comparison test. Both technological tests are suitable for determination of optimized parameters in forging. The risk of material failure can thus be eliminated or minimized.

1 Introduction

Developing an open-die forging process for poor-workability alloys has always been a challenge. This paper presents a rapid method that meets this purpose. As such, it can be readily employed in industrial forge shops. A simplified workability test was used to find the workability limits. The risk of workpiece failure can thus be avoided or minimized [1].

Among the various process conditions, the coefficient of friction has a major impact on the forging process. It would be very convenient to have a simple method of managing friction variations during certain stages of the forging process. To assess the coefficient of friction, the Spike test was designed. When the material deformation behavior and the process conditions are known, it significantly helps to control the process and quality. The both test results further provide input into inverse calculations of reverse engineering method. Validated simulation model is main goal for optimizing technologies [2],[3].

2 Workability testing

In terms of metallurgical workability, two aspects are of prime importance. The first is the cohesion of the material which prevents failure or local cracking during forming. The second aspect relates to the microstructure: optimal thermomechanical conditions can lead to the desired properties within a particular volume. This paper places particular emphasis on the first aspect. It means that the plasticity limit is explored (when the first cracks begin to form). Workability depends on four groups of factors: metallurgical condition of steel (chemical composition, microstructure, thermally-activated processes), thermomechanical factors (forming temperature, speed and amount of deformation), stress state and processing conditions. The success of the processing procedure depends on the appropriate combination of these conditions [4],[5],[6].

Workability can be assessed by various standardized tests or by tests specifically designed to characterize this property. Standardized hot tests include: tensile test, compressive test, torsion test, hot impact test and wedge rolling test. Tensile test is characterized by tensile stress and by the absence of friction. Tensile stress strongly responds to formation of cracks. Elongation, reduction of area, engineering stress and true stress are evaluated. Compressive test is particularly suitable for the determination of flow stress. Torsion test is good for evaluating large strains without effects of external friction. Hot impact testing is appropriate for assessing the effects of strain rate by comparison. Wedge rolling test involves rolling of a wedge-shaped specimen. Its purpose is to evaluate a wide range of strains across side notches. It is clear from the above descriptions that the tests are not generally applicable to any forming process. In some cases, the tests must be combined to yield the desired information, or a specific test must be designed for a particular forming technology. Standardized tests provide indirect workability assessment, through elongation or reduction of area [4],[5],[6].

3 Workability testing for open die forging

The choice of workability tests for open-die forging depends on the structural condition of the particular ingot. Ingots tend to possess non-uniform structure, which means that their workability varies depending on location. Ordinarily, intergranular cracks develop in segregated regions and cracking occurs as a result of coarse grain (severe stress concentration). Therefore, workability assessment involves special-geometry specimens (figure 1 and figure 2), such as deformed flat (figure 3), and deformed conical specimens (figure 4). Their sizes are not defined because they will be derived from the grain size or the size of a region of interest in the ingot. Using these tests, the behavior of material can be studied at various temperatures, strain rates and strain magnitudes (these vary with changes in the specimen geometry). One can therefore identify safe conditions under which failure does not occur. In addition, the share of recrystallized structure can be determined [7],[8].

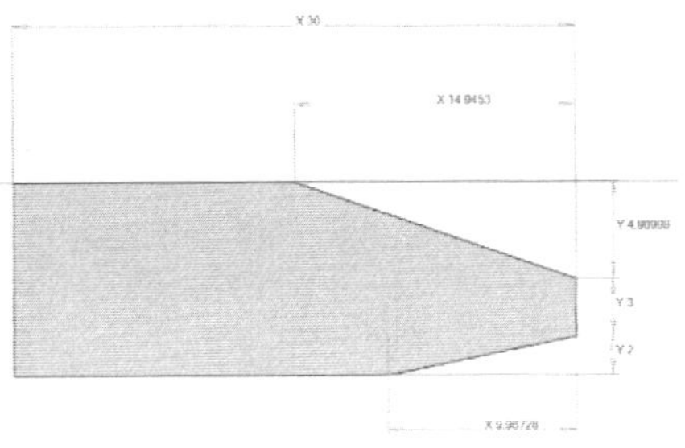

Figure 1. Flat sample. [1]

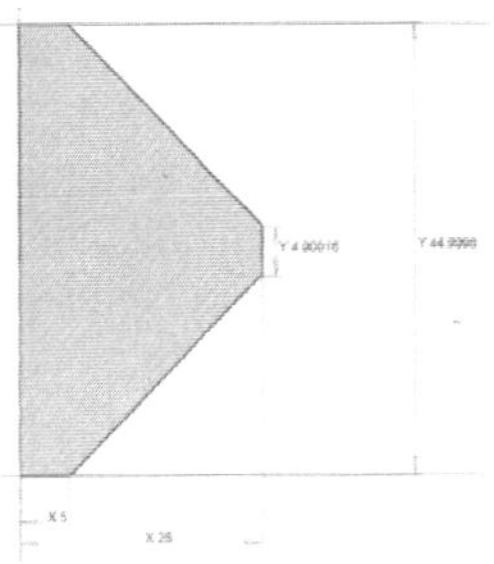

Figure 2. Conical sample. [1]

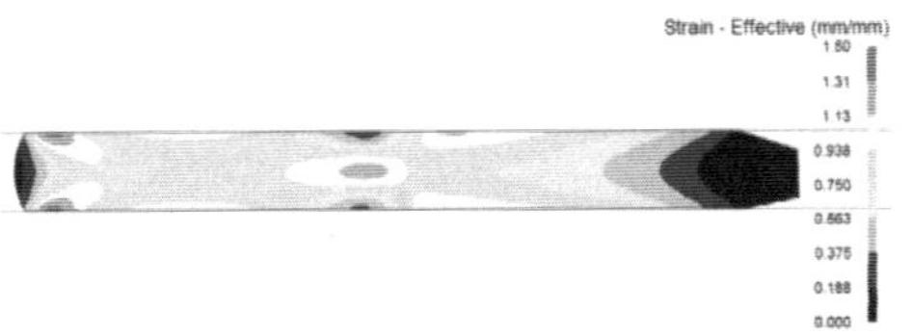

Figure 3. Strain distribution in deformed flat sample. [1]

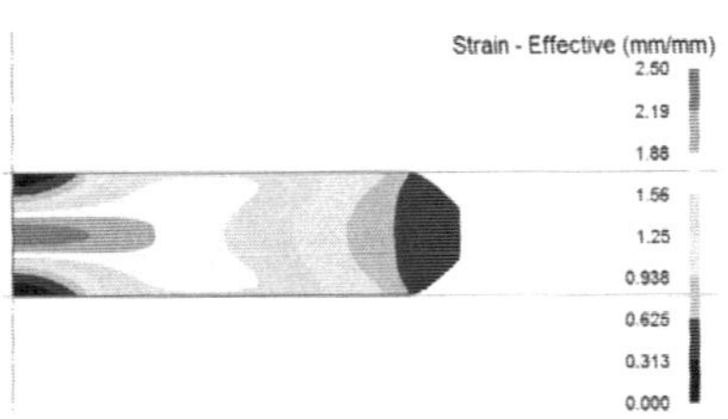

Figure 4. Strain distribution in deformed conical sample. [1]

4 New concept of workability testing

The above-listed workability tests were inadequate for our purpose. The objective was to perform a workability test outside the laboratory and as part of ordinary operation.

The design of the shape of the deformed specimen was dictated exclusively by the desired stress state for assessing formability at the chosen temperature when compared to a reference standard. An example of the new specimen upon deformation is shown in figure 5. The test specimens are plain cuboids. The dies are specially-shaped. The shaped die imparts various amounts of strain to the material. Various test temperatures are used and modes of failure are explored in relation to test conditions. Findings from actual experiments can be compared with the results of numerical modelling. This provides for their refinement and better accuracy. Based on the data, one can optimize forming operations and reductions at appropriate temperatures and stress states [1],[9].

4.1 Failure criteria

The right shape of the test specimen was sought with the aid of finite element modeling. In these calculations, the failure description (1) was based on general continuum mechanics calculations. They involved triaxiality, a parameter which is governed by the stress state.

$$TF = \frac{\sigma_h}{\sigma_{eqv}} = \frac{{}^{1}/_{3}(\sigma_1+\sigma_2+\sigma_3)}{\frac{1}{\sqrt{2}}\sqrt{(\sigma_1-\sigma_2)^2+(\sigma_2-\sigma_3)^2+(\sigma_3-\sigma_1)^2}} \quad (1)$$

Where:

TF = triaxiality
σ_h = hydrostatic stress
σ_{eqv} = equivalent von Mises stress
σ_1, σ_2, σ_3 = principal stress

4.2 Numerical simulation of new test

The shaped upper die produces the desired strain and stress distribution in the specimen. The sequence of images figure 6 – figure 9 shows the parameters of interest as computed by FE analysis. For the sake of shorter computing time, only one quarter of the specimen was considered. One can therefore easily monitor the parameter fields not only on the surface but also in the centre of the specimen. The material chosen for the calculation had an ordinary formability and little susceptibility to surface cracking [1],[9].

When stress states are monitored in terms of von Mises equivalent stress or effective engineering strain, it is not certain whether surface cracking would occur.

However, an in-depth analysis involving maximum tensile stresses and triaxiality may offer insights in the form that the locations with the most severe compression are likely to develop cracks, particularly in difficult-to-form materials.

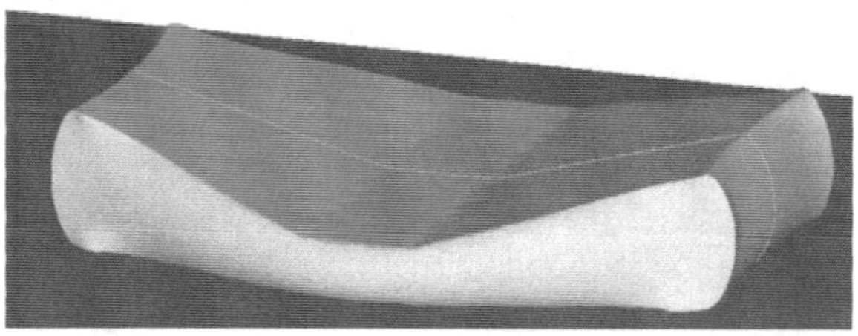

Figure 5. New test sample after deformation.

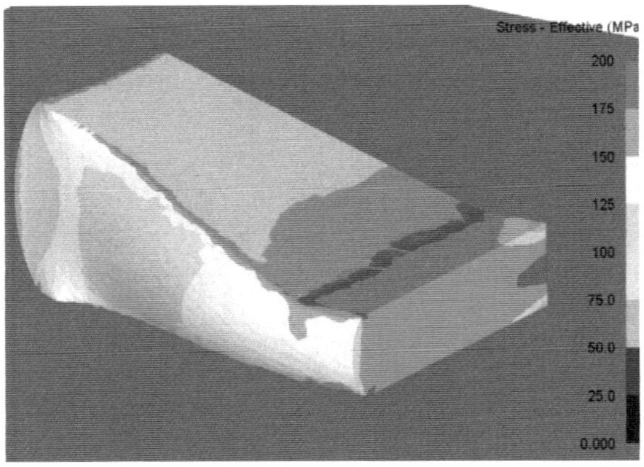

Figure 6. Distribution of von Mises stress.

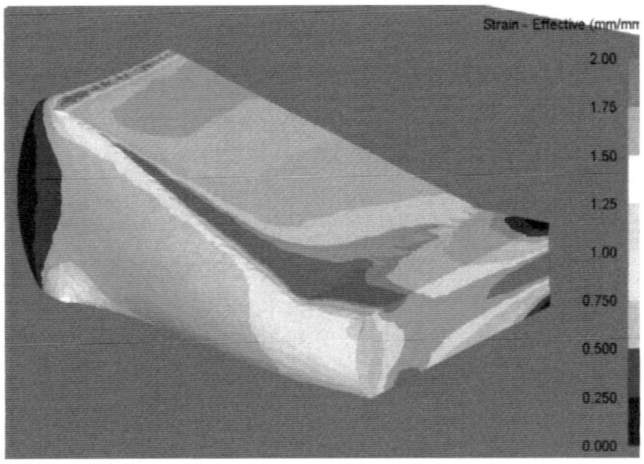

Figure 7. Strain distribution.

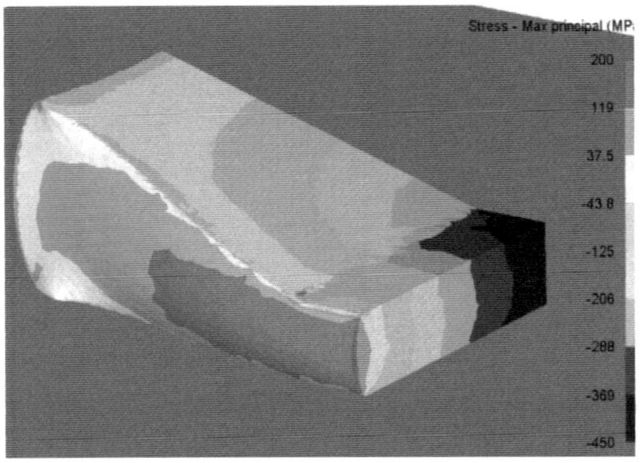

Figure 8. Distribution of max. principal stress.

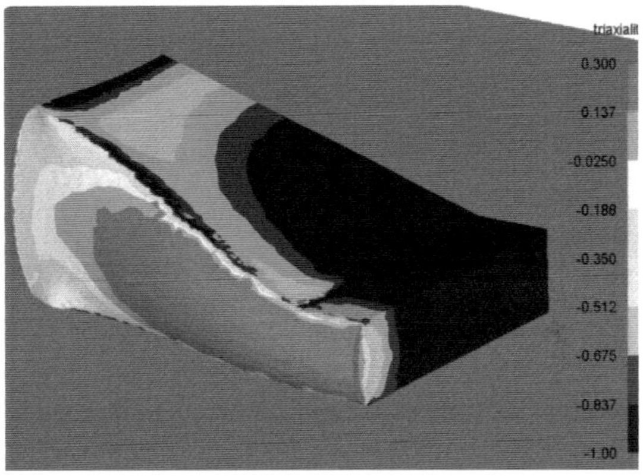

Figure 9. Distribution of triaxiality.

5 Design and optimization of friction testing method

A correctly chosen lubricant for bulk forming facilitates the filling of die and mould cavities, reduces the tool wear, and thus improves its life. Numerical simulations of these processes depend on the friction coefficient value. Correct values make the entire numerical model and the prediction of the material flow more accurate [9]. These are the reasons behind the development of the spike test, which is a simple technological test. The test uses a specially-designed die, the action of which strongly depends on the lubricant used. The height (h) and radius (r) of the specimen after the test are measured and the values are used for determining the friction coefficient or for direct comparison between various lubricants, temperatures and materials [2],[3].

5.1 Design and optimization of die shape

The shape of the test die was designed and optimized with the aid of numerical simulation conducted in the DEFORM software. The test specimen was made from C45 steel (ČSN 12050). It was a cylinder, 40 mm in diameter and 25 mm in height. The forming temperature was 900°C. The speed of movement of the top die was constant: 10 mm/s [3]. A drawing of the test die (figure 10 a) presents the parameter ranges explored. In the tests of various die geometries, the main parameters varied were the edge radius, the central cavity angle α and the die face angle β [2],[3].

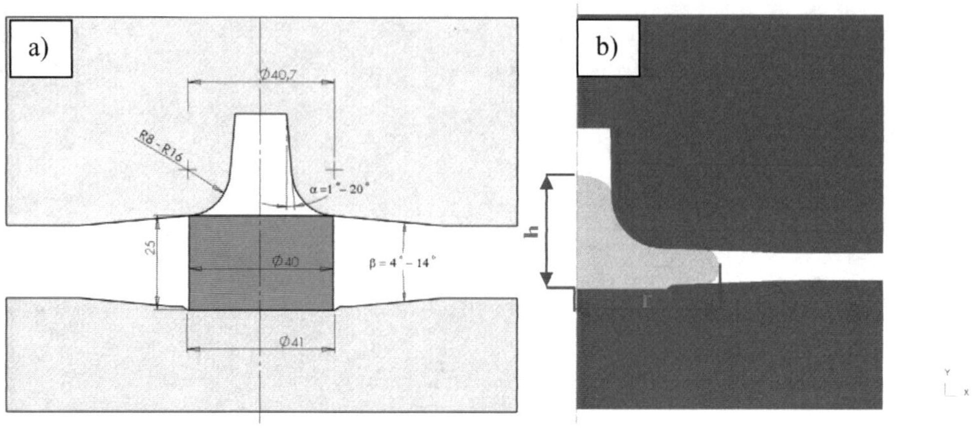

Figure 10. a) Die geometry b) Measured dimensions of specimens. [3]

The dimensions shown in figure 10 b are measured after the top die descended, stopping 16 mm below its initial position. The following graph (figure 11) contains the r/h values calculated in individual simulation runs. It is decisive for selecting the appropriate die shape with respect to the friction coefficient. The higher the r/h ratio, the greater is the sensitivity of the test to friction.

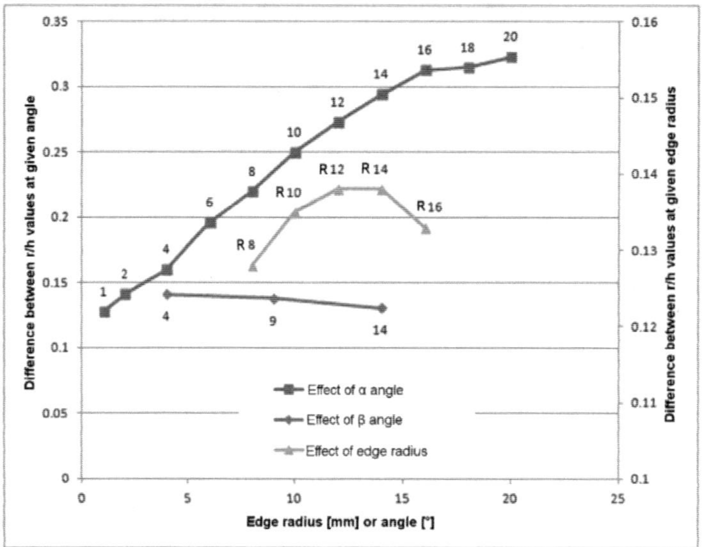

Figure 11. Plot of results for effects of α and β angles and edge radius. [3]

First, the die edge radius was varied and its effect was studied. Then, the effects of the β and α angles were explored. The radius of the top die edge does not have a substantial impact on the friction conditions. Another variant of the die shape involved an edge radius of 12 mm, as shown in the graph in figure 11. Smaller angles of the die face proved more suitable for demonstrating the effects of friction. Hence, only the angle of 4° was used in all subsequent test runs. The graph shows that the top die cavity angle has a profound effect on the friction conditions. The larger the angle, the larger is the difference between the final specimen shapes for various friction coefficients. At angles above 16°, the difference between specimen shapes obtained at two different friction coefficients begins to decrease, when compared to angles smaller than 16°. This is why the angle of 16° was chosen for the subsequent test runs. The top die cavity angle of 16° which is measured on one half of the die in the 2D model translates into an angle of 32° of the top die cavity [3].

5.2 Experimental verification

The trials of the real die were carried out in the PZSK 2500 tf universal forging press. The material of the forged specimen was 38MnVS6 steel. The specimen was heated in a furnace to 1100 °C and then placed in the die and compressed. The transfer of the specimen from the furnace to the die took 12 seconds. The top die speed was kept constant at 80 mm/s. In this experiment, the effect of the graphite lubricant Berulit 913 on the flow of the material was explored. Three specimens were processed with each set of parameters.

The average value of the d/h ratio in specimens formed without lubricant was 2.24. In the specimens formed with lubricant, it was 2.84. The difference between these values confirms that the sensitivity of the die is sufficient for its purpose. An optimized testing method was validated using the top die velocity of 80 mm/s. The final shape of the specimen does not substantially differ from the shape calculated for the lower velocity of 10 mm/s.

This test was therefore adopted as an in-house technological test by the company COMTES FHT a.s. The value of the friction coefficient is determined by substituting the measured values in a numerical model created in the DEFORM program.

6 Conclusion

A new workability testing procedure was developed. It enables materials with similar behaviours to be classified in terms of their workability. The benefits are simple testing and evaluation. The testing sample is block which is possible to make only by cutting.

A spike test procedure was developed and optimized to investigate the effects of friction conditions. The geometry of test specimens was proposed with regard to the ease of manufacture (Ø40x25 mm). It enables materials with certain conditions to be easy classified in terms of their friction coefficient. Data from both the tests can be used as input into inverse calculations of reverse engineering method. Validated simulation is used as main optimization tool for technological processes.

Acknowledgment

This paper was created under the project no. FV10019 „Progressive Material-Technological Increase in Efficiency of the Turbine Blades Production".

References

[1] Palán J Kubec V Urbánek M and Duchek M 2018 Damage criteria´s of material in context of open die forging technology *Kovárenství* vol 66 pp 22-25

[2] Duchek M Kubec V Poláková I and Urbánek M 2016 Optimalizace technologických zkoušek pro určení koeficientu tření a životnosti nástrojů *Kovárenství* vol 57 pp 3-7 ISSN 1213-9289

[3] Duchek M Poláková I 2015 Development of Spike Test Method to Improve Numerical Model Accuracy *Material Sci Eng*

[4] Schnider I and Kawulok P 2013 Deformační chování materálů *VŠB Ostrava*

[5] Žídek M 1995 Metalurgická tvařitelnost ocelí za tepla a za studena ISBN 80-85341-45-X

[6] Dieter E G Kuhn A H Semiatin L S 2003 Handbook of workability and Process Design *ASM INTERNATIONAL* ISBN 0-87170-778-0

[7] Maleček L Zemko M Šteller M and Hodek J 2015 Numerické simulace a programové kování – nástroje pro optimalizaci procesů volného kování Kovárenství vol 53 pp 34-39

[8] Kubec V Fedorko M Zemko M and Maleček L 2015 Možnost využití numerických simulací pro zápustkové kování *Kovárenství* vol 53 pp 40-43

[9] Poláková I Kubina T 2013 Flow stress determination methods for numerical modelling *METAL 2013* TANGER pp 273 ISBN 978-80-87294-41-3

IOP Conf. Series: Materials Science and Engineering **723** (2020) 012016 doi:10.1088/1757-899X/723/1/012016

Design and optimization of a closed die forging of nickel-based superalloy turbine blade

Jakub Kotous[1a], Václav Kubec[1b], Pavel Salvetr[1c] , Michal Duchek[1d] , Miroslav Majer[2e]

[1]COMTES FHT a.s., Dobřany 995, Dobřany, 334 41,
[2]CZECH PRECISION FORGE a.s., Tylova 1/57, Plzeň, 301 00,

E-mail: [a]jkotous@comtesfht.cz, [b]vkubec@comtesfht.cz, [c]psalvetr@comtesfht.cz, [d]mduchek@comtesfht.cz, [e]Miroslav.majer@cpforge.com

Abstract. The nickel-based superalloys belong to widely used materials for most demanding industrial applications. The design and the experimental verification of manufacturing technology of NIMONIC 80A turbine blade is presented in this paper. A finite element (FEM) simulation was exploited for the closed die forging technology optimization. Based on the precision material model and boundary conditions, the deformation behaviour in the range of hot working temperatures was studied. The process conditions including the strain rates were preset according to the industrial scale practise. Based on the FEM simulation results the necessary tools were manufactured and the experimental closed die forging of turbine blades was performed. Subsequently, a heat treatment of forged blades was carried out. The minimum of 1300 MPa tensile strength was achieved. A metallographic survey was carried out to verify the structure homogenity.

1. Introduction

Due to its ability to withstand very high temperatures, nickel-based superalloys is ideal for use in aircraft parts and gas turbine components such as turbine blades, for instance, where the pressure and heat are extreme. Nickel based alloys show worse workability than it has steel. This behaviour often results in initiations of cracks during forming operations therefore it is necessary to follow technological procedures [1].

The design manufacturing technology of turbine blade is presented in this paper. First it is necessary to created material model which will be validated according real conditions. In numerical simulations, the plastic behavior of material is characterized by flow stress data. In the first iteration, this data is calculated from results of tensile and compressive tests. Tensile test data typically cover only small strains. Compressive testing can provide data for larger strains. Real-world manufacturing processes, however, often involve larger strains than those achievable in a compressive test. In such cases, the flow stress data must be extrapolated. Based on the data, one can optimize manufacturing procedures by using numerical modelling [2],[3].

Main goal of this paper was design and the experimental verification of manufacturing technology of NIMONIC 80A turbine blade [4],[5],[6]. Requirement of final products was ultimate strength over 1300 MPa. A finite element (FEM) simulation was exploited for the closed die forging technology optimization. Based on the precision material model and boundary conditions, the deformation behavior in the range of hot working temperatures was studied. Based on the FEM simulation results the experimental closed die forging of turbine blades was performed. Subsequently, a heat treatment of forged blades was carried out. In order to attain the desired strength, the workpiece must undergo sufficient mechanical working prior to heat treatment. At small strains, these limit values cannot be achieved [7]. The resultant properties were investigated by metallographic analysis and mechanical testing.

2. Material model

At this stage, the purpose was to develop a mathematical description of the material behavior for numerical modelling based on experimental data. The extent to which these models are accurate is decisive for the outcome of this exercise.

Basic characteristics, such as yield strength YS, ultimate tensile strength UTS, elongation El and reduction of area RA were determined from tensile tests. Material was in solution annealed condition as supplied. The values are given in table 1. The test specimens were 10 mm in diameter, had a gauge length of 50 mm, and were provided with M16 threads. The force-displacement data in table 2 were measured in a compressive test at a particular temperature and strain rate. Flow stress data was then calculated and input into DEFORM simulation software. The data was calibrated using the simulation tool.

Table 1. Results of tensile tests at various temperatures.

Specimen	**Temp.** [°C]	**YS** [MPa]	**UTS** [MPa]	**El** [%]	**RA** [%]
N20C_1	20	464	915	37.9	46.0
N20C_2	20	463	923	42.3	41.2
N1000C_11	1000	74	82	92.7	96.8
N1000C_12	1000	69	81	113.0	97.0
N1100C_13	1100	38	45	125.9	95.5
N1100C_14	1100	42	48	121.2	97.6
N1200C_15	1200	24	29	155.0	98.9
N1200C_16	1200	25	29	127.9	97.8

Table 2. Results of compressive tests (Samples Ø8x12 mm) with various strain rate (e).

Specimen	**T** [°C]	**e** [s^{-1}]	**Kf_0** [MPa]	**Kf_{max}** [MPa]
N_20C_0	20	1	460	1394
N_20C_1	20	29	520	1425
N_20C_2	20	31	510	1450
N_20C_12	20	34	520	1489
N_20C_13	20	38	510	1495
N_1000C_9	1000	1	177	312
N_1000C_3	1000	21	191	521
N_1000C_4	1000	22	201	532
N_1000C_14	1000	21	200	545
N_1100C_10	1100	1	149	214
N_1100C_5	1100	24	185	399
N_1100C_6	1100	23	180	389
N_1200C_11	1200	1	45	134
N_1200C_7	1200	24	151	280
N_1200C_8	1200	24	178	282

Compressive tests were conducted across ranges of strain rates and temperatures corresponding to the forging process. The resulting data was a basis for calibrating the material model for the simulation.

Comparisons between measured and calculated forces after calibration are given in the graphs below (figure 1). Graphs for different temperatures include plots for two strain rates (figure 2). In each case, the curve for the slower rate (1.5 s^{-1}) is below that for the faster rate (30 s^{-1}).

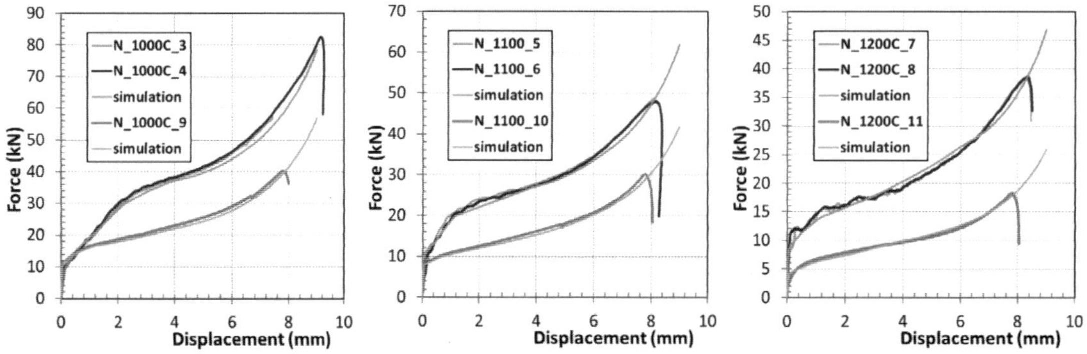

Figure 1. Comparison between the measured and the calculated data, temperature 1000, 1100 and 1200 °C, strain rates 1.5 and 30 s^{-1}.

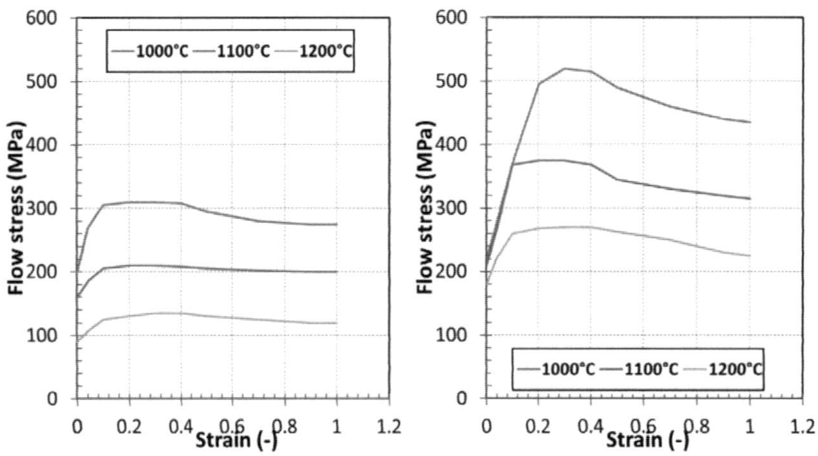

Figure 2. Flow stress curves for various temperatures and strain rates 1.5 s^{-1} (left) and 30 s^{-1}(right).

3. Numerical simulation

Using DEFORM 2D/3D, a numerical model of forging of a blade from a round stock was created. Numerical simulations were based on process planning data, including relevant forming temperatures, lubrication, cooling of the dies, and the handling times as determined by the company CPF where the experimental production of blades took place. The calculation was performed in 3D environment. The plastic material model was considered - involving plastic deformation only. The size of the input stock: Ø50 x 300 mm, initial forming temperature 1200 °C, dies preheated at 150 °C. Friction: shear 0.33, constant. Closed-die forging was performed in three steps in a screw-hydraulic press. Energy: 1st upsetting – 80 kJ, 2nd blow – 240 kJ, 3rd blow – 350 kJ.

Several simulation runs were performed to optimize the closed-die forging process. The final optimized process sequence with the most favorable strain and stress distributions is shown in the following figure 3 and figure 5.

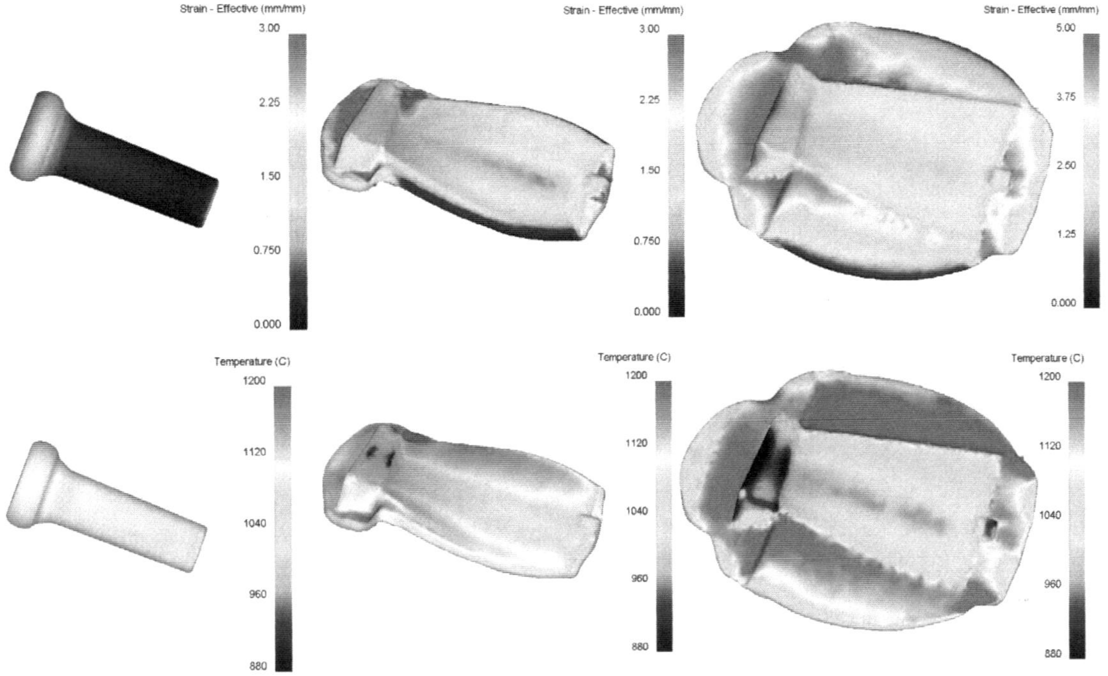

Figure 3. Optimized deformation and temperature distribution after different strokes.

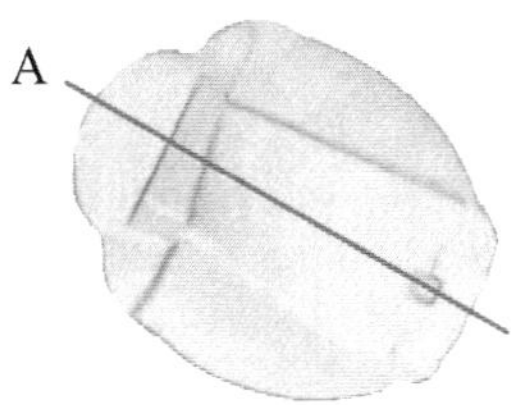

Deformation distribution after different strokes in the middle cross-section (figure 4) follow in figure 5.

Figure 4. Cross-section A.

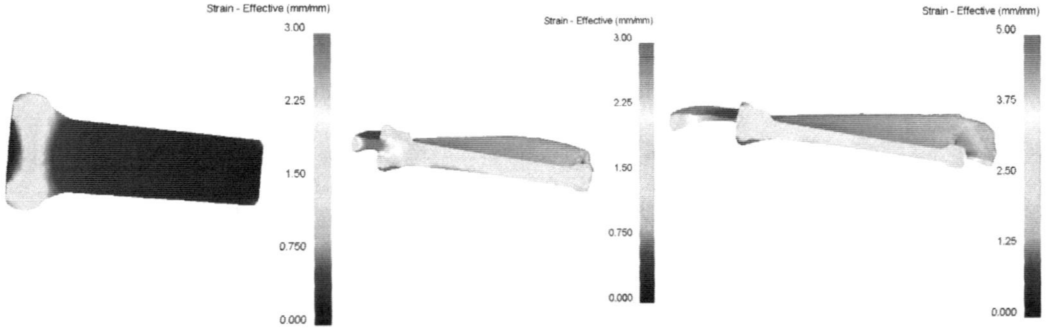

Figure 5. Deformation distribution in the cross-section A.

4. Heat treatment

Optimizing of heat treatment was in precipitation hardening. Solution annealing for all regimes was same: 1080 °C/ 2 hours/ air cooling. The hardening experiment consisted of variable temperature and time. Results you can see in table 3. The hardening regime 700 °C with 16 hours holding was chosen for final treatment, because others properties are better than the previous regime with UTS 1323 MPa.

Table 3. Hardening experiment.

Sequence	YS [MPa]	UTS [MPa]	El [%]	RA [%]
600°C/ 10 hrs	992	1152	15.2	24.8
600°C/ 12 hrs	1000	1152	14.3	23.4
600°C/ 14 hrs	1009	1164	14.3	23.4
600°C/ 16 hrs	1026	1173	13.8	23.4
600°C/ 18 hrs	1037	1195	12.5	24.0
600°C/ 22 hrs	1024	1199	12.1	21.9
650°C/ 10 hrs	1038	1258	10.7	18.6
650°C/ 16 hrs	1100	1282	9.9	18.7
650°C/ 22 hrs	1119	1323	10.4	18.4
700°C/ 16 hrs	1132	1312	12.0	19.8
750°C/ 16 hrs	1024	1255	16.4	21.0

5. Experimental forging and results

5.1. Experimental forging

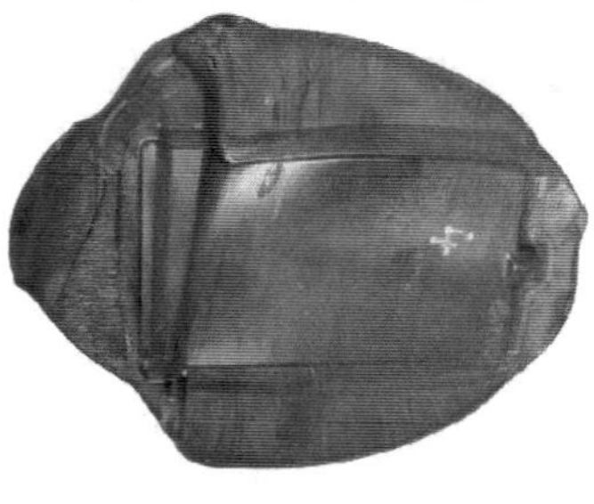

Figure 6. Forged turbine blades.

The experimental turbine blades were close die forged in CPF company. The heating of input material Ø50 x 300 mm was carried out in medium-frequency induction equipment. The second heating after one stroke was in electrical furnace. It was made 160 pcs of prototype turbine blades from NIMONIC 80A. The surface quality was perfect, see figure 6. There wasn´t any cracks. Microstructure was homogenous. It was consisted of recrystallized grain in matrix and intermetallic phases. It can be for example carbides of Ti, Cr and Mo. The chemical composition of forged blades are in table 4.

Table 4. Chemical composition of forged blades from NIMONIC 80A.

C	Mn	P	S	Si	Cr	Ti	Fe	Al	Co	Ni
0,007	0,02	0,001	0,003	0,10	20,83	2,53	0,74	1,05	0,01	Rest.

5.2. Final results

The forged blades were heat treated. The solution annealing was at 1080 °C, hold 2 hours and air cooled. The final hardening annealing was at temperature 700 °C, hold 16 hours and air cooled. The evolution of microstructure is seen in figure 7. The results of tensile test are in table 5. It is in agreement with heat treatment experiment.

Table 5. Final mechanical properties of turbine blades.

Sequence	YS [MPa]	UTS [MPa]	El [%]	RA [%]
700°C/ 16 hrs	1129 ± 12	1319 ± 16	15.2 ± 0.2	24.8 ± 0.6

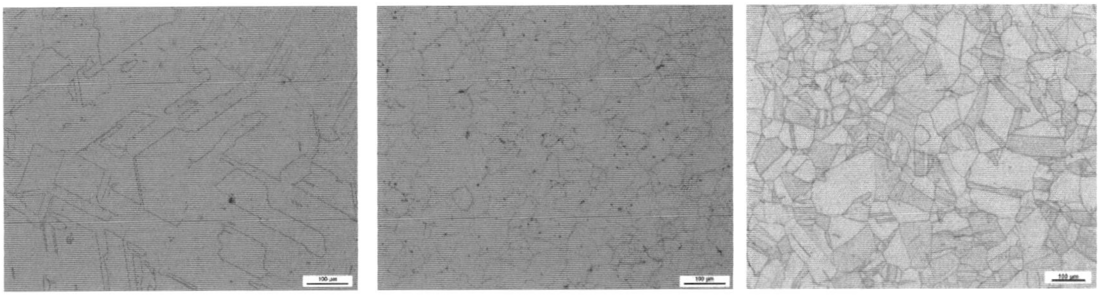

Figure 7. Microstructure of input material (left), after close die forging (middle), after heat treatment (right), NIMONIC 80A.

6. Conclusion

The valid material model of NIMONIC 80A was created. Then design and optimizing of close die forging was performed by numerical simulation. Closed die forging process has been developed with co-operation of CPF company. The top and bottom die were made after evaluation of best simulation. Then experimental turbine blades were produced. It was made 160 pcs of prototype turbine blades. Heat treatment experiment consisted of solution annealing and optimizing of hardening annealing. Final heat treatment (hardening at 700°C/16 hrs) were performed on the forged blades to obtain a strength over 1300 MPa.

Acknowledgment

This paper was created under the project no. FV10019 „Progressive Material-Technological Increase in Efficiency of the Turbine Blades Production".

References

[1] Srinivasa N Prasad Y V R K 1995 Hot working characteristics of nimonic 75, 80A and 90 superalloys: a comparison using processing maps *Journal of Materials Processing Technology* vol 51 pp 171-192

[2] Kubec V Fedorko M Zemko M and Maleček L 2015 Možnost využití numerických simulací pro zápustkové kování *Kovárenství* vol 53 pp 40-43

[3] Poláková I Kubina T 2013 Flow stress determination methods for numerical modelling *METAL 2013* TANGER pp 273 ISBN 978-80-87294-41-3

[4] Kumar R R and Pandey K M 2017 Static structural and modal analysis of gas turbine blade *IOP Conference Series: Materials Science and Engineering* vol 225 pp 1-9 ISSN 1757-8981

[5] Torabi S H R Alibabaei B Bonab B B Sadeghi M H and Faraji G 2017 Design and oplimization of turbine blades preform forging using RSM and NSGA II *Journal of Inteligent Manufacturing* vol 28 (6) pp 1409-1419 ISSN 0956-5515

[6] Wright D C and Smith D J 2013 Forging of blades for gas turbines *Material Science and Technology* vol 2 (7) pp 742-747 ISSN 0267-0836

[7] Special Metals Corporation 2004 NIMONIC alloy 80A [online] [seen 7. august 2019] available from http://www.specialmetals.com/assets/smc/documents/alloys/nimonic/nimonic-alloy-80a.pdf

Effect of Mn addition on serrated plastic flow behaviour in high-strength multiphase steels with retained austenite

Aleksandra Kozłowska, Barbara Grzegorczyk, Marcin Staszuk, Adam Grajcar

Silesian University of Technology, Department of Engineering Materials and Biomaterials, 44-100 Gliwice, Konarskiego 18a Street, Poland

aleksandra.kozlowska@polsl.pl

Abstract. The effect of manganese addition (from 3 to 5%) on the serrated plastic flow behavior in medium manganese TRIP steels was investigated in static tensile tests performed in a temperature range of 20-200°C. The Portevin-Le Chatelier (PLC) effect was observed in a steel containing the higher Mn content. The effect of deformation temperature on the critical strain for the serrated flow was noted. The relationships between the manganese content, deformation temperature, mechanical properties and the appearance of serrated flow were characterized.

1. Introduction

Steels are still the most commonly used materials in automotive industry, despite the increasing tendency to replace them by lightweight materials such as aluminum alloys, magnesium alloys, composite materials and polymers. Therefore, many scientific research are focused on the advanced high strength steels (AHSS). In comparison to the first generation AHSS, such as dual phase (DP) steels and transformation induced plasticity (TRIP) steels, and the second generation AHSS, such as twinning induced plasticity (TWIP) steels, the third generation of AHSS provides attractive mechanical properties as combination of high strength, relatively high ductility at relatively low cost. These properties result from the transformation of the metastable retained austenite into martensite during deformation [1-6].

The occurrence of the serrated flow in tensile curves results in the appearance of technological problems during plastic deformation; thus it should be avoided. Several reports [7, 8] concerning the PLC effect in steels, confirm that the chemical composition and deformation temperature influenced both the intensity and critical stress initiating the serrated flow behavior. In case of the TRIP-aided steels, the deformation temperature also affects the stability of retained austenite.

Plastic deformation of medium manganese steels with a multiphase microstructure is a complex process. The deformation of these steels is related to the intensity of martensitic transformation and appearance of PLC effect. Hence, the aim of the present study was to understand the simultaneous effects of Mn content and deformation temperature on the microstructure, mechanical properties and the appearance of serrated flow behavior in TRIP steels with different manganese contents.

2. Materials and experimental methods

In order to determine the influence of Mn addition on the serrated flow behaviour, two medium manganese TRIP steels with different Mn contents were chosen. The chemical compositions of the investigated steels are listed in table 1. Both steels were produced by the thermomechanical processing and controlled cooling was applied to obtain a microstructure consisting of bainite and retained austenite. Detailed information on the processing parameters can be found in [9, 10].

Series of tensile tests carried out at a constant strain rate of 10^{-3} s^{-1} were conducted in a temperature range 20-200°C to determine the influence of deformation temperature on the appearance of PLC effect and resulting microstructure and mechanical properties.
The microstructural details were revealed with a scanning electron microscope Zeiss SUPRA 25 operating at 20 kV. A surface topography of the investigated samples was characterized using the atomic force microscope (XE-100 Park Systems). The SEM and AFM methods were applied to compare the morphological details of the steels in the initial state (non-deformed) and deformed at different temperatures. Specimens for microstructural observations were prepared using standard metallographic procedures and etched in 3% nital to reveal the microstructure. A fraction of retained austenite was measured by using XRD method.

Table 1. Chemical composition of investigated steels, wt.%.

steel type	C	Mn	Al	Si	Mo	S	P	N
3Mn	0.17	3.3	1.7	0.22	0.23	0.014	0.010	0.0043
5Mn	0.16	4.7	1.6	0.20	0.20	0.004	0.008	0.0039

3. Results and discussion

3.1. Mechanical properties

Results of static tensile tests performed at temperatures: 20°C, 100°C, 140°C and 200°C are presented in figure 1. Obtained results confirm that a deformation temperature affects mechanical properties of the investigated steels and the occurrence of serrated flow phenomenon. Both specimens deformed at 20°C show relatively high strength properties and ductility (figure 1a) when compared to higher deformation temperatures. It is related to the highest effectiveness of the TRIP effect at this deformation temperature. No evidence of PLC effect was observed at 20°C. An increase in deformation temperature to 100°C resulted in a decrease in UTS, YS and TE values for both investigated steels (figure 1b). Luo et al. [1] also observed the decrease in elongation in a temperature range 20-100°C in 0.12C-0.3Si-7Mn TRIP steel. Moreover, the characteristic serrations were observed on tensile curve of 5Mn steel. The type B serrations were observed in the range of post-uniform deformation. The UTS and TE value of specimens deformed at 140°C (figure 1c) are higher when compared to the samples deformed at 100°C (figure 1b). The deformation temperature in a range 20-140°C affects less the value of yield strength; similar results have been reported by Feng et al. [11] in a QP steel. Increasing the deformation temperature to 140°C results in a higher intensity of serrations and a lower critical strain value in the steel containing 5% Mn (figure 1c). The serration type was classified as B+C type. Serrations were not observed in the 3Mn steel. Increasing the deformation temperature to 200°C increased mechanical properties, especially in 5Mn steel (figure 1d). Both steels do not show a serration behavior at this deformation temperature (figure 1d). The steel characterized by the higher Mn content shows higher mechanical properties and lower ductility than 3Mn steel. It is due to a smaller austenite amount at the initial state in case of 5Mn steel (~11%), when compared to 3Mn steel (~16%) [9]. It is related to the fact, that Mn addition lowers C solubility in γ phase which results in the lower stability of retained austenite.

The plastic instability phenomenon observed in Fe-Mn alloys by other authors [1, 7, 8] was explained based on a DSA (dynamic strain ageing) theory. Manganese and carbon affect the serration behavior due to formation of Mn–C couples, which interact dynamically with dislocations during deformation. During the DSA, mobile dislocations are pinned and depinned by diffusing solute atoms, e.g. C atoms and Mn–C atom pairs, resulting in the occurrence of characteristic serrations on tensile curves. It was reported in literature [1, 7], that the DSA is also a temperature-dependent effect. The decrease in temperature results in a longer reorientation time of C-Mn complex, which suppresses the DSA. In present study, serrations were observed only in 5Mn steel deformed at 100 and 140°C (figures 1b and c), while no serrations were observed at 20 and 200°C (figures 1a and d). Some researchers observed a similar effect, that DSA occurs in a specific range of temperature and strain. Min et al. [7] reported that at low temperatures where serrations do not occur, dislocation unpinning occurs very rapidly relative to dislocation pinning. At high temperatures, where serrations do not occur, solute diffusion is fast enough to reduce the pinning force on dislocations. Kipelova et al. [8] noted that serrations observed in 0.13C-8.6Cr-3.2Co-1.2W-0.9Mo steel in a temperature range 200-300°C were present in a post uniform elongation range. It was explained due to the carbide precipitation effect which increased the critical strain activating the DSA to a value larger than the uniform strain of the steel. Hence, the PLC effect could not be observed before necking. Moreover, they reported that at the deformation temperature higher than 350°C serrations disappeared due to carbides precipitation, which resulted in reduced concentration of interstitial atoms in solid solution, terminating the DSA. The very similar situation took place in the present 5Mn steel.

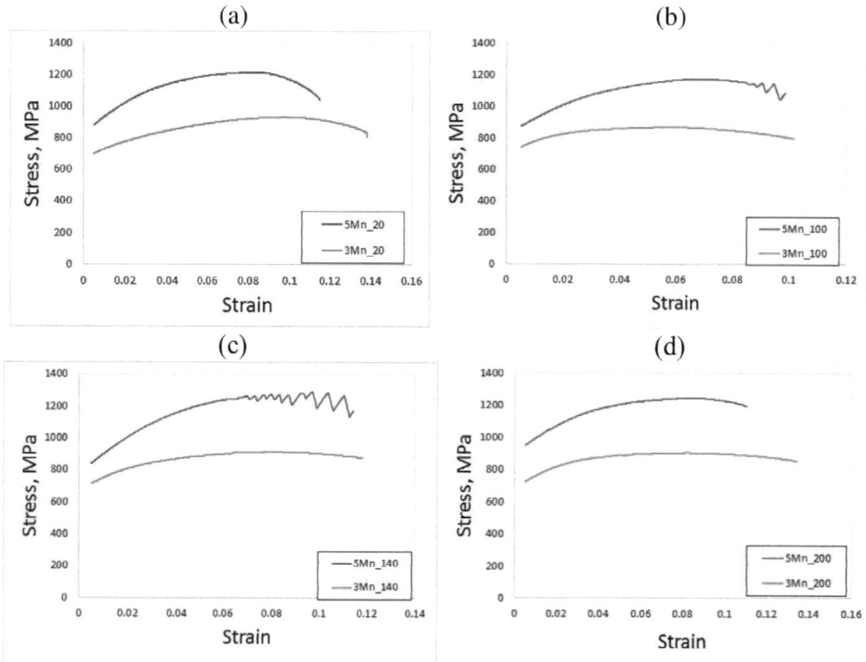

Figure 1. Tensile curves of investigated steels deformed at: 20°C (a), 100°C (b), 140°C (c) and 200°C (d).

3.2. Microstructure characterization

Figure 2 shows the SEM micrographs of investigated steels in the initial state. The matrix consists of fine lath bainitic ferrite and some fraction of martensite. The amount of martensite is higher in 5Mn steel (figure 2b). It is related to the lower carbon enrichment and therefore the lower stability of γ phase [12]. Retained austenite can be observed in the microstructure as thin layers located between bainitic laths or as martensitic-austenitic areas. In case of 5Mn steel, small precipitates characterized by a different shape located inside the bainitic areas can also be identified (figure 2b). The AFM maps in figure 3 show the location of the microstructural constituents in the initial state. The AFM surface morphology maps indicate that the height of austenite is higher than that of bainite or martensite. An evidence of martensite can be obseved in the steel containg 5% Mn (figure 3b). The decrease in height within retained austenite is characteristic for the martensitic areas.

SEM micrographs of the investaigated steels deformed at different tempertures are shown in figure 4. The microstructure of the 3Mn steel (figure 4a) deformed in a static tensile test at 20°C is characterized by the presence of small fraction of retained austenite located between bainitic ferrite or as MA areas. Some evidences of retained austenite can be observed in 5Mn steel (figure 4b) in a form of very thin layers. Increasing the deformation temperature to 100 and 140°C resulted in an increase in the stability of retained austenite (figures 4c and d). The amount of bright areas represented by γ phase becomes higher. However the fraction of RA in 5Mn type steel is lower than in 3Mn steel. 3Mn steel deformed at 200°C (figure 4e) possesses a lower fraction of RA than specimens deformed at 140°C. It is related to higher intensity of TRIP effect presumably due to initiation of precipitation processes and resulting decrease in the stability of retained austenite (possessing a smaller C content). The microstructure of 5Mn steel deformed at 200°C (figure 4f) is similar to the microstructure of the specimen deformed at 140°C. Results of microstructural observations correspond to obtained mechanical properties. An increase in γ phase stability together with increasing temperature (not the case at 200°C) is related to the higher SFE value [1]. The AFM maps in figure 5 show the surface height of the 5Mn steel deformed at 100°C. It was observed that martensite formed during deformation divided the austenitic areas into smaler parts. The freshly formed martensite is located between retained austenite laths and it is characterized by a lower height AFM profiles than the remaining austenite.

(a) (b)

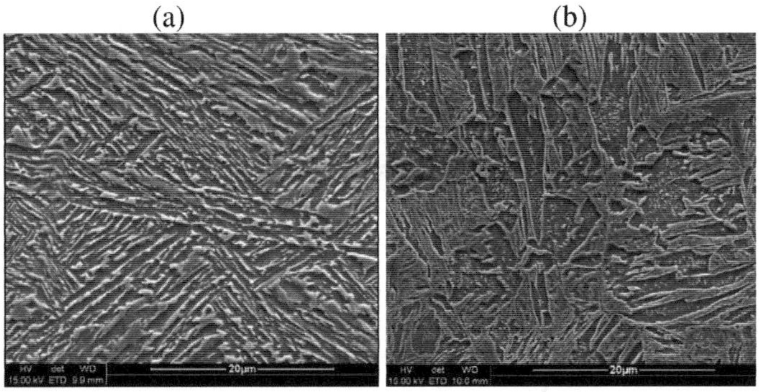

Figure 2. SEM micrographs of the 3Mn (a) and 5Mn (b) steels at the initial state (non deformed).

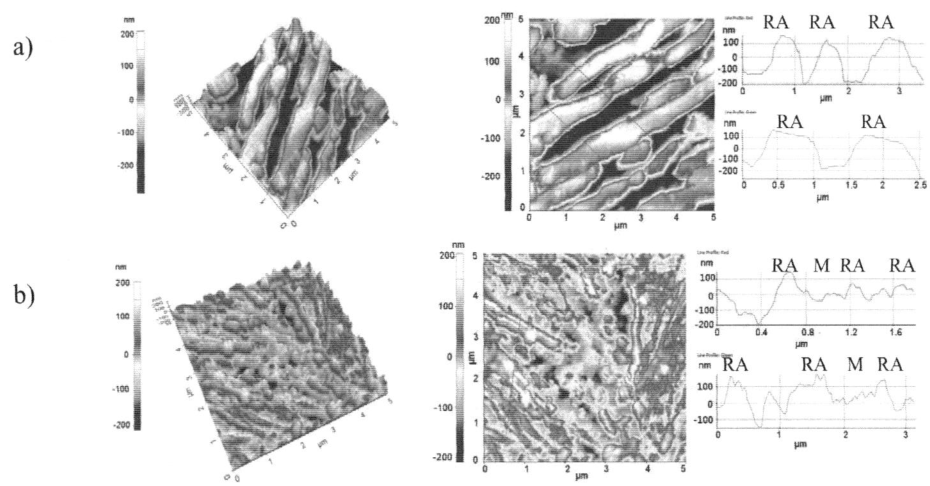

Figure 3. The surface height images registered in AFM: 3Mn (a) and 5Mn (b) steels in the initial state and corresponding height profiles.

(a) (b)

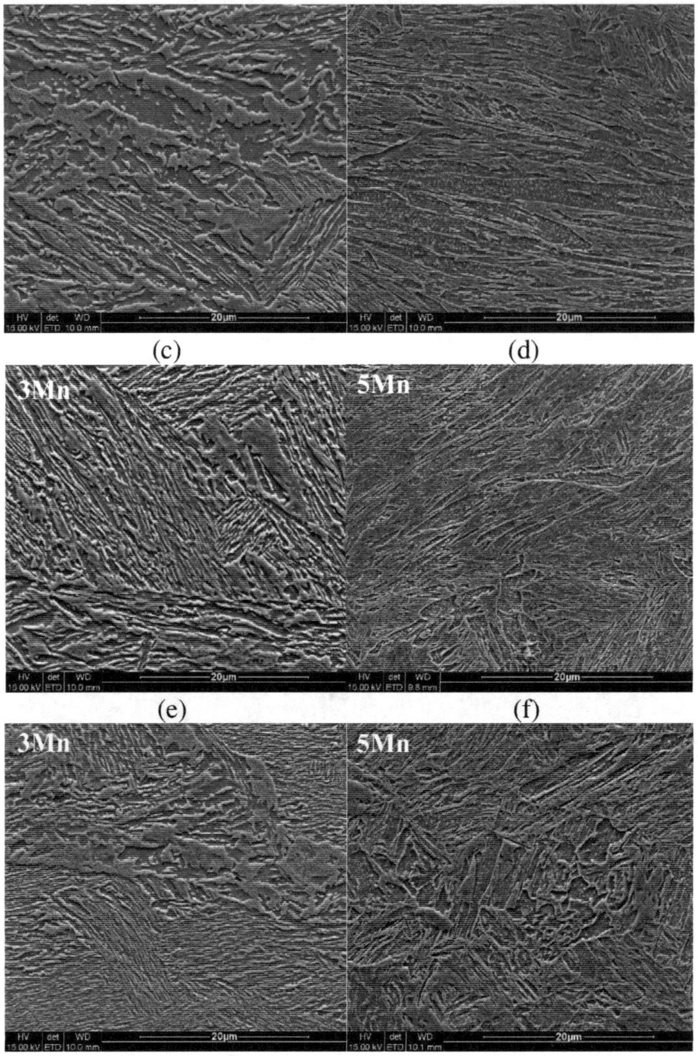

Figure 4. SEM micrographs of the 3Mn and 5Mn steels deformed at 20°C (a,b), 140°C (c,d) and 200°C (e,f).

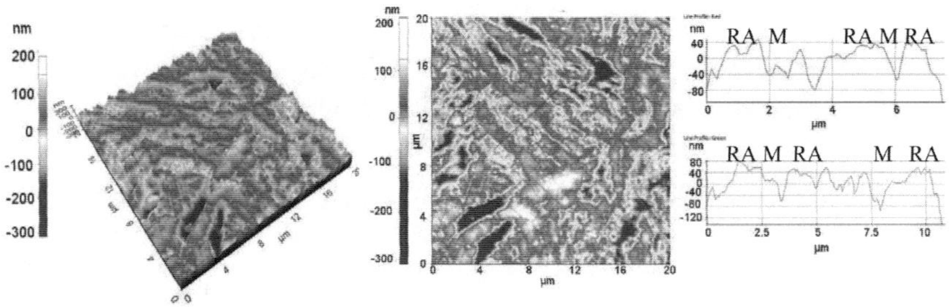

Figure 5. The surface height images registered in AFM of 5Mn steel deformed at 100°C and corresponding height profiles.

4. Conclusions

The study concerns the characterization of relationships between the manganese content, deformation temperature, mechanical properties and appearance of serrated flow phenomenon. It was found that manganese addition and deformation temperature affect the appearance of serrated flow behaviour. The critical strain value activating the PLC effect and obtained mechanical properties are also temperature-dependent. The serrated flow phenomenon was observed in the steel with a higher Mn content at deformation temperatures 100 and 140°C. It is due to the higher amount of C-Mn pairs which interact dynamically with dislocations during deformation. The disappearance of serrations at 200°C can be related to the increased diffusion intensity which reduces the pinning force on dislocations or precipitation process reduces the concentration of intestinal atoms in solid solution. The 3Mn steel is free of the PLC effect due to a lower amount of C-Mn pairs.

Acknowledgement

The work was supported by the scholarship fund of the Silesian University of Technology in 2019.

References

[1] Luo L, Li W, Li W, Zhou S and Jina X 2017 Tensile behaviors and deformation mechanism of a medium Mn-TRIP steel at different temperatures *Mater. Sci. Eng. A.* vol 682 p **698-703** doi: 10.1016/j.msea.2016.11.017

[2] Jirkova H, Kucerova L, Aisman D and Masek B 2014 Optimization of the Q-P process parameters for low alloyed steels with 0.2% C *Arch. Metall. Mater.* vol 59 p **1205-12**10 doi: 10.2478/amm-2014-0209

[3] Kuc D, Hadasik E, Niewielski G, Schindler I, Mazancova E, Rusz S and Kawulok P 2012 Structural and mechanical properties of laboratory rolled steels high-alloyed with manganese and aluminium *Arch. Civ. Mech. Eng.* vol. 12 p **312-317** doi:10.1016/j.acme.2012.06.008

[4] Jablonska M B 2014 Mechanical properties and fractographic analysis of high manganese steels after dynamic deformation tests *Arch. Metall. Mater.* vol 59 p **1193-1197** doi:0.2478/amm-2014-0207

[5] Gorka J and Stano S 2018 Microstructure and properties of hybrid laser arc welded joints (laser beam-MAG) in thermo-mechanical control processed S700MC steel *Metals* vol 8 doi:10.3390/met8020132

[6] Kvackaj T and Mamuzic I 2006 Development of bake hardening effect by plastic deformation and annealing conditions *Metalurgija* vol 45 p **51-55**

[7] Min J, Hector Jr L G, Zhang L, Sund L, Carsley J E and Lin J 2016 Plastic instability at elevated temperatures in a TRIP-assisted steel *Mater. Des.* vol 95 p **370-386** doi: 10.1016/j.matdes.2016.01.113

[8] Kipelova A, Kaibyshev R, Skorobogatykh V and Schenkova I 2010 Portevin–Le Chatelier effect in an E911 creep resistant steel with 3% Co additives *J. Phys. Conf. Ser.* vol 240 doi: 10.1088/1742-6596/240/1/012100

[9] Grajcar A, Skrzypczyk P and Kozłowska A 2018 Effects of temperature and time of isothermal holding on retained austenite stability in medium-Mn steels. *Appl. Sci.* vol 8 doi:10.3390/app8112156

[10] Grajcar A, Skrzypczyk P and Wozniak D 2014 Thermomechanically rolled medium-Mn steels containing retained austenite *Arch. Metall. Mater.* Vol. 59 p **1691-1697** doi:10.2478/amm-2014-0286

[11] Feng W, Wu Z, Li W and Speer J G 2013 Effect of testing temperature on retained austenite stability of cold rolled CMnSi steels treated by quenching and partitioning process *Steel Res. Int.* vol. 84 p **246-252** doi:10.1002/srin.201200129

[12] Grajcar A, Kwasny W and Zalecki W 2015 Microstructure-property relationships in TRIP aided medium-C bainitic steel with lamellar retained austenite *Mater. Sci. Technol.* vol 31 p **781-794** doi:10.1179/1743284714Y.0000000742

Effect of soaking temperature on the microstructure and mechanical properties of heat-treated Al-Si-Nb TRIP steel

L Kučerová[1], K Burdová[1], A Stehlík[1]

UWB in Pilsen, Univerzitní 8, 306 14 Plzeň, Czech Republic

E-mail: skal@rti.zcu.cz, kburdova@rti.zcu.cz, stehlika@rti.zcu.cz

Abstract. Two-step heat treatment with bainitic hold at 425 °C was applied to low alloyed high strength steel with partial substitution of silicon by aluminium. The steel was further micro-alloyed by niobium. Various soaking temperatures in the range of 750-1250 °C were used to analyse the effect of soaking on mechanical properties and microstructure of the steel. According to calculations in JMatPro, the temperatures below 900 °C should lie in an intercritical (tow phase) region, while the temperature of 900 °C should be the first one to ensure full austenitization of this steel. To dissolve different portion of niobium into solid solution, higher soaking temperatures were also used. Tensile strengths of 780-1069 MPa and total elongation of 22-46 % were achieved.

1. Introduction

Ongoing development of materials with increased tensile strength, ductility and formability are fuelled particularly by the demands of an automotive and vehicle industry. Increasing requirements on CO_2 reduction and lower fuel consumption coupled with increasing safety requirements could be satisfied by application of materials with higher strengths which will enable weight reduction without compromising on the safety of the constructions.

Several grades of advanced high strength steels (AHSS) have been developed in recent years, with various combinations of tensile strength and ductility [1-5]. TRIP (transformation induced plasticity) steels belong to the first generation of AHSS, using the principle described by Zackay [6] on a novel lean alloying concept. Modern TRIP steels are low alloyed steels with carbon contents between 0.2-0.4 % [1, 7-11], but most frequent are carbon contents around 0.2 %. The steel is alloyed by 1-2 % of manganese which increases stability of retained austenite and originally, between 1-2 % of silicon was used to prevent carbide precipitation during the treatment. As those high silicon steels possessed low surface quality, which presents problems during galvanising, attempts were made to find suitable substitution [12]. The most promising possibility seems to be partial substitution of silicon by aluminium, which has the same effect on carbide precipitation hindering and solid solution strengthening as silicon [13, 14]. However, the effects of aluminium are weaker than those of silicon for the same contents of both elements. This is the reason why a full substitution of silicon is not convenient, as it results in a decrease of tensile strengths [15-18]. Other minor alloying elements were also investigated in the attempts to improve mechanical or technological properties of TRIP steels. In this work, micro-alloying by niobium was used, as it was shown to have a beneficial effect on sensitivity to pearlite formation [5]. Carbide precipitation and pearlite formation are undesirable in TRIP steels, as they consume carbon which is needed for proper stabilization of retained austenite. TRIP steels contain typically multi-phased microstructures with either ferrite or bainite matrix, always containing bainite, retained austenite and often also small amounts of either martensite or so-called M-A constituent. M-A constituent are larger bulk islands of remaining austenite that were not stable enough during the final cooling to the room temperature and therefore partially transformed to martensite (mainly at the edges of the islands). Retained austenite in TRIP steels can be of either lath of bulky morphology, the laths are mainly part of the bainitic microstructure, smaller islands of irregular shapes are often part of globular bainite while larger bulky islands are typically found at the edges of bainitic blocks [3, 5, 11]. The volume fraction, morphology and distribution of phases and structural components have a significant effect on mechanical properties of TRIP steels [9]. In this work, the effect of soaking temperature on the final microstructure and mechanical properties of Al-Si-Mn-Nb TRIP steel.

2. Experimental program

2.1. *Material*

Low carbon low alloyed high strength steel with partial substitution of silicon by aluminium and micro-alloying by niobium was used in this work (table 1).

Table 1. Chemical composition of used steel in weight %.

C	Mn	Si	P	S	Cr	Ni	Cu	Al	Nb
0.2	1.5	0.5	0.008	0.005	0.008	0.072	0.058	1.5	0.06

TTA (Time–temperature-austenitization) diagram of this steel was calculated using JMatPro software (figure 1 a). According to the diagram, the austenitization of used experimental steel starts around 740 °C and full austenitization can be expected above the temperature of 980 °C.

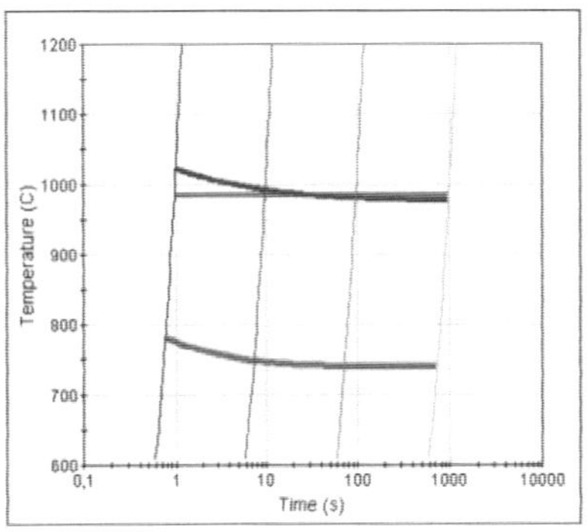

a)

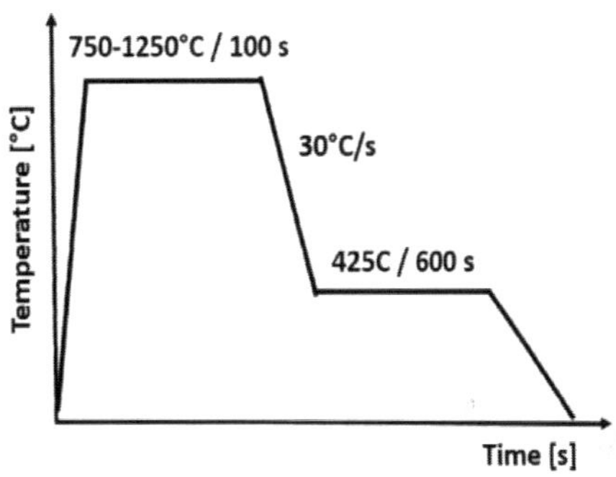

b)

Figure 1. a) TTA diagram. b) Scheme of used two-step heat treatment.

2.2. *Heat treatment*

Experimental steel was cast in induction vacuum furnace, the ingot was cut into four parts and forged at 1150 °C into bars with the diameter around 25 mm. The bars were cooled in the cold furnace after the forging used for the preparation of samples for subsequent heat treatment.

Heat treatment was carried out at thermo-mechanical simulator, which enables the application of precisely defined and controlled thermal and deformation parameters. High-frequency electrical resistance heating is used to heat the specimen. With a maximum power of 3 kW, the specimens can be heated in a controlled manner at rates in excess of 200 °C/s, depending on the material and shape of the test piece. The active part of the sample is cylindrical with 8 mm diameter and length of 16 mm.

Two-step heat treatment typical for processing of TRIP steels was designed for this work (figure 1 b). Various soaking temperatures in the region of 750-1250 °C were applied to the steel, the other processing parameters were kept the same and chosen based on previous experiments [10,13]. All the samples had the same 100 s hold at the soaking temperature, followed by cooling to the second

600 s hold at 425 °C, where bainite transformation took place. Constant cooling rate of 30 °C/s was used. Lower soaking temperatures of 750-950 °C should be in the two-phase intercritical temperature interval, according to a JMatPro calculation and therefore a varying fraction of polygonal ferrite should remain in the resulting microstructure. The temperature of 1050 °C should already ensure a full austenitization of the prior microstructure and only proeutectoid ferrite should appear in the final microstructure. The highest temperatures of 1230 and 1250 °C were chosen to observe the effect of niobium dissolved in solid solution on the mechanical properties, as niobium particles were reported to start dissolving in these steels at around 1050 °C [19].

2.3. Characterisation

Processed samples were cut along the longitudinal axes for metallographic analysis and the metallographic sections were prepared in a standard way. All samples were etched in 3 % Nital to reveal the microstructure, which was observed by an Olympus BX 61 light microscope and a Zeiss Crossbeam scanning electron microscope. The volume fraction of the retained austenite was determined by X-ray diffraction phase analysis using an AXS Bruker D8 Discover automatic powder diffractometer with a HI-STAR detector and Co lamp ($\lambda K\alpha = 0.1790307$ nm). The measurement was carried out in the central part of the samples and spectra were taken in the range of 2 ϑ from 25° to 110°. The integrated intensities of (200) ferrite peak and (111), (002) and (022) austenite peaks were used for evaluation. Mechanical properties were determined by tensile tests of sub-size flat samples with a gauge length of 5 mm and a cross-section of 2 x 1.2 mm. Two samples from each treatment were tested and average values are given in table 2.

3. Results and discussion

A significant amount of polygonal ferrite was found in the samples with soaking temperatures of 750 and 770 °C (figure 2 a). The sample soaked at 800 °C possessed a mixture of polygonal ferrite grains and proeutectoid ferrite placed at prior austenite boundaries (figure 2 b). Interestingly, polygonal ferrite was missing in sample soaked at 850 °C despite JMatPro prediction. Only large islands of bainitic ferrite could be observed at light micrographs (figure 2 c).

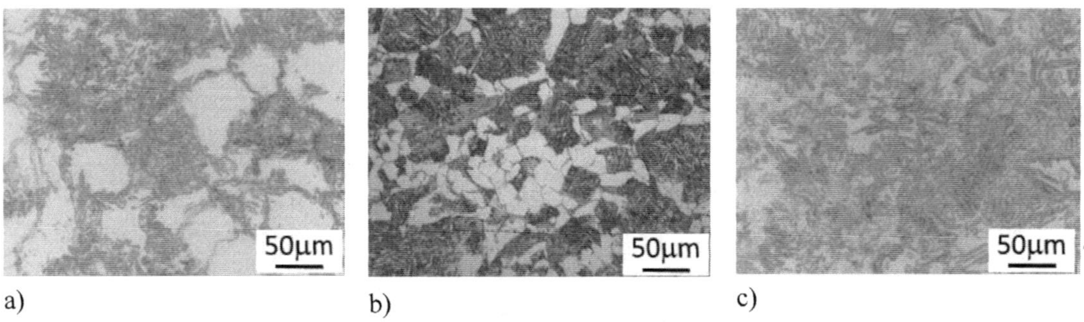

Figure 2. Microstructures of samples soaked at a) 770 °C, b) 800 °C, c) 850 °C.

Very fine lines of proeutectoid ferrite could be still seen also in the microstructure obtained by soaking at 900 °C (figure 3 a), while higher soaking temperature resulted in the bainitic microstructures (figure 3 b, c) with various bainite morphologies.

Detailed microstructure analysis (figure 4) showed changing bainitic microstructure. For lower soaking temperatures, the bainite was predominantly granular, with just occasional area of lath bainite. Retained austenite in bainite obtained at soaking temperature of 750 °C (figure 4 a) possessed relatively uniform size and by far the largest amount of retained austenite (23%) was measured in this microstructure by an X-ray diffraction phase analysis (table 2). Around 50% of ferrite in the microstructure could have contributed to this extensive retained austenite stabilization, as more carbon was, therefore, present in the remaining austenite leading to more significant chemical stabilization of austenite.

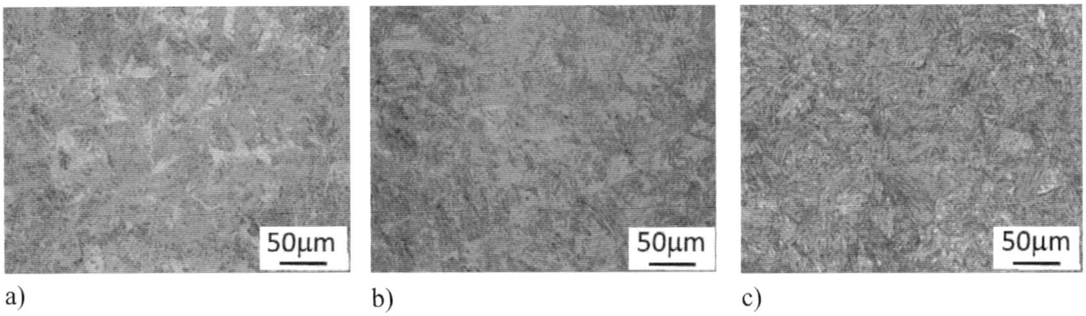

Figure 3. Microstructures of samples soaked at a) 900 °C, b) 1050 °C, c) 1250 °C.

Table 2. Mechanical properties (YS - yield strength, UTS – ultimate tensile strength, EL – total elongation) and retained austenite volume fraction (RA)

Soaking temperature [°C]	YS [MPa]	UTS [MPa]	EL [%]	RA [%]
750	521±13	786±10	41±0	23
770	633±6	781±8	46±0	14
800	561±10	789±2	45±0	12
850	614±3	822±2	46±0	10
900	582±1	836±1	43±1	14
950	658 ±4	843±3	43±1	12
1050	897±1	1030±5	24±0	11
1230	960±5	1069±6	22±0	11
1250	946±6	1054±1	22±0	12

Even this high retained austenite content, however, did not result in an increased elongation in comparison with following samples with 10-14 % of retained austenite. It could be therefore expected that about half of this austenite didn't contribute to the TRIP effect. Relative thin austenitic phase was detected at grain boundaries of proeutectoid ferrite obtained in the sample with a soaking temperature of 800 °C (figure 4 b).

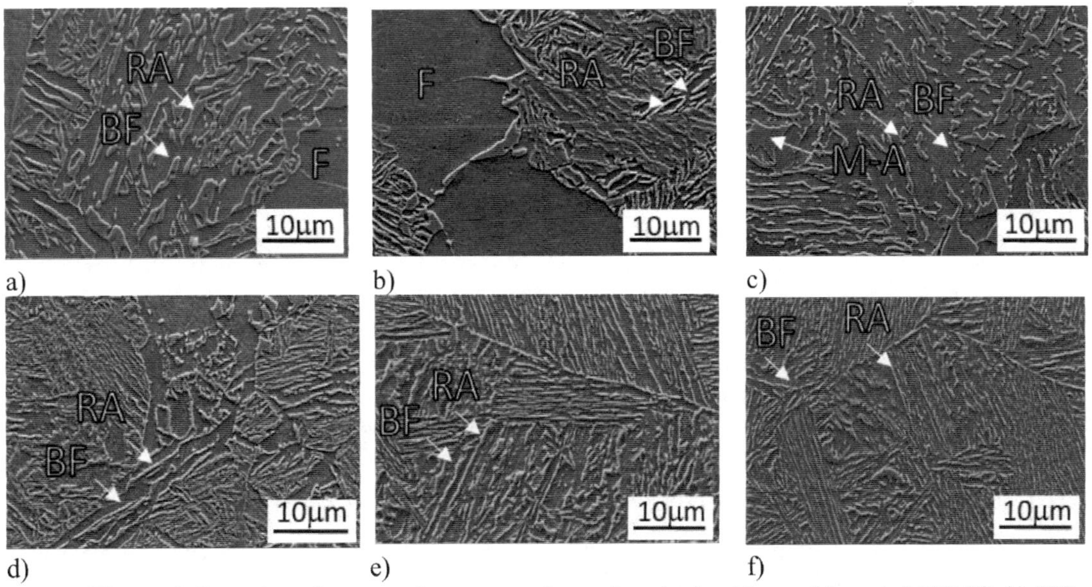

Figure 4. Scanning electron microscopy of samples obtained by soaking at a) 750 °C, b) 800 °C, c) 900 °C, d) 1050 °C, e) 1230 °C, f) 1250 °C.

The samples with soaking temperature in the two-phase region (750-800 °C) possessed very similar mechanical properties, reaching tensile strength of 780-790 MPa with total elongations above 40 %. Bainitic microstructure without polygonal ferrite was obtained in the sample soaked at 900 °C (figure 4 c), consisting of bainitic ferrite plates with sparely distributed areas of retained austenite. The mixture of granular and lath bainite was found, with granular bainite prevailing. Larger islands of M-A constituent were observed in the microstructure. The best combinations of tensile strengths and total elongations were obtained for samples obtained by soaking at temperatures of 850-950 °C. These samples reached tensile strengths of 820-840 MPa with elongations of 43-46 %.

With increasing soaking temperature, the bainite became denser and more of a lamellar type, with thinner laths of retained austenite and bainitic ferrite (figure 4 d-f). These laths are however slightly growing in length with increasing soaking temperature. Fine grains of proeutectoid ferrite can be found occasionally in the microstructure obtained by soaking at 1050 °C (figure 4 d). The mixture of lamellar upper bainite and granular bainite could be found in the microstructures obtained by soaking at the highest temperatures above 1200 °C (figure 4 e, f). The highest soaking temperature of 1250 °C resulted in the thinnest bainitic laths. The microstructures obtained by soaking at 1050-1250 °C have the same volume fraction of retained austenite (11 %), regardless of the morphology of the bainite. Similarities in the microstructures are mirrored by their total elongations of 22-24 % and high ultimate tensile strengths of 1030-1070 MPa.

It can be assumed that these high strengths were caused by thin bainitic laths, prevailing lath bainite could be also responsible for the drop of elongation. Lower elongation could be also assigned to the very high stability of retained austenite in these samples, which caused by extreme thickness of retained austenite laths (size stabilization). Nearly 9 % of retained austenite were determined by X-ray diffraction phase analysis at the neck area of fractured tensile sample processed by soaking at 1250 °C. This suggests that only a very small amount of retained austenite actually transformed during the deformation and TRIP effect, therefore, could not be utilized.

4. Conclusions

Two-step heat treatment with annealing hold at 425 °C was applied to low carbon low alloyed TRIP steel. Nine heat treatments with various soaking temperatures in the region of 750-1250 °C were carried out to determine the effect of soaking temperature on the final microstructure and mechanical properties. Samples with soaking temperature in a two-phase (ferrite-austenite) region possessed the lowest tensile strengths around 785 MPa due to the presence of polygonal ferrite in the final microstructure. Their total elongation was above 40%. The soaking temperature of 850 °C already produced final microstructure without polygonal ferrite despite JMatPro calculations.

Samples with soaking temperatures around 900 °C produced predominantly bainitic microstructures with mainly granular morphology. They reached the best combinations of tensile strengths around 835 MPa with total elongations about 43 %.

The last group of samples was processed by the highest soaking temperatures of 1050-1250 °C, where the increasing dissolution of niobium particles could be expected. Lath bainite was becoming the dominant morphology with increasing soaking temperature. These samples obtained the highest strengths around 1050 MPa with the lowest elongations around 22 %. It was found out that the retained austenite in these samples was over stabilized and did not transform to martensite during tensile loading, probably due to very thin laths of retained austenite in the bainitic microstructures.

Acknowledgements

The present contribution has been prepared with the support of the student grant competition of University of West Bohemia in Pilsen, SGS 2018-048 Design of heat and thermo-mechanical treatment of high strength steels of the third generation. The project is subsidised from specific resources of the state budget for research and development.

References

[1] De Cooman B C 2004 Structure-properties relationship in TRIP steels containing carbide-free bainite *Current Opinion in Solid State and Materials Science*, Vol. 8, pp. 285–303.

[2] Kucerova L, Aisman D, Jirkova H 2009 Optimization of Q-P process parameters with regard to final microstructure and properties, Annals of DAAAM for 2009 & proc. of the 20th international DAAAM symposium (Annals of DAAAM and Proceedings vol 20) pp 1035-1036
[3] Pereloma E V, Timokhina I B, Hodgson P D 1999 Transformation Behaviour in Thermomechanically Processed C–Mn–Si TRIP Steels with and without Nb *Mater. Sci. Eng. A.*273–275 pp 448–452
[4] Masek B, Jirkova H, Kucerova L 2012 Rapid Spheroidization and Grain Refinement Caused by Thermomechanical Treatment for Plain Structural Steel *Materials Science Forum* vol 706-709) pp 2770-2775
[5] Kucerova L, Opatova K, Kana J 2017 High versatility of Niobium alloyed AHSS *Archives of Metallurgy and Materials* vol 62 Issue: 3 pp 1485-1491
[6] Zackay V F, Parker E R, Fahr D, Busch R 1967 *Trans. ASM.* vol 6 pp 252-259.
[7] Kucerova L 2017 The Effect of Two-Step Heat Treatment Parameters on Microstructure and Mechanical Properties of 42SiMn Steel *Metals* vol 7 Issue: 12 AN: 537)
[8] Shen Y F, Qiu, L N, Sun, X 2015 Effects of retained austenite volume fraction, morphology, and carbon content on strength and ductility of nanostructured TRIP-assisted steels *Mater. Sci. Eng. A.* vol 636 pp 551-564
[9] Grajcar, A. 2014 Thermodynamic analysis of precipitation processes in Nb–Ti-microalloyed Si–Al TRIP steel *J Therm Anal Calorim* Vol. 118 (2), pp 1011-1020
[10] Kučerová L, Bystrianský M 2017 Comparison of thermo-mechanical treatment of C-Mn-Si-Nb and C-Mn-Si-Al-Nb TRIP steels *Procedia Engineering* pp 1856-1861
[11] Tirumalasetty G K 2012 Deformation-Induced Austenite Grain Rotation and Transformation in TRIP-Assisted Steel *Acta Materialia* 60, 3 pp 1311–1321
[12] Bhattacharyya T 2013 An Assessment on Coatability of Transformation Induced Plasticity (TRIP)-Aided Steel *Surface and Coatings Technology* 235 pp 226–34
[13] Kučerová L, Jandová A, Jeníček Š 2018 Effect of Thermo-Mechanical Treatment with Various Annealing Hold Conditions on Al Alloyed TRIP Steel *Manufacturing Technology* 18/5 pp 764-768
[14] Li L, De Cooman B C, Wollants P, He Y, Zhou X 2004 Effect of Aluminum and Silicon on Transformation Induced Plasticity of the TRIP Steel, *J. Mater. Sci. Technol* vol 20 pp 135-138
[15] Chiang J, Boyd J D, Pilkey AK 2015 Effect of microstructure on retained austenite stability and tensile behaviour in an aluminum-alloyed TRIP steel, *Mater. Sci. Eng. A.* vol 638 pp 132-142
[16] Li L 2004 Effect of Aluminum and Silicon on Transformation Induced Plasticity of the TRIP Steel *Journal of Materials Science & Technology* 20, 2 pp 135–38
[17] Monsalve A, Guzmán A, De Barbieri F 2016 Mechanical and Microstructural Characterization of an Aluminum Bearing Trip Steel *Metallurgical and Materials Transactions* A 47(6)
[18] Timokhina I B, Enomoto M, Miller M K, Pereloma E V 2012 Microstructure-Property Relationship in the Thermomechanically Processed C-Mn-Si-Nb-Al-(Mo) Transformation-Induced Plasticity Steels Before and After Prestraining and Bake Hardening Treatment *Metallurgical and Materials Transactions* A 43A(7)) pp 2473-2483
[19] Hong S Ch, Lim S H, Hong H S, Lee K J, Shin D H Lee K S 2003 Effects of Nb on strain induced ferrite transformation in C-Mn steel *Mater. Sci. Eng. A.*Vol. 355 pp 241-248

Effect of revolutions number on mechanical properties of HPT processed copper

D Melzer[1a], I Smirnov[2b], A Evstifeev[2c], S Rzepa[1d], P Konopík[1e]

[1]COMTES FHT a.s., Prumyslova 995, 334 41 Dobrany, Czech Republic

[2]Saint Petersburg State University, Universitetskaya nab. 7/9, St. Petersburg, 199034, Russia

[a]daniel.melzer@comtesfht.cz, [b]si.v.smirnov@spbu.ru, [c]ad.evstifeev@gmail.com, [d]sylwia.rzepa@comtesfht.cz, pavel.konopik@comtesfht.cz

Abstract. This paper studies the effect of high-pressure torsion (HPT) method at ambient temperature on mechanical properties of material. The aim is to compare copper subjected to HPT for 2, 10 and 30-revolutions with coarse grain structured copper in a cold-rolled state. Miniaturized tensile tests were performed to evaluate anisotropy of mechanical properties within a disc product of HPT process. Also the strain rate sensitivity was examined. The results of mechanical tests demonstrate that increasing shear strain leads to ultra-fine grain structure (UFG) which resulted in increasing of material strength. The plasticity of material decreases correspondingly to that. With respect to character of HPT process, discs are known as non-homogenous products, where shear strain effect increases in radial direction from the centre section to the edge. Results show that with different number of HPT revolutions the anisotropy of mechanical behaviour is changing. With increasing number of revolutions the anisotropy within a single disc is increasing.

1 Introduction

Mechanical properties of materials are key factor for their successful application. Achieving the optimal combination of strength and ductility is the main challenge in material processing. One of the way how to enhance mechanical properties of material is to use high pressure torsion (HPT) which belongs to severe plastic deformation (SPD) techniques in order to achieve suitable combination of high strength and ductility. SPD processes create ultra-fine grain (UFG) structure with high angle grain boundaries (HAGB) via imposing severe shear strains [1 – 2]. UFG materials may have strength exceeding that of coarse grained (CG) and even alloyed metals [3]. Grain size distribution is in the sub-micrometer or nanometer scale [4]. However, SPD techniques have their limitations. With increasing number of HPT revolutions the strength of material increases but regarding to that the ductility of material decreases [5]. This results in the application limitation of material. For reliable application the material must be characterized by a good combination of strength and ductility. It means rather limitation of number of revolutions processing via HPT, or application of post-processing. Volume of material, which can be HPT processed is insufficient to produce standard specimens. Thus, only small specimen test techniques are possible to use.

In this study, cylindrical rods of pure copper in an as cold-rolled state were subjected to HPT processing at 2, 10 a 30-revolutions. Set of miniaturized tensile tests were performed at room temperature. The effect of number of revolutions on mechanical properties was discussed as well as dependency of material anisotropy.

2 Experimental material and procedure

Resulting properties after HPT are dependent on the initial material state. The base pure copper (99.9%) of M1 grade with coarse-grained structure was delivered in the form of cold-rolled rods with 20 mm in diameter. This material state is referred as the Coarse Grained (CG). Chemical composition is presented in Table 1.

Table 1. Chemical composition of pure copper M1.

As	Bi	Fe	Ni	O_2	Pb	S	Sb	Sn	Zn	Cu+Ag
<0.001	<0.001	<0.001	<0.001	0.04	<0.001	<0.001	<0.001	<0.001	<0.001	base

The rod of CG material of diameter 20 mm and 2.5 mm thickness were cut perpendicular to the rolling direction. The UFG structure of the material was produced by SPD using the HPT method [6]. HPT was conducted on Walter-Klement GmbH HPT-07 press at room temperature, presented on Figure 1a). Schema of material processing also is presented, see Figure 1b).

The HPT processing is defined by applied force F and total strain ε expressed by equation (1) [7], wherein N is revolution number, t is the specimen thickness and r is the distance from the centre of the disc to the edge. The equation clearly shows that with increasing distance from the centre and increasing number of revolutions, imposed strain increases.

$$\varepsilon = 2\pi rN/t\sqrt{3} \tag{1}$$

The billets were compressed under required pressure (6 GPa), then the torsion with a given number of revolutions (2, 10 and 30) was conducted under conditions of hydrostatic pressure and the speed of 1 rev/min. The absence of slippage was checked by the coincidence of the drawn strips on the sample surfaces after torsion. As a result, the load capacity and anvil dimensions allow to produce the discs with diameter of 20 mm and thickness of ~1.9 mm shown on Figure 1c). Two discs of UFG copper after 2 revolutions, 10 revolutions and one disc after 30 revolutions were produced in total. Equivalent strain in the middle of the radius was 4.2, 5.9 and 7 for 2 revolutions 10 revolutions and 30 revolutions, respectively.

a)

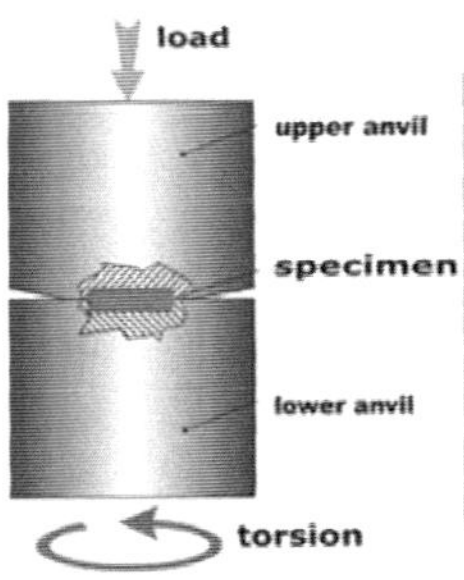

b)

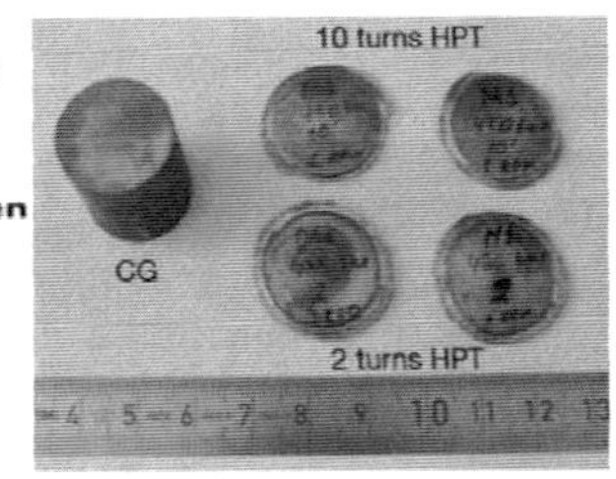

c)

Fig. 1. a) Walter-Klement GmbH HPT-07 press, b) Schema of HPT process, c) CG and HPT processed copper.

Evolution of ultrafine-grained structure for different revolution number was described in many studies [8-10]. It was shown, that for smaller number of revolutions the differences in grain size from the center of disc to the edge are smaller, since the recrystallization and limited grain growth occurred. These effects act against refining process and lead to creation of coarse grains. For higher number of revolutions no significant recrystallization areas were observed in the edge area. Nevertheless, in the center area of the disc, equivalent strain for certain revolution number may exceed value, when recrystallization may appear.

3 Miniaturized tensile test (MTT)

MTT is widely used testing technique on different fields of mechanical science nowadays. [11-13]. To examine mechanical properties of HPT discs processed by different number of revolutions, sets of miniaturized tensile tests (MTT) were performed. Examination of mechanical properties of UFG material via MTT was successfully used in different studies [14-17]. Three testing specimens with known location were extracted from each produced disc. MTT specimens in longitudinal (Y) and perpendicular (X) orientation were also produced from CG copper. Specimens were extracted using a wire electro-discharge grinding machine. To keep the same dimensions, specimens extracted from discs were polished to final thickness t = 0.5 mm. Extracted specimens and their geometry are presented on figure 2 a) – b).

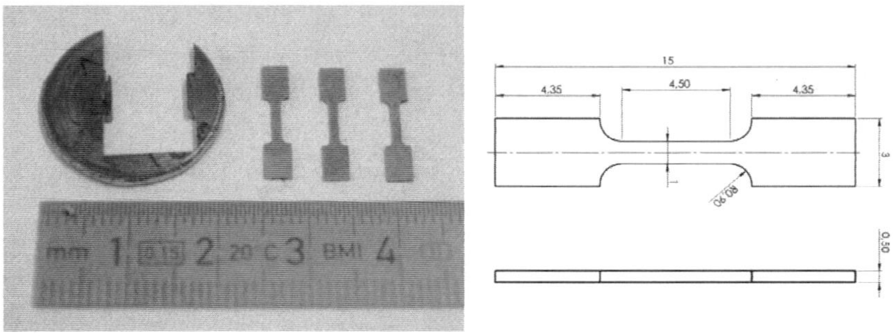

a) b)

Figure 2. a) MTT specimens extracted from disc, b) MTT specimens geometry.

Effect of absolute size or state of material microstructure can significantly influence reliability of measured results. State of microstructure, especially grain size has major effect on plastic deformation and Yield strength (YS). With changing of grain size, the number of grains in the specimen cross section is also changing. The main mechanism of plastic deformation for steels is slip of dislocations. With increasing number of grains, the number of dislocations and number of its interaction with grain boundaries increases too. A phenomenon of dependency of YS on number of grains is well known and described by Hall-Patch equation (2) [18]. With respect to miniaturized specimen the predominant dimension is thickness of specimen. The ratio t/d, where t is specimen thickness and d is grain diameter, is the decisive parameter, whether the microstructure affects the tensile results or not. Nevertheless, in case of UFG materials, the number of grains located in the cross-section of a specimen is large due to their small size. Therefore, tensile test results are more comparable to standard specimens.

$$YS = \sigma_0 + \frac{k_y}{\sqrt{d}} \quad (2)$$

MTT were carried out on testing machine with linear drive with the loading capacity of 5kN. Deformation was measured using optical extensometer of Mercury RT system. All the tests were performed at ambient temperature. Cross-head speeds of 0.25 mm/min and 10 mm/min were set up to examine the strain rate effect.

4 Tensile tests results

For all test specimens tensile curves were evaluated. Stress – strain curves of tested MTT specimens are presented on Figure 3. For better clarity and readability, only average curve of each tested batch is shown on graph. In case of specimens subjected to 30 revolutions, all results are presented. Table 2 present average results and standard deviation of mechanical parameters for each tested group.

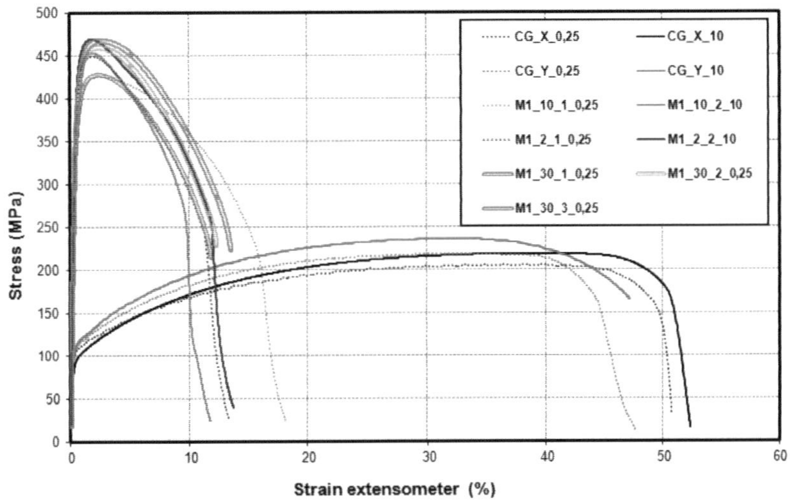

Figure 3. Average engineering stress-strain curves of tested MTT in CG and UFG state.

Table 2. Sumarization of average values of mechanical parameters of tested MTT.

Specimen	Value	$R_{p0,2}$ MPa	R_m MPa	A_g %	A_5 %	Z %	Cross-head mm/min	copper state
CG_X_0.25	Average	**99.2**	**207.7**	**35.5**	**47.0**	**74.2**	0.25	Coarse Grain
	St. Dev.	6.7	2.0	0.8	2.2	5.7		
CG_X_10	Average	**91.8**	**218.4**	**39.1**	**53.1**	**72.6**	10	
	St. Dev.	5.6	0.5	0.7	2.4	0.2		
CG_Y_0.25	Average	**104.1**	**220.4**	**33.2**	**45.3**	**69.2**	0.25	
	St. Dev.	4.1	2.7	1.0	1.2	0.2		
CG_Y_10	Average	**106.2**	**235.5**	**31.6**	**46.4**	**74.3**	10	
	St. Dev.	1.0	3.0	0.2	0.2	6.7		
M1_2_0.25	Average	**353.8**	**444.2**	**1.6**	**11.1**	**72.2**	0.25	HPT Processing 2 rev.
	St. Dev.	18.1	6.2	0.4	0.8	3.0		
M1_2_10	Average	**401.7**	**475.8**	**1.7**	**11.8**	**72.7**	10	
	St. Dev.	7.7	6.1	0.2	0.4	9.1		
M1_10_0.25	Average	**334.2**	**427.1**	**2.0**	**15.2**	**80.5**	0.25	HPT Processing 10 rev.
	St. Dev.	15.9	10.4	0.0	1.7	3.6		
M1_10_10	Average	**390.3**	**446.7**	**1.5**	**10.2**	**75.1**	10	
	St. Dev.	23.0	15.1	0.2	1.0	3.9		
M1_30_0.25	Average	**367.3**	**452.1**	**1.9**	**12.4**	**75.7**	0.25	HPT Processing 30 rev.
	St. Dev.	20.9	21.3	0.2	0.8	1.6		

4.1 Tensile parameters comparison

As can be seen from presented graph, CG copper structure was characterized by ultimate tensile strength (UTS) about 210 MPa, YS = 100 MPa and elongation of 50 %. Thanks to as-rolled state with no heat treatment processing slight anisotropy of base material was present. Considerable difference of tensile properties is visible for HPT processed copper with UFG structure. Results for copper subjected to 2 revolutions show more than 3-times and 2-times higher value of YS and UTS, respectively. Correspondingly to UTS and YS increase, there is significant decrease of material ductility, when uniform elongation changes from 40 % to 2 %. This difference has significant influence on applicability of UFG material. For copper processed by 10 revolutions, UTS and YS values are slightly lower compared to 2-revolutions specimens.

4.2 Strain rate sensitivity
Effect of strain rate to UFG behavior was examined on CG, 2 and 10 revolutions specimens. Strain rates $\dot{\varepsilon}$ = 0.0009 and 0.03 s^{-1} were applied. For all three material states the increase of UTS and YS is visible. Nevertheless, changes are more noticeable for HPT processed copper after 2 and 10 revolutions. This can be effect of history of deformation, since for increasing number of revolutions the strain sensitivity slightly increases, too.

4.3 Anisotropy of mechanical properties
The last stage of experimental program was to produce UFG copper after 30 revolutions. Three MTT specimens were extracted and tested with applied strain rate of 0.0009 s^{-1}. The results show, that for higher number of revolutions, the changes of tensile properties are almost negligible. Nevertheless, with increasing number of revolutions there is more significant anisotropy of mechanical properties within the produced disc. For discs produced by 2 and 10 revolutions, good repeatability of tests is clearly visible, regardless to the area of extraction. For each tested group of three specimens the results are in good agreement with relatively small standard deviation. On the other hand, in case of specimens subjected to 30 revolutions, lower values of tensile parameters are noticeable for specimens extracted from the center of the disc, compare to specimens from sides of disc. This clearly demonstrates, that although increasing number of revolutions doesn't affect the tensile properties, continuous shear deformation affects the anisotropy of discs. Described behavior can be a result of recrystallization of structure in the center of disc, whereas on the edge of the disc, imposed strain causes structure's homogeneity. Comparison of average values and standard deviation of YS and UTS for HPT processed copper is presented of Figure 4 a) and b).

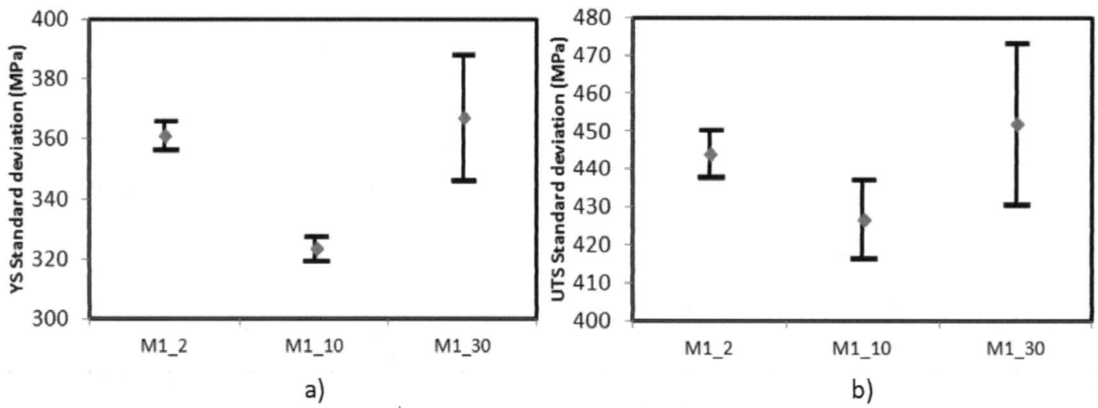

Figure 4. Average and standard deviation comparison of a) Yield stress, b) Ultimate tensile stress.

5 Conclusion
In the presented work, the effect of different revolutions on creation of ultrafine-grained structure was examined. Copper in 4 different structure states was investigated. Base coarse grain material in the cold-rolled state and HPT processed copper after 2, 10 and 30 revolutions were produced. Series of miniaturized tensile tests were performed on extracted specimens to determine differences of processed material. There were found following differences of material behavior.

- With increasing number of revolutions, the tensile parameters, such as YS and UTS increased till saturated state was achieved.
- UFG structures show higher strain rate sensitivity in comparison to CG structure.

- For smaller number of revolutions (2 – 10) anisotropy of mechanical properties within one disc was not observed.
- For 30 revolutions, there is noticeable dependency of mechanical parameters on location of specimen extraction.

References

[1] Gleiter H 1989 Nanocrystalline materials, *Prog. Mater. Sci.* **33** pp 223 – 315
[2] Valiev R, Alexandrov I, Zhu Y, & Lowe T 2002 Paradox of strength and ductility in metals processed by severe plastic deformation *J. Mater. Res.* **17** pp 5-8
[3] Kawasaki M, Lee HJ., Ahn B, Zhilyaev AP and Langdon TG 2014 Evolution of hardness in ultrafine-grained metals processed by high-pressure torsion *J. Mater. Res. Technol.* **3** pp 311-8
[4] Jamalian M, Hamid M, Vincentis N D, Buck Q, Field D P and Zbib HM 2019 Creation of heterogeneous microstructures in copper using high-pressure torsion to enhance mechanical properties *Mater. Sci. Eng.* A **756** pp 142-8
[5] Trojanova Z, Halmesova K, Dzugan J, Palcek P, Minarik P and Lukac P 2018 Influence of strain rate on deformation behaviour of an AX52 alloy processed by equal channel angular pressing (ECAP) *Mater. Lett.* **8** pp. 517-23.
[6] Zhilyaev AP and Langdon TG 2008 Using high-pressure torsion for metal processing: Fundamentals and applications *Prog. Mater. Sci.* **53** pp 893-979
[7] Jonas JJ, Gosh C and Toth LS 2014 The equivalent strain in high pressure torsion *Mater. Sci. Eng.* A **607** pp 530 - 5
[8] Al-Fadhalah KJ, Alhajeri SN, Almazrouee AI and Langdon TG 2013 Microstructure and microtexture in pure copper processed by high-pressure torsion *J. Mater. Sci.* **48** pp 4563–72
[9] Cengeri P, Kerber MB, Schafler E, Zehetbauer MJ and Setman D 2019 Strengthening during heat treatment of HPT processed copper and nickel *Mater. Sci. Eng.* A **742** pp 124-31
[10] Kral P, Stanek J, Kuncicka L, Seitl F, Petrich L, Schmidt V, Benes V and Sklenicka V Microstructure changes in HPT-processed copper occurring at room temperature *Mater. Charact.* **151** pp 602-11
[11] Dzugan J, Seifi M, Prochazka R, Rund M, Podany P, Konopik P and Lewandowsk, JJ 2018 Effects of thickness and orientation on the small scale fracture behaviour of additively manufactured Ti-6Al-4V Mater. Charact. 143 pp 94-109.
[12] Konopik P, Farahnak P, Rund M, Dzugan J and Rzepa S 2018 Applicability of miniature tensile test in the automotive sector IOP Conf. Ser.: Mater. Sci. Eng. **461**
[13] Dzugan J, Sibr M, Konopik P, Prochazka R and Rund M 2017 Mechanical properties determination of AM components IOP Conf. Ser.: Mater. Sci. Eng. **179**
[14] Mathis K, Kover M, Straska J, Trojanova Z, Dzugan J and Halmesova K 2018 Micro-tensile behavior of Mg-Al-Zn alloy processed by equal channel angular pressing (ECAP) *Materials* **11**
[15] Hilser O, Rusz S, Tanski T, Snopinski P, Dzugan J and Kraus M 2017 Mechanical properties and structure of AZ61 magnesium alloy processed by equal channel angular pressing IOP Conf. Ser.: Mater. Sci. Eng. **179**
[16] Hilser O, Rusz S, Szkandera P, Cizek L, Kraus M, Dzugan J and Maziarz W 2018 Study of the microstructure, tensile properties and hardness of az61 magnesium alloy subjected to severe plastic deformation *Metals* **8**
[17] Trojanova Z, Dzugan J, Halmesova K, Nemeth G, Minarik P, Lukac P and Bohlen J 2018 Influence of accumulative roll bonding on the texture and tensile properties of an AZ31 magnesium alloy sheets *Materials* **11**
[18] Strnadel B and Brumek J 2013 The Size Effect in Tensile Test of Steels AMSE *PVP* **6B**

Coiling simulations of medium-Mn sheet steels using dilatometry

M. Morawiec, A. Skowronek, W. Pakieła, A. Grajcar

Department of Engineering Materials and Biomaterials, Silesian University of Technology, ul. Konarskiego 18a, 44-100 Gliwice, Poland

Corresponding author: E-mail address: mateusz.morawiec@polsl.pl

Abstract. The work presents the results of coiling simulations of two medium manganese steels containing 3 and 5% Mn. The steels were subjected to the heat treatment including an austenitization at 1000°C for 300s and next isothermal holding at temperatures of 750, 700, 650 and 500°C for duration of 5 hours. The results of dilatometric analysis showed that in case of the 3Mn steel the ferritic transformation occured during the isothermal holding at 750 and 700°C. The amount of ferrite created during this step at 750°C was smaller compared to 700°C. Lowering the temperature to 650°C led to a transformation lack during the holding time. At 500°C a bainitic transformation occurred. Increasing the manganese content resulted in prolonging the incubation time before any transformations. For the 5Mn steel for all isothermal holding temperatures no transformation occurred within 5 hours. The conclusion was that manganese shifted significantly the ferritic and bainitic regions to longer times.

1. Introduction

Still growing demands for weight reduction and crashworthiness properties in automotive industry are the driving force for new steel grades and advanced heat-treatment cycles. Medium Mn steels, classified as the III generation of AHSS, are characterized by indirect properties of the two previous generations - high strength combined with very good ductility. Those features may be gained with relatively low cost due to reduced content of expensive alloying elements [1-3]. The main point in production process of medium manganese steel is the heat treatment [4]. During this stage performed in an intercritical region the redistribution of carbon and manganese in steel should occur. In the intercritical annealing two phases coexist – ferrite and austenite. Ferrite due to its low carbon solubility, displaces it, allowing austenite to enrich in this chemical element. Both carbon and manganese stabilize the austenite causing a decrease in its M_s temperature. When it is lower than room temperature, after cooling the material, a structure contains retained austenite. This phase exhibits a TRIP effect – during deformation metastable austenite grains transform to martensite causing very intense strain strengthening. This phenomenon in combination with a very fine grain provides great properties of medium manganese steels.

There are two different processing routes for the medium manganese steels. One performs the intercritical annealing which takes place after plastic forming (hot, warm or cold) from room-temperature microstructure (martensite in the majority of medium Mn steels). Recent researches have been mostly focused on this common route for obtaining considerable amount of retained austenite (20-40%) to realize better mechanical properties through the TRIP effect. On the other side, the different route of thermomechanical processing starting from hot rolling phase – austenite – can be applied from an economical viewpoint, since a conventional equipment in the steel industries can be used [5,6]. After finishing the hot rolling process, material is cooled to coiling temperature. After this step coils of medium manganese sheet steel cool in the air. During first part of this stage material stays in the intercritical region allowing the chemical elements redistribution. From this viewpoint, the present study aims to clarify a microstructure evolution during physical simulation of coiling medium manganese steels with different content of Mn at various temperatures.

2. Materials and methods

Chemical compositions of the investigated 3 and 5% Mn steels given in Table 1 were designed from the viewpoint of maximization of retained austenite and obtaining carbide-free bainite by high-Al low-Si concept (susceptibility to galvanizing) [7,8]. The steels are characterized by high metallurgical purity connected to low concentration of phosphorus and sulphur (Table 1). Steel was melted in a vacuum furnace (Balzers VSG-50). Casting occurred in Ar atmosphere. Ingots with a mass of 25kg after austenitization at 1200°C within 3 hours were forged at a temperature range from 1200 to 900°C to a final width of 160 mm and a thickness of about 22 mm [9]. A last stage of preparing material to tests were cutting the dilatometric samples with dimensions ⌀4x10mm.

First investigation step was the austenitization at 1000°C for 300s. This stage was followed by cooling the samples to an intercritical temperature in the range 500-750 and holding them for 5 hours, which simulates a coiling process. The last step was cooling the material to room temperature at a cooling rate of 10°C/s. Investigations was performed using a dilatometer DIL805, which allows to determine time and temperatures of phase transformations during the simulation. Summarized heat treatment cycles simulating steel coiling are shown in Figure 1.

Table 1. Chemical composition of the medium Mn steels used in the present study.

	Chemical composition [wt%]						
Steel	C	Mn	Al	Nb	Mo	Si	M_s [°C]
3Mn	0,17	3,1	1,6	0,04	0,22	0,22	384
5Mn	0,16	4,7	1,6	-	0,20	0,20	320

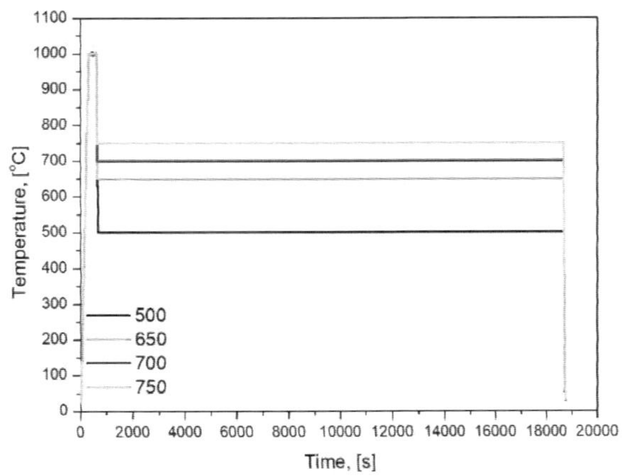

Figure 1. Heat treatment cycles applied during investigations.

For light optical microstructural investigations samples were grinded, polished and etched with 5% nital etchant. More detailed microstructural characterizations were carried out by scanning electron microscopy (SEM) on Zeiss SUPRA 35 microscope.

3. Results and discussion

3.1. 3Mn steel

Figure 2A shows the isothermal holding stage for 3Mn samples. In the case of samples tested at 650°C, the function is linear, which indicates the absence of phase transitions. This fact is confirmed by the fully martensitic structure shown in microscopic observations. At the initial endurance phase the 3Mn500 sample shows a significant increase in the relative length, indicating the start of phase transformation, as shown in Figure 2B. Microscopic observations (Figure 3) showed the presence of bainitic structures among the martensitic matrix. The decrease in length after this stage indicates that the precipitation processes in the material have started due to the long soaking time. Samples 3Mn700 and 3Mn750 (Figure 2C) show the slow start of transformation during the endurance step. A small proportion of the ferritic phase was obtained in both samples (Figure 4). The increase in elongation during this stage for the steel annealed at a lower temperature is twice as large, which indicates the more intense transformation kinetics, resulting in a larger ferrite content. In addition, it can be seen that at a lower temperature the transformation started immediately after reaching the isothermal temperature. Increasing the temperature by 50°C increased the incubation time needed to initiate the transformation. It is concluded that the transformation from austenite to ferrite is quite slow above A_{r1} temperature in the tested steel whereas bainitic transformation initiates quickly at temperatures lower than 600°C.

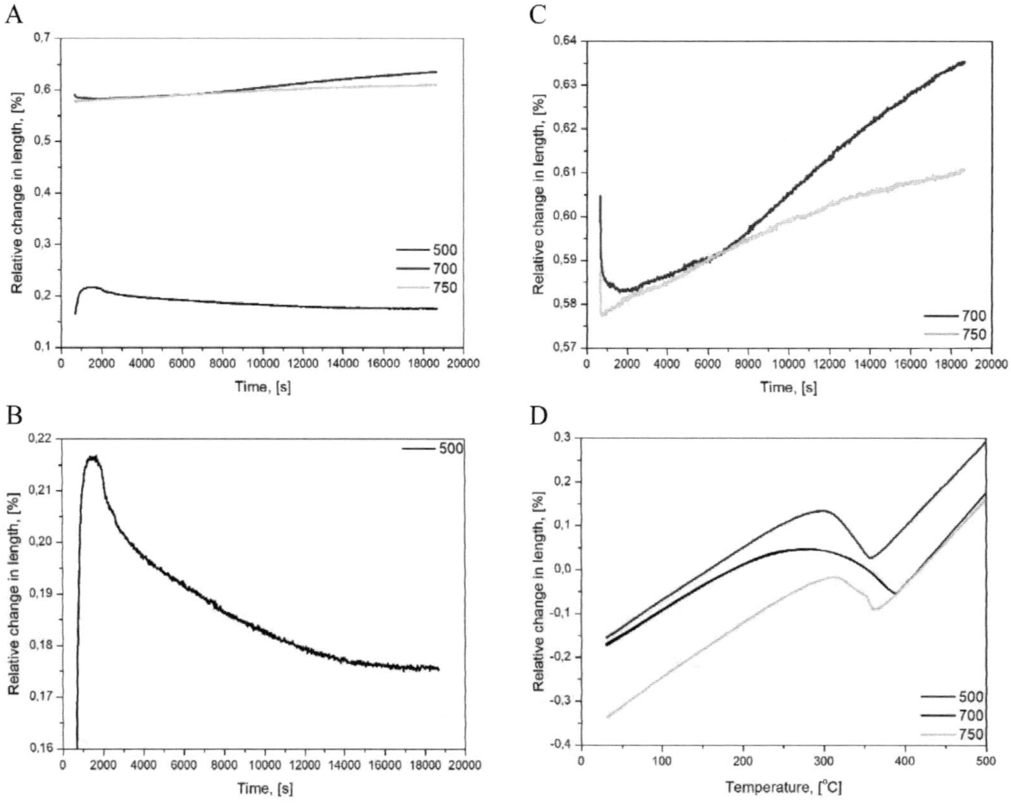

Figure 2. Dilatometric curves of 3Mn samples; A – holding stage for all samples; B – enlarged graph of holding stage for samples tested at 700 and 750°C; C – enlarged graph of holding stage for the sample tested at 500°C; D – final cooling stage.

Figure 2D shows the stage of final sample cooling. For samples annealed at different temperatures, there are slight differences in the temperature at the beginning of the phase transition, coinciding with the results of the hardness measurement (Table 2). Lower M_s temperatures indicating some redistribution of alloying elements correlate well with the reduced hardness of the samples, which indicates the generation of non-martensite phases in the material.

Table 2. Dilatometry-determined M_s temperatures and hardness for 3Mn steel.

Sample	M_s, [°C]	Hardness, [HV]
3Mn500	393	438
3Mn650	392	421
3Mn700	374	379
3Mn750	385	348

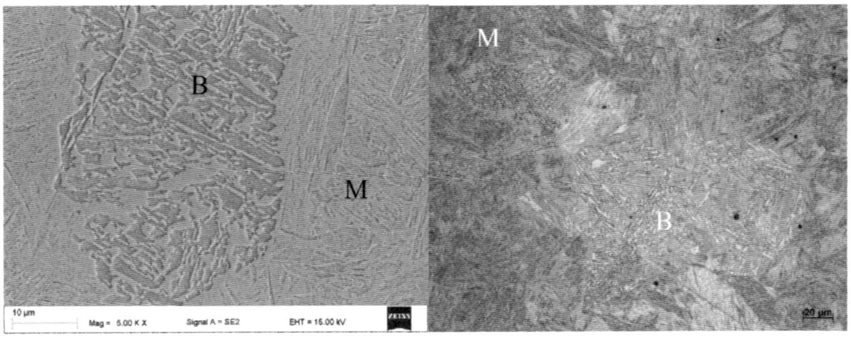

Figure 3. Microscopic images for 3Mn500 sample; M – martensite; B – bainite.

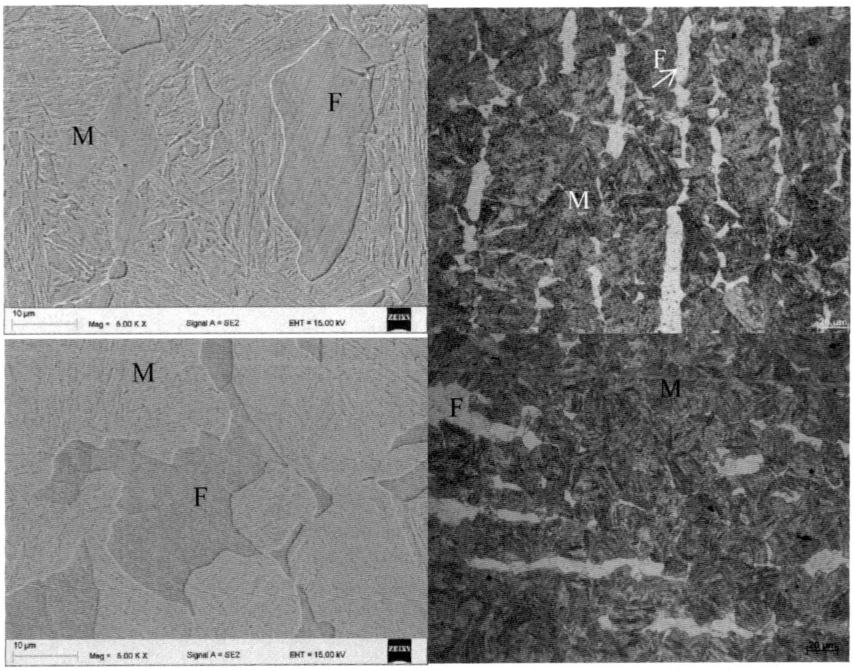

Figure 4. Microscopic images for 3Mn samples; on the top 3Mn750 sample images; on the bottom 3Mn700 sample images; M – martensite; F – ferrite.

3.2. 5Mn steel

The linearity of the function during the isothermal holding stage (Figure 5A) for 5Mn steel samples indicates the absence of phase transformation. This in turn suggests that 5Mn steel is not susceptible to an applied method of heat treatment. Figure 5B shows the stage of final sample cooling. The dysfunction that indicates the start of the martensitic transformation occurs for all samples at a similar moment, which clearly indicates the lack of redistribution of alloying elements during the previous stage. Microscopic images (Figure 6) confirmed the assumptions about achieving a full martensitic structure.

It should be emphasized that ferrite transformation in conventional low-carbon steels containing 1.5-2 wt% Mn is commonly very quick. The increase of Mn content to 5 wt% significantly delayed the ferritic transformation of the tested steel. This fact has been proven in earlier studies [6,10].

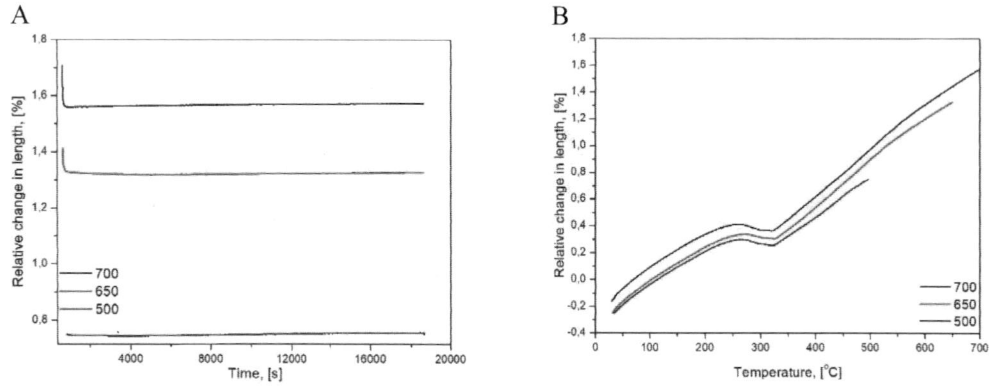

Figure 5. Dilatometric curves of 5Mn samples; A – holding stage; B – final cooling stage.

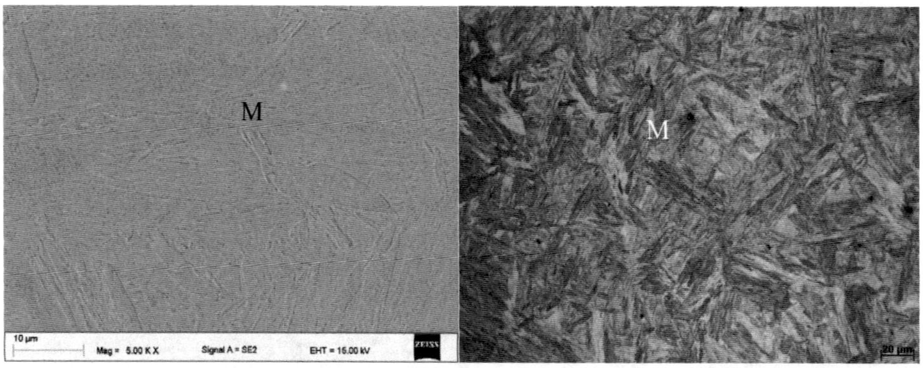

Figure 6. Microscopic images for 5Mn sample; M – martensite.

4. Conclusions

Transformation kinetics and microstructure during coiling simulation of medium manganese steels have been investigated in two steels containing 3 and 5 wt% Mn using the dilatometry approach. The major results obtained are as follows:

- The occurrence of ferritic transformation in 3Mn steel led to a slight decrease in the M_s temperature, which indicates the redistribution of carbon during the simulated coiling process.

- In the case of 3Mn steel, lowering the holding temperature below 650° C leads to entering the area of bainitic transformation, during which the precipitation process prevents the enrichment of austenite in carbon.
- The isothermal holding of 3Mn steel at temperatures of 700 and 750° C enabled the start of ferritic transformation, which is confirmed by dilatometric and microscopic tests and the results of hardness measurements. However, the time was too short to generate a sufficient part of ferrite to stabilize retained austenite by redistributing carbon and manganese.
- Along with the increase in manganese content from 3 to 5%, the transformations occurring during the cooling of steel are significantly delayed, preventing their start even during several hours of annealing.
- Samples of 5Mn steel annealed in the temperature range of 500-750° C do not show phase transformation of the austenite to ferrite/bainite. This is demonstrated by the full martensitic structure.
- Under the experimental conditions used, the simulated coiling without a previous deformation can not be an alternative to intercritical heat treatment of medium-manganese steel. The further research will be focused on coiling process following plastic deformation. A positive effect can be brought by the use of plastic deformation before holding, which requires further research.

Acknowledgement
The work was supported by the scholarship fund of the Silesian University of Technology in 2019.

References

[1] Matlock D and Speer J 2009 Third generation of AHSS: microstructure design concepts *Microstructure and texture in steels* (London, UK) p 185 – 205
[2] Kučerová L, Opatová K, Káňa J and Jirková H 2017 High versality of nobium alloyed AHSS *Archives of Materials Science and Engineering* vol **62** p 1485 – 91
[3] Mazancova E, Wyslych P and Mazanec K 1995 Physical metallurgy of granular bainite *Kovove Materialy-Metallic Materials* vol **33** p 94 – 104
[4] Grajcar A, Skrzypczyk P and Woźniak D 2014 Thermomechanically rolled medium-Mn steels containing retained austenite *Archives of Metallurgy and Materials* vol **59** p 1691 – 7
[5] Rusz S, Kubina T and Schindler I 2013 Comparison of methods for physical determination of phase transformations temperatures *Metalurgija* vol **52** p 525 - 8
[6] Ito A, Shibata A and Tsuji N 2017 Thermomechanical processing of medium manganese steels *Mater. Sci. Forum* vol **879** p 90 – 4
[7] Kuziak R, Kawalla R and Waengler S 2008 Advanced high strength steels for automotive industry *Archives of Civil and Mechanical Engineering* vol **8** p 103 – 17
[8] Kvackaj T 2000 The review of steel and steel products flat-rolled production *Metalurgija* vol **39** p 185 – 9
[9] Grajcar A, Skrzypczyk P, Kuziak R and Gołombek K 2014 Effect of finishing hot-working temperature on microstructure of thermomechanically processed Mn-Al multiphase steels *Steel Research International* vol **85** p 1058 – 69
[10] Hou F, Bai Y, Shibata A and Tsuji N 2018 Microstructure evolution during thermomechanical processing in 3Mn-0.1C medium-Mn steel *Mater. Sci. Technol.* vol **34** p 1 – 8

IOP Conf. Series: Materials Science and Engineering **723** (2020) 012021 doi:10.1088/1757-899X/723/1/012021

Flash pulse phase thermography for a paint thickness determination

L Muzika[1] and M Švantner[1]

[1] University of West Bohemia, New Technologies Research Centre, Plzen, Czech Republic

muzika@ntc.zcu.cz

Abstract. The contribution describes a fast contactless measurement of a paint thickness non-uniformity using flash pulse thermography. Specimens sprayed by a paint were thermally excited by a flash lamp and temperature responses were recorded by an infrared camera. The recorded sequences were post-processed with Fast Fourier Transform to obtain phase angles. Differences in the resulting images showed phase differences which corresponded to a paint thickness non-uniformity. Furthermore, the phases were correlated with the thickness by means of calibration curve so that the paint thickness could be determined with flash pulse phase thermography measurement. The method showed a promising potential in the contactless evaluation of the paint thickness. Average error of the thickness determination was less than 10 % for samples with paint thickness from 41 to 74 μm on AISI 304 substrate. Advantages, disadvantages and limitations of described method were discussed.

1 Introduction

Paint thickness is one of the key parameter when dealing with paints. The thickness can influence both decorative and functional properties. Among decorative flaws belong blistering, blooming, sagging and many more. Paints used in industry usually have some function for example a protection against corrosion, heat, acids. The problem isn't just insufficient amount of a paint which causes that required properties (e.g. anticorrosion) are not met but also a surplus of paint which leads to increase of weight of painted part, economic losses (less paint cost less) and in some cases decrease of adhesion.

The paint thickness is therefore often tested in industry. Nondestructive thickness measurement can be performed by many different techniques. One of the most used are [1]: measurement with magnetic gauges, ultrasound testing, measuring with devices based on magnetic induction or based on eddy current. Some more advanced and more expensive techniques can be also used as beta-backscatter measuring method [2] and X-Ray fluorescence [2]. They measure thickness but also provide some additional information. The reason for existence of that many techniques is that just limited number of materials can be measured with each of them. For example, with magnetic induction method the thickness can be determined just for non-magnetizable paint on magnetizable substrate. For eddy current (amplitude sensitive) the substrate must be from electrically conductive material. All mentioned techniques take thickness readings from a point. When many readings need to be acquired, the measurement process can be time demanding. Because of that a technique which would provide readings from the whole surface could speed up the thickness measurement. In this contribution we introduce technique for a planar thickness determination based on flash pulse phase thermography [3].

2 Thickness determination using flash pulse phase thermography

Flash pulse thermography (FPT) is an inspection method which belongs to infrared nondestructive testing [3]. The technique is based on excitation of tested subject by a flash lamp with one short pulse (few ms). The temperature response to this pulse is recorded by an infrared camera. This response can show changes in heat transfer caused by defects. Those defects can be for example cracks, delaminations but also uneven thickness of a coating (paint). The result of FPT measurement is a sequence of thermograms (images with temperature distribution). This sequence is then post-processed with an algorithm to enhance defect detectability. To the most popular post-processing techniques belongs pulse phase thermography [4]. This technique is based on post-processing with Fast Fourier Transform. The result is a new sequence where images are in phases or in amplitudes and time is transformed to frequency. In general phase images are more used, especially due to less sensitivity to uneven heating. The thickness of coating can be retrieved via calibration curve phase-thickness as was demonstrated in [5] or [6] for thermal barrier coatings. The principle is that the phases are dependent on thickness of coating, when no additional damages or thermal properties changes are present. There were also some other attempts to determine thickness with FPT as is shown in table 1. The thickness determination was mostly tested on thermal barrier coatings (TBC) with different error in thickness determination (from 5.35 to 59 %). Detection of paint thickness inhomogeneity was published in [7], but a quantification of thickness measurement results was not performed. A goal for this contribution was to determine if it is possible to use PPT for estimating thickness of paint even for thicknesses which are around several tens of μm.

Table 1. Summary of FPT for thickness determination.

Method	Camera wavelength	Camera framerate (fps)	Coating	Thickness (mm)	Error in thickness determination (%)	Source
FPT	Not specified	Not specified	TBC (ZrO2)	0.25-1.5	15.69	[8]
FPT	Long (7.5-13 um)	50	TBC (ZrO2)	0.1-0.6	5.35	[6]
FPT simulation	Not specified	Not specified	TBC (ZrO2)	0.1-0.6	2.35	[5]
FPT	Long (8-9 um)	60	TBC (ZrO2)	0.15-0.3	59	[9]
FPT	Middle (2.5-5 um)	100	TBC (SiC)	0.04-0.12	8.2*	[10]
FPT	Long (8-14 um)	60	Paint	0.01-0.25	Not specified	[7]
FPT	Long	1000	Polymer coating	0.1-2	4.1	[11]
FPT	Long	70	TBC (ZrO2)	0.05-0.13	10	[12]

* calculated by us from data

3 Description of experiment

Two specimens from AISI 304 steel with dimensions 100 x 50 x 2 mm were painted with a black paint (LabIR HERP-LT-MWIR-BK-11). The paint was unevenly sprayed on the surface of specimens. Experimental measurement was performed on both samples. IR camera FLIR SC 7650 was used for thermographic measurement. The framerate was 250 Hz and 1000 frames were recorded. Specimens were excited by 6000 J flash lamp. Center of the tested specimen was in the center of the flash lamp during measurements. The flash lamp was 38 cm from the tested specimen and was perpendicular to the

IOP Conf. Series: Materials Science and Engineering **723** (2020) 012021 doi:10.1088/1757-899X/723/1/012021

specimen. Special measures had to be taken considering synchronization of the IR camera and the flash lamp. This is crucial for the thickness estimation. Without synchronization it is possible to detect inhomogeneity in thickness but the phases differ even for repeated measurement. This is caused by capturing frames in different times after excitation. The used device allowed that the IR camera and the lamp were synchronized with precision of +/- 2 μs, which was sufficient for the thickness estimation. The result of PPT is a sequence. Phase values differ in each frame. Therefore, it is necessary to choose a frame which will be further analyzed. We chose frame which corresponded to frequency 2.25 Hz because it showed highest differences in phases in the area of specimens.

During measurement we encountered a problem which can significantly influence results. With a high framerate (in this case 250 Hz) the camera sometimes skipped some frames. Several measurements were performed for one specimen and all of them contained frames drops. This is a problem for calculating phases because data for FFT have to be evenly spaced. If they are not, phases can differ when important frames (especially those after excitation) are missing. Therefore, the raw data was fitted with an artificial function. The coefficients of this function were retrieved. Fitted temperature response was calculated for evenly spaced time with time difference of 1/250 (frequency of the camera). The equation of fitting function is shown in equation 1.

$$f(t) = K_0 + K_1 \cdot \log t + K_2 \cdot (\log t)^2 + K_n \cdot (\log t)^n \qquad (1)$$

, where K_0-K_n are constants and t is time.

With this fitting noise were also suppressed. FFT was then applied to fitted data and the frame which corresponded to frequency 2.25 Hz was used for further analysis. To show difference between raw and fitted data we presented both results.

Specimen 1 was used for creating a calibration curve. Thickness of several points of specimen 1 had been measured with thickness meter (combination of eddy current and magnetic induction principle). The aim was to find points which cover the whole thickness range of sprayed paint. Phase values of those points were then obtained from the phase image of specimen 1. Therefore, it was possible to assign thickness values to phase values. Curve fitting App from Matlab were used to get an equation of calibration curve (equation 2). This equation allows estimation of paint thickness based on the phase.

$$z_c = C_1 \cdot \Phi^3 + C_2 \cdot \Phi^2 + C_3 \cdot \Phi + C_4 \qquad (2)$$

, where z_c is thickness, C_1-C_4 are constants and Φ is the phase angle.

4 Results

Figure 1 shows position of points used for creation of calibration curves. These points covered thickness range from 41 to 74 μm. Phase values are in table 2.

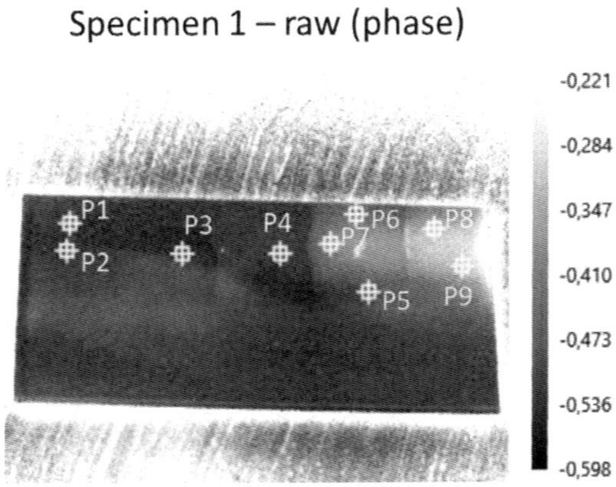

Figure 1. Position of points for calibration curve (phase image).

Table 2. Calibration points.

Point	P1	P2	P3	P4	P5	P6	P7	P8	P9
Thickness (μm)	41	47	52	54	56	60	66	73	74
Phase Raw (rad)	-0.584	-0.519	-0.516	-0.506	-0.502	-0.431	-0.409	-0.332	-0.266
Phase Fitted (rad)	-0.521	-0.493	-0.486	-0.479	-0.459	-0.431	-0.359	-0.272	-0.211

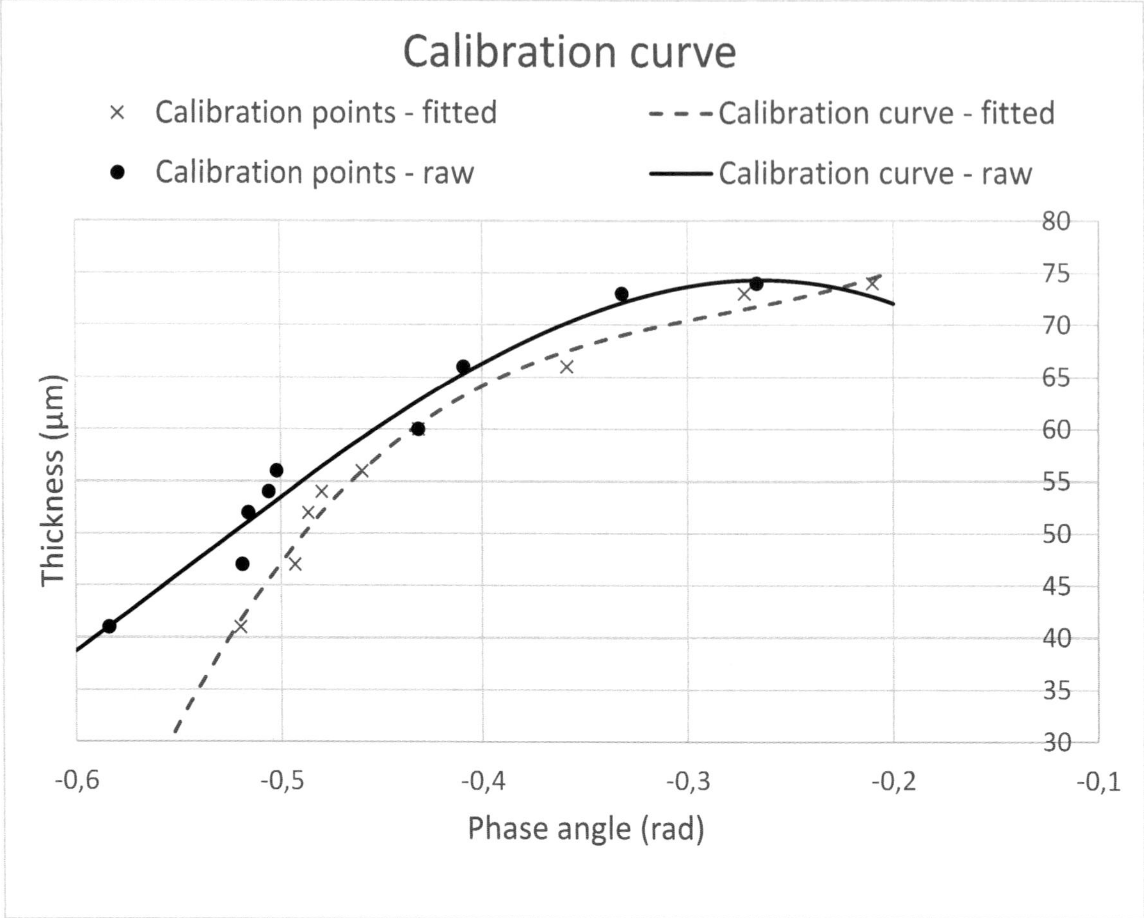

Figure 2. Calibration curve.

Figure 2 shows calibration curves for raw and fitted data. The graph shows good match between measured data (points in the graph) and calibration curves (obtained via equation 2, with constants C_1-C_4 for raw data -590.8, -981.5, -394.6, 27.67 and for fitted data 1515, 1284, 401.4, 116.2). The curves were valid for interval 41-74 µm. The dependency of phase angle on thickness cannot be predicted out of this interval because the functions were designed just for the inside of this region. We can see this effect on the calibration curve obtained from raw data, where phase went down after reaching 74 µm but in reality it should continue to grow. It should be pointed out, that more data for calibration can provide better determination of thickness. Different functions can be also used. The important thing is to have as many pairs (phase-thickness) as possible for the precise fit. The phase values for fitted and raw data differed for the same thicknesses. This was caused by noise suppression.

Phase images were converted via calibration curve (equation 2 with corresponding constants) to images of paint thickness. Thickness maps based on phase images are shown in figure 3. Results obtained from raw data and fitted data showed approximately the same thickness. Nonetheless the raw data were little bit noisier.

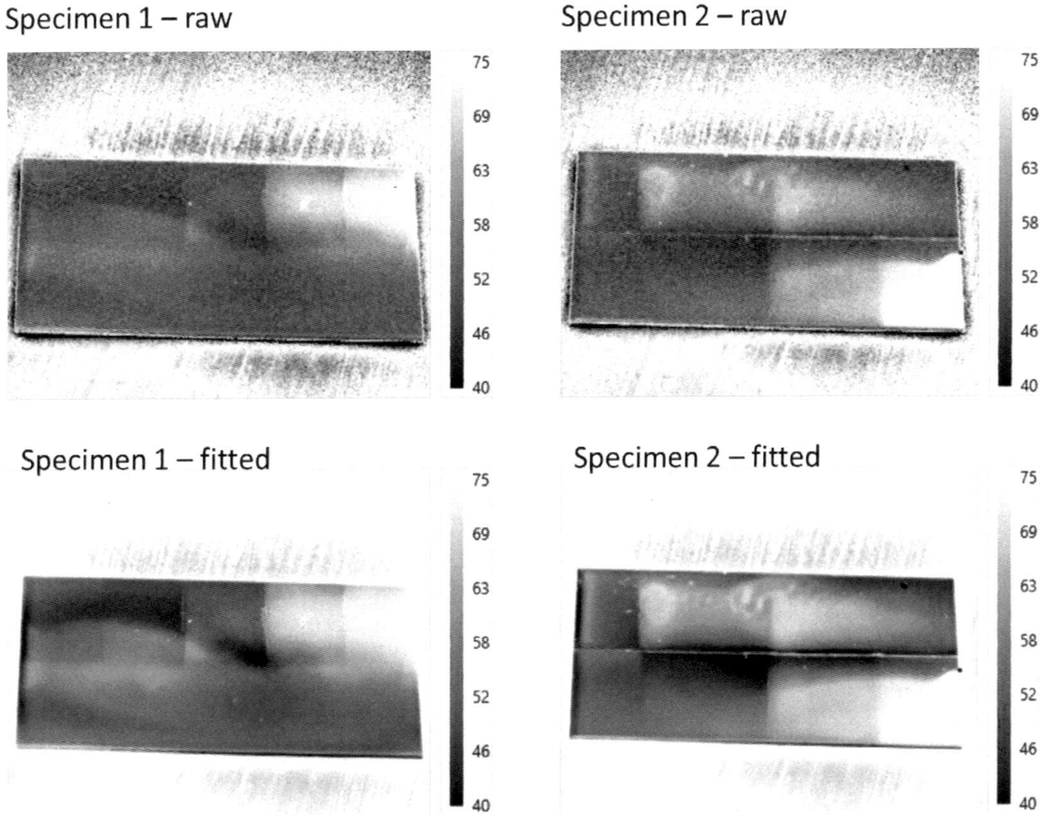

Figure 3. Results, the range is in μm.

For determination how well the thickness was estimated, both specimens were measured with thickness meter. Specimens were divided in ten zones from which the average thickness was measured. Nine measurements were performed in each zone (3 x in the top part,3 x in the middle and 3 x in the bottom part). The zones can be seen in figure 4. Correspondingly, the average of the zones was obtained from those zones from results by PPT.

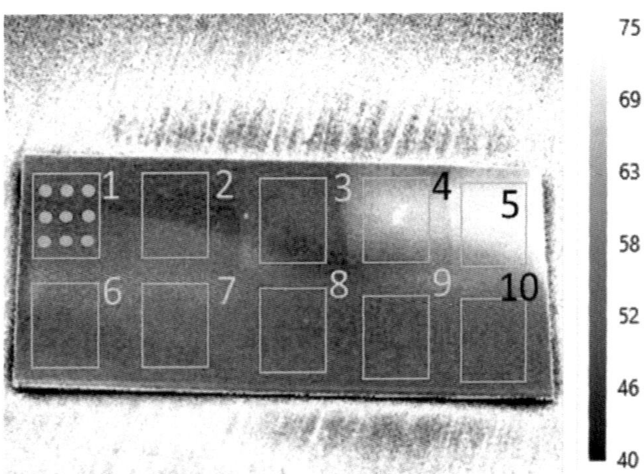

Figure 4. Placement of zones, white dots in zone 1 represents points of measurement.

The results are in table 3. Thickness determination for specimen 1 was worse than for specimen 2. Average relative error for specimen 1 for raw data was 4.9 % for fitted data 4.3 % and for specimen 2 for raw data 3.6 % and for fitted data 2.8 %. As expected with fitted data we obtained more precise results. The difference was not big but for some application it can be crucial. We expected that the thickness estimation would be better for sample on which the calibration (specimen 1) was performed but that was not confirmed. The probable cause for that was that non-uniform heating influenced phase values in the center of the lamp (zone 3 and 8) due to higher energy (energy was the biggest in the center of the lamp). Those phase values were slightly shifted and therefore the thickness differed. This was especially problem for specimen 1 in the zone around the center of the lamp due to steep calibration curve around 56 μm. Specimen 2 had thickness around 65 μm in the center of the lamp and the effect of a non-uniform heating was masked due to not that steep calibration curve around this thickness. Basically the thickness was less phase sensitive around 65 μm. We should point out that even with highest relative error 8.9 % the absolute thickness difference between measured thickness by thickness meter and estimated by PPT was less than 6 μm, which is acceptable for many of industrial applications.

Table 3. Results of estimated thicknesses

	Specimen 1										
	Zone	1	2	3	4	5	6	7	8	9	10
Average thickness from the zone (um)	Thickness meter	47	52	56	67	73	52	55	56	55	54
	FFT phase raw	51	51	51	65	71	53	54	51	51	52
	FFT phase fitted	49	50	51	66	71	53	56	52	51	52
Relative error (%)	FFT phase raw	8.5	1.9	8.9	3.0	2.7	1.9	1.8	8.9	7.3	3.7
	FFT phase fitted	4.3	3.8	8.9	1.5	2.7	1.9	1.8	7.1	7.3	3.7

	Specimen 2										
	Zone	1	2	3	4	5	6	7	8	9	10
Average thickness from the zone (um)	Thickness meter	54	64	65	60	56	48	51	64	65	74
	FFT phase raw	52	59	63	58	53	51	51	61	65	73
	FFT phase fitted	52	61	65	59	53	51	51	62	66	73
Relative error (%)	FFT phase raw	3.7	7.8	3.1	3.3	5.4	6.3	0.0	4.7	0.0	1.4
	FFT phase fitted	3.7	4.7	0.0	1.7	5.4	6.3	0.0	3.1	1.5	1.4

5 Conclusion

Flash pulse phase thermography measurement was used for the paint thickness estimation on AISI 304 substrate. The thickness was determined from phases via the calibration curve thickness-phase with satisfactory results. The average relative error was less than 3.6 %. The maximum difference between the thickness measured by the standard thickness meter and obtained with PPT was less than 6 μm in thickness range from 41 to 74 μm. It should be pointed out that the thickness obtained and thus the measurement error with PPT greatly depends on the calibration curve, which is also a big disadvantage of the measurement procedure. The calibration curve depends on phase values, which can differ for example due to used energy, energy distribution of a lamp, distance between a lamp and a specimen, a used paint. Therefore, when phase values changes due to a measurement setup, calibration curve has to be redetermined. That means that calibration curve is valid for a fixed measurement setup. The calibration curve doesn't have a linear characteristic and therefore thickness sensitivity and thus relative error can differ in different thickness ranges. The effect of non-uniform heating can negatively influence the precision of thickness determination. This effect was negligible for the tested samples. The biggest advantage is in the speed of the measurement. Thickness values of whole samples (100 x 50 mm) were measured in 4 s, which corresponded to the measurement time of two points by a standard thickness meter. Another advantage is that the measurement is non-contact and therefore no surface damage can be induced (cracks, indents, etc.). The results suggest that flash pulse phase thermography could be a viable tool for an estimation of a paint thickness on planar parts with a high speed.

Acknowledgements

The work has been supported by the Technology Agency of the Czech Republic within the project no. TE01020068 and by the Ministry of Education, Youth and Sports of the Czech Republic within the OP RDI program, CENTEM project, no. CZ.1.05/2.1.00/03.0088, co-funded by the ERDF; and National Sustainability Programme I., CENTEM PLUS project, no. LO1402 and by project SGS-2019-008.

References

[1] Mezghani S, Perrin E, Bodnar J L, Marthe J, Cauwe B, Vrabie V and Samples A 2015 Evaluation of Heterogeneity of Paint Coating on Metal Substrate Using Laser Infrared Thermography and Eddy Current [online]. 9(5), 837–842. ISSN 1359-6640. doi:10.1039/B919800H
[2] Palomo S and Mills B 2015 Comparison of Beta Backscatter and X-ray Fluorescence Methods to Measure Coating Thickness. In: STAR (STEM Teacher and Researcher) Presentations.
[3] Yang R and He Y 2016 [1] R. Yang, Y. He, Optically and non-optically excited thermography for composites: A review, Infrared Phys. Technol. [online]. 26-50. doi: 10.1016/j.infrared.2015.12.026
[4] Waugh R C, Dulieu-Barton J M and Quinn S 2014 Modelling and evaluation of pulsed and pulse phase thermography through application of composite and metallic case studies. NDT and E International [online]. 66, 52–66. ISSN 09638695. doi:10.1016/j.ndteint.2014.04.002
[5] Shrestha R and Kim W 2018 Evaluation of coating thickness by thermal wave imaging: A comparative study of pulsed and lock-in infrared thermography – Part II: Experimental investigation. Infrared Physics and Technology [online]. 92(May), 24–29. ISSN 13504495. doi:10.1016/j.infrared.2018.05.001
[6] Shrestha R and Kim W 2018 Evaluation of coating thickness by thermal wave imaging: A comparative study of pulsed and lock-in infrared thermography – Part I: Simulation. Infrared Physics and Technology [online].92, 24–29. ISSN 13504495. doi:10.1016/j.infrared.2018.05.001
[7] Aoyagi M, Hiraguri T and Ueno T 2014 Nondestructive analysis of uneven paint coatings by dynamic heat conduction following flash heating. Journal of Coatings Technology and Research [online].11(3), 311–318. ISSN 15470091. doi:10.1007/s11998-013-9537-8
[8] Shrestha R, Chung Y and Kim W 2016 Thermal Behavior Variations in Coating Thickness Using Pulse Phase Thermography. Journal of the Korean Society for Nondestructive Testing [online], 36(4), 259–265. ISSN 1225-7842. doi:10.7779/jksnt.2016.36.4.259
[9] Zhao S, Zhang C, Wu N and Wang H 2011 Quality evaluation for air plasma spray thermal barrier coatings with pulsed thermography. Progress in Natural Science: Materials International [online]., 21, 301–306. ISSN 1002-0071. doi: 10.1016/S1002-0071(12)60061-6
[10] Liu J Y, Tang Q J, Wang Y, Gong J L a Qin L 2015 Investigation on Coating Uniformity of High-Temperature Alloy with SiC Thermal Barrier Coating Using Pulsed Infrared Thermographic Technique. International Journal of Thermophysics [online]. 36(5–6), 1252–1258. ISSN 0195928X. Dostupné z: doi:10.1007/s10765-015-1834-2
[11] Altenburg S J, Weber H and Krankenhagen R. 2016 Thickness determination of semitransparent isolated solids using the flash method and an analytical model., 71–78.
[12] Chiwu B, Qingju T, Yuanlin L, Fengyun Y, Chen M aND Yawei Z 2016 Quantitative detection of thermal barrier coating thickness based on simulated annealing algorithm using pulsed infrared thermography technology. Applied Thermal Engineering [online].99, 751–755. ISSN 13594311. doi:10.1016/j.applthermaleng.2016.01.143

IOP Conf. Series: Materials Science and Engineering **723** (2020) 012022 doi:10.1088/1757-899X/723/1/012022

CFD simulation of the multiphase heat transfer during the quenching process

J Novosád[1*] P Peukert[1] N Pomp[1] P Klouček[1]

[1]VÚTS, a.s., Svárovská 619, Liberec, Czech Republic

*jan.novosad@vuts.cz

Abstract. The paper presents the results of the CFD (Computational Fluid Dynamics) simulation of quenching process for the ring shape sample. The aim of the work is to develop and validate the methodology for multiphase CFD simulation including the boiling during the quenching process. CFD simulation is provided in ANSYS Fluent. The Lee model is used for modelling the phase change during the quenching process. The first step consider the simulation of cooling of the ring sample when the correct model parameters will be found. Validation of results is performed by comparison with experimental data. Experimental was realized inside the own designed quenching bath device filled with quenching polymer. The general description of the experimental setup is included in the paper. The CFD results are cooling curves, i.e. variation of solid temperature on time. The Lee model parameters especially the evaporation frequency was tuned. Thin polymer film on the solid surface was considered to bring the results closer to the experimental data. The comparison between experiment and CFD shows very good agreement for higher temperatures, which covers the boiling stage. On the other hand for lower temperatures worse match of results was found caused probably by the sensitivity on the inlet velocity profile settings. Some recommendations for future work were defined.

1 Introduction

In mechanical engineering, the parts production process contains a heat treatment and quenching used to improve physical and chemical properties of a material, e.g. strength, toughness, durability etc. The quenching process uses the quenchant as a medium for controlled cooling of the parts.

The heat transfer [1] during quenching gradually passes three stages [2] (illustrated in figure 1):

- Vapor film stage.
- Boiling stage.
- Convection stage.

Vapor film stage is represented by vapor barrier formation on the surface of the part. The barrier is cause by the thin vapor film with low thermal conductivity, i.e. low heat transfer rate.

Boiling stage begins by the vapor film collapse. Then the nucleate boiling, characterized by high heat transfer rate, continues until the temperature of the surface falls below the boiling point of the liquid.

Convection stage, i.e. convection heat transfer, is the final stage of the quenching process. The heat transfer rate is the slowest.

As the quenchant different kind of liquid, solid or gas could be used. In our case the polymer quenchant is used as the solution of the organic polymer in water. The using of polymer gives some advantages [2] comparing to using of oil quenchants. The quenching speed could be maintained by varying the concentration. The uniform polymer film avoids the soft spots and reduces surface thermal gradients and residual stresses caused by non-uniformity in heat transfer. Finally there is no smoke and fume occurred during quenching which improves the environment quality.

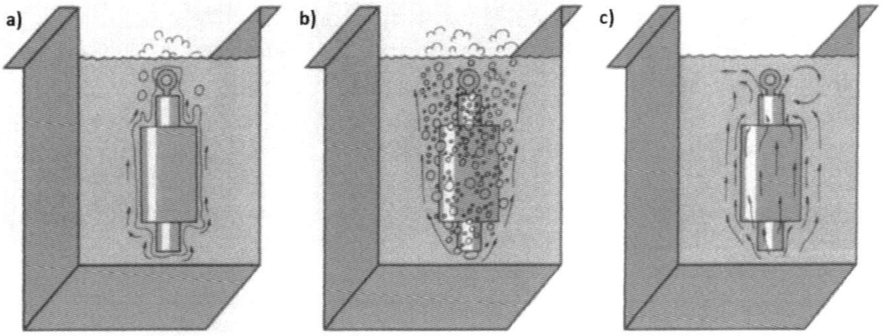

Figure 1. Quenching stages [1] – a) vapor film, b) boiling stage, c) convection.

The experimental investigation of the thermal field inside the quenched part during the industrial process could be very difficult. However, many authors deal with the investigation of quenching process using the CFD (Computational Fluid Dynamics [3]) analysis. In paper [4] MacKenzie have described the abilities of CFD in 'real world' applications of quenching. Then some design study about affecting the flow of quenchant were performed, e.g. by Kobayashi et al. [5]. The main advantage of modelling is the coupling between CFD and FEA (Finite Element Analysis) analysis, described e.g. by Banka et al. [6]. These coupled studies can give as the results not only the fluid and thermal fields, in addition the chemical and mechanical parameters, i.e. the austenite decomposition, internal stress state and distortion of the finished part.
The aim of our study is to develop and validate the methodology for CFD analysis. For phase change the Lee [7] model will be used. The validation of the CFD results will be based on the experimental results. The measurement will be provided inside the quenching tank with our design.

2 Experimental part

First of all the experimental quenching device (see figure 2) was designed and built in VÚTS. The main part is the welded steel tank with the filling volume approx. 1.4 m^3. The circulation of polymer PVP (polyvinylpyrrolidon) solution is performed by two pumps on the sides of the tank. In the bottom of the tank there are bent sheets for directing the flow. The main flow inside the tank is vertical from bottom to top. Quenched parts are situated on the grid which is submerged in the tank.

Figure 2. Quenching tank.

2.1 Samples for experiment
For the experiment two samples with different shapes (shown in figure 3) were prepared. Thermocouples (TC) for temperature measurement were installed in both samples. Positions of TCs were chosen so as not to affect the flow and to prevent damage of the thermocouples during measurement. Positions are illustrated in figure 3 (red crosses). Samples prepared for measurement, including mounting of thermocouples, is shown in figure 4.

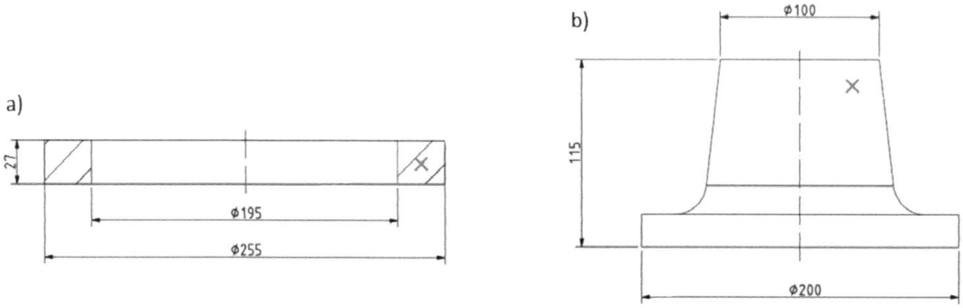

Figure 3. Sketch of samples – a) ring, b) bell-shaped.

Figure 4. Samples with thermocouples.

2.2 Setup and preparation
Temperature measurement was performed using the data logger Ahlborn ALMEMO MA5690-2 with two thermocouples Ahlborn FTA05L0500 for measuring temperature of samples. In addition, two thermocouples were connected to measure temperature of the quenching bath and inside the furnace for heating up the samples. Temperatures were measured throughout the whole process with the sample rate 1 Hz.

At first both samples were heated up to 850 °C and maintained at that temperature for several tens of minutes (30 min for ring, 90 min for bell-shape). Then the first sample was putted on the grid and submerged into the quenching bath maintained at 30 °C. The manipulation time does not exceed 15 s. The bath was mixed by pumps. Quenching was stopped when the sample temperature reaches the value under 50 °C. After that the whole process was repeated for the second sample.

3 CFD analysis

The main goal of our work was to develop the methodology for CFD simulation of the quenching process and validate that by the experimental results. The simulation have to be based on the same conditions as for experiments, i.e. geometry and initial temperature of samples, temperature of bath etc. For the first stage of simulations the ring shape sample data will be used.

3.1 Geometry and mesh

Based on the geometry of samples (see figure 3) the CAD model of the ring was created respecting the dimensions. On the other hand some elements were neglected for simplification, e.g. screws and holes for thermocouples. Then symmetry was used to create final 2D axisymmetric CAD model. Geometry for simulation is only the cutout around the ring, not the whole quenching bath.

Computational grid is composed as two domains, one for fluid, second one for solid. The quad grid with several triangles was created. The grid was refined near the interface. Total number of elements is approx. 10 000 for ring.

3.2 CFD model

The simulated problem could be considered as the transient multiphase flow with boiling on the hot wall. The mixture model was used for multiphase flow of two fluids, i.e. polymer solution and its vapor. The Lee model [8] was used for evaporation (boiling). The influence of the Lee model parameters, i.e. tune the evaporation and condensation frequency is a part of our work.

The properties of fluids were temperature dependent. The values of density, thermal conductivity and viscosity were used from the company material database respecting the solution concentration.

The calculation is transient with variable time step respecting the flow Courant number $Co < 2$. Final time is the same as was in the experiment, approx. 300 s for ring.

3.3 Boundary and initial conditions

The boundaries of the model are consistent with the experiment. For simplification the symmetry of the models was applied. Boundary conditions of the models are illustrated in figure 5. Symmetry is applied to create the axisymmetric model. At the inlet the velocity was given as 0.4 m/s in accordance with the previous CFD study of quenching bath. Moving wall is applied to simulate the far field, hence the velocity value and direction is the same as for inlet. The volume fraction of vapor is 0 at the inlet. Outlet is set as pressure outlet boundary.

At the beginning the initial temperature of solid is set to 850 °C. Then during the calculation the temperature at point P (red crosses in figure 5) is reported for future evaluation per each 0.1 s.

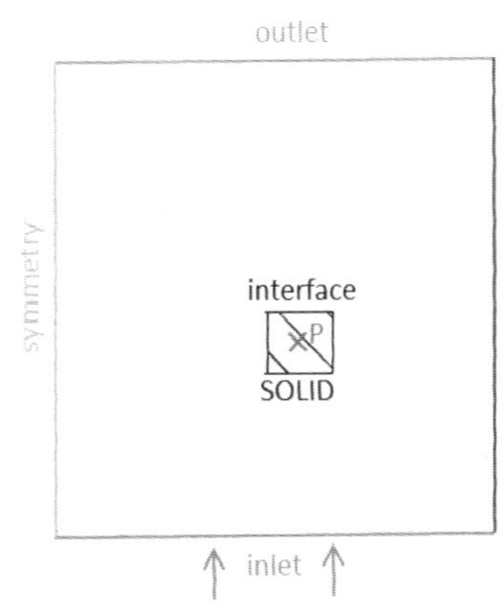

Figure 5. BCs of the computational domain – ring.

4 Results

The results show the comparison of temperature versus time from the simulation of the ring with different settings of flow parameters. Several calculations were provided to tune up the Lee model parameters. The best fitted parameters were used for the final calculation.

4.1 Ring

The cooling curves obtained from the experiment and three CFD cases are shown in figure 6. The first CFD case describes the quenching by natural convection, i.e. the inlet velocity is set as 0. The flow is driven by buoyant forces. During the boiling stage the flow is speed up by the water vapor. It is clear that the vapor film slows down cooling. In comparison with the experimental data less slope of the curve shows slower cooling than during the experiment.

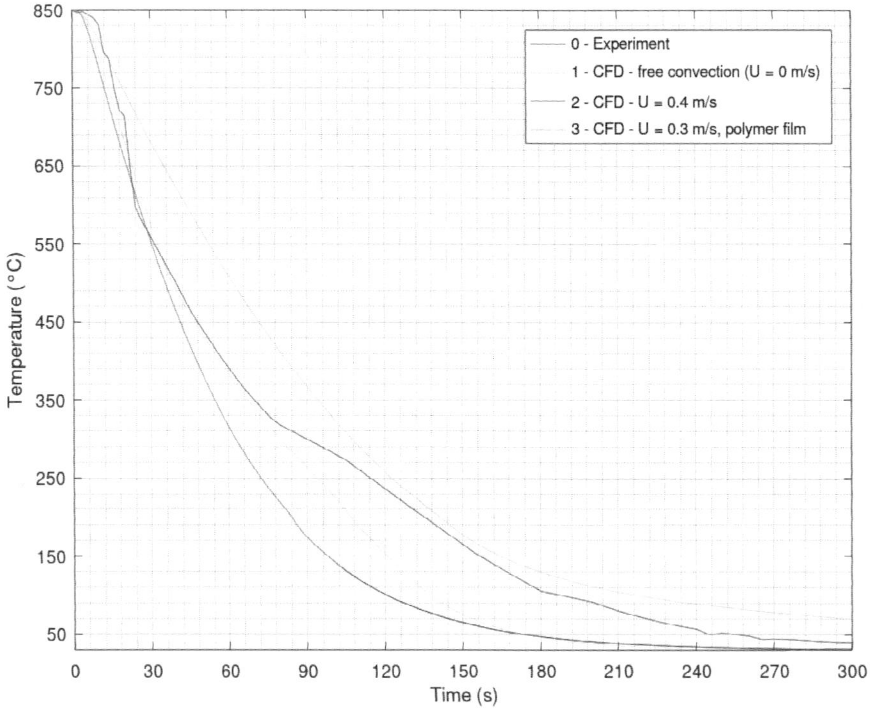

Figure 6. Ring – cooling curves - experimental and CFD results.

The inlet velocity in the second case was set to value 0.4 m/s. Thus represents the nominal value in the quenching tank. The good agreement is observed for temperatures above 500 °C. For lower temperatures the curve extends below experimental data that means the cooling is faster. These two CFD cases results were used for the tuning of the coefficients and „more real“ boundary conditions for the final case.

The final case was modified in several ways. The inlet velocity was reduced to only 0.3 m/s. This should reduce the non-reality of the uniform velocity profile at the inlet of the model. The second modification covers the boiling process when the thin polymer film was defined on the solid surface. Finally the Lee model parameters were tuned to minimize the difference between CFD and experimental results during the boiling stage. It could be said that the good agreement is obtained for the temperatures higher than 300 °C. For lower temperatures the velocity field was probably inhomogeneous during the experiment therefore the velocity apparently was not constant. This could lead to decreasing of the cooling speed.

5 Conclusion

The aim of our work was to develop and validate the methodology for CFD simulation of the quenching process. Several cases were solved to set up and tune the multiphase flow model including evaporation. The experimental data was used for validation. The comparison with CFD data for particular cases shows (see figure 6), that the inlet velocity value significantly affects the results. The Lee model parameters especially the evaporation frequency was tuned. The inclusion of a thin polymer film on the solid surface was also necessary to bring the results closer to the experimental data.

To conclude, the comparison between experiment and CFD shows very good agreement for higher temperatures, which covers the boiling stage. At this stage the main influence on cooling is by heat transfer caused by evaporation, i.e. the correctness of the results depends on the parameters of the boiling model. This part was fully done which is confirmed by the match of results. On the other hand the convection stage is more affected by the fluid flow. In this part worse match of results is probably caused by the insufficiently accurate inlet velocity specification because no experimental data was available.

Finally, the measured flow field data as the input for CFD could improve the quality of results for the convection stage. This should be done before the CFD methodology will be tested for the bell shape sample. In the future it is assumed to use the methodology for the industrial cases.

Acknowledgments

This publication was supported by the Ministry of Industry and Trade (MPO) within the framework of institutional support for long-term strategic development research organization - provider MPO, recipient VÚTS, a. s.

References

[1] Lienhard J H IV, Lienhard J H V 2018 A heat transfer textbook, 4th Ed. (Cambridge Massachusetts: Phlogiston Press)
[2] MacKenzie D S 2003 Advances in quenching *Houghton International, Inc.*
[3] Ferziger J H, Peric M 2002 Computational methods for fluid dynamics (Berlin: Springer-Verlag) ISBN 3-540-42074-6
[4] MacKenzie D S 2010 Application of computational fluid dynamics to understand quenching problems *18th Congress International Federation for Heat Treatment and Surface Engineering*
[5] Kobayashi K, Nakamura O and Haragucchi Y 2016 Water quenching CFD (Computational Fluid Dynamics) simulation with cylindrical impinging jets *Nippon steel & sumitomo metal technical report* 111
[6] Banka A, Franklin J, Li Z, Ferguson L B and Aronov M 2008 CFD and FEA used to improve the quenching process *Heat treating progress* September 2008 pp 50-56
[7] Lee W H 1979 A Pressure Iteration Scheme for Two-Phase Modeling (Los Alamos: Scientific Laboratory Report) LA-UR 79-975
[8] Ansys Fluent theory guide *Ansys, Inc.*

IOP Conf. Series: Materials Science and Engineering **723** (2020) 012023 doi:10.1088/1757-899X/723/1/012023

Materials analyses of composite coatings with selected fillers

Jan Novotný, Štefan Michna, Martin Jaskevič

J. E. Purkyně University in Ústí nad Labem, Faculty of Mechanical Engineering, Ústí nad Labem, 400 01, Czech Republic

jan.novotny@ujep.cz, stefan.michna@ujep.cz, martin.jaskevic@ujep.cz

Abstract. This article focuses on the selection of fillers for composite coatings according to their key properties, sample composition and particle morphology. Properties that we want to improve with coatings are especially durability, abrasion resistance and heat resistance. TiO_2, Al_2O_3, WC a W were selected as suitable materials. XRD and XRF analysis was performed on selected fillers to determine the exact composition and crystallographic structure. Furthermore, the samples were milled in a ball mill, sieved according to the size of the fraction and SEM analysis was performed for each fraction to determine the particle morphology.

1. Selected materials and basic properties

Coating of metallic and non-metallic materials with composite coatings is intended to improve their mechanical properties and thereby increase their service life and utility value. It is therefore very important to choose the right matrix and functional filler.

PTFE emulsion, which has already proven itself in previous research [1], [2], geopolymer substances and PMMA [3], [4], will be used as the matrix of composite coatings.

The functional filler in the form of powder was chosen according to its properties for a particular type of use of the given component (material) in order to achieve an improvement of the required properties (improved abrasion resistance, increased temperature resistance, reduced thermal conductivity…). Four materials were selected as a filler.

- TiO_2 – this metal oxide has proven itself in previous research for its advantageous properties such as UV resistance, hydrophobic and antistatic properties, abrasion resistance, etc., high hardness [4], [5].
  Examples of possible applications include: increasing component life, improving abrasion resistance, improving surface non-adhesion, improving UV resistance.
- Al_2O_3 – heat resistant, hard, low thermal conductivity.
  Examples of possible applications: increasing the service life of the component, improving wear resistance, improving the mechanical properties.
- WC – heat resistant, hard, good electrical conductivity [2].
  Examples of possible applications are: increasing component life, improving abrasion resistance, increasing heat resistance (heat shields).
- W – high melting point, high hardness, heat resistance.
  Examples of possible applications include: increasing component life, improving abrasion resistance, increasing heat resistance (heat shields).

By a suitable combination of matrix and functional filler, it is possible to form a composite that improves the desired property. It is also possible to apply multi-layer composite systems (for example, a high temperature undercoat and a wear resistant topcoat).

Another important variable affecting the resulting properties of the composite coating is the weight concentration of filler particles in solution. In previous researches, concentrations of tenths of percent to units of percent were chosen [1], [2], [3], [4]. If the concentration is too high, the individual particles are not firmly anchored in the coating and are dropped by rubbing at very low pressure (swipe). Conversely, if the concentration is too low, the particles in the coating are not sufficient to exhibit their beneficial properties (further research will be conducted with appropriate concentration of filler particles).

The choice of particle fraction size (and their morphology) is another variable affecting the resulting properties, see chapter 3.

There is also a choice of many methods of applying a composite coating to a component, which also affects the resulting surface quality and properties. The coating can be applied, for example, by dipping, brushing, sputtering, etc.

2. XRD a XRF analysis

XRD analysis was performed on a Panalytical X´Pert PRO diffractometer. The measurement was performed in a symmetrical Bragg-Brentan arrangement, which is used for standard powder diffraction. A Cu cathode X-ray lamp was used. The measured diffractograms were evaluated using HighScore Plus software with PDF-2 database. The result of XRD analysis is presented as a diffraction pattern with commentary.

XRF analysis was performed on a WDXRF instrument Rigaku ZSX Primus 4. Data were evaluated by a non-standard method of fundamental parameters, which allows the determination of elements in the range of F-U in concentrations from hundredths ppm to 100 %. The result of the XRF analysis is presented in the table (due to the large amount of other elements and oxides contained in the samples at very low concentrations, only those with a proportion greater than 0.2 % were selected for the tables).

- TiO_2

Table 1. XRF analysis of TiO_2.

No.	Component	Result	Unit	Det. Limit	El. line	Intensity	w/o normal
1	**TiO_2**	99.8	%	0.02068	Ti-KA	784.3756	97.4513

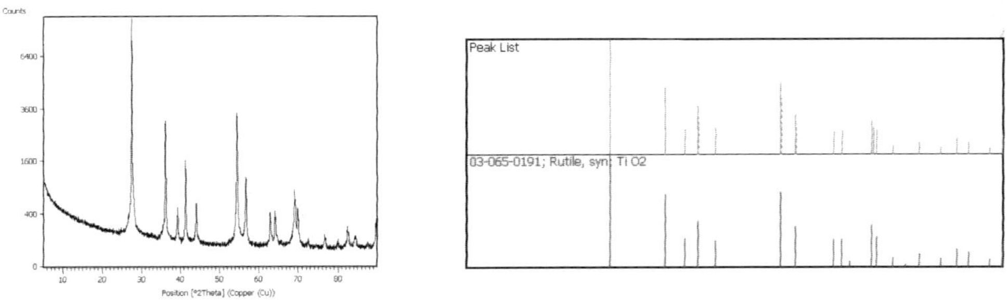

Figure 1. XRD analysis of TiO_2.

The sample contains TiO_2 in the form of rutile.

- Al_2O_3

Table 2. XRF analysis of Al_2O_3.

No.	Component	Result	Unit	Det. Limit	El. line	Intensity	w/o normal
1	**Al_2O_3**	88.6	%	0.04815	Al-KA	90.4043	71.4467
2	**TiO_2**	4.23	%	0.04847	Ti-KA	1.3127	3.4149
3	**SiO_2**	3.34	%	0.01194	Si-KA	1.4998	2.6945
4	**CaO**	1.30	%	0.00755	Ca-KA	1.0570	1.0468
5	**Fe_2O_3**	0.990	%	0.00913	Fe-KA	2.9819	0.7982
6	**MgO**	0.425	%	0.02924	Mg-KA	0.1741	0.3427
7	**K_2O**	0.310	%	0.00871	K-KA	0.2393	0.2501

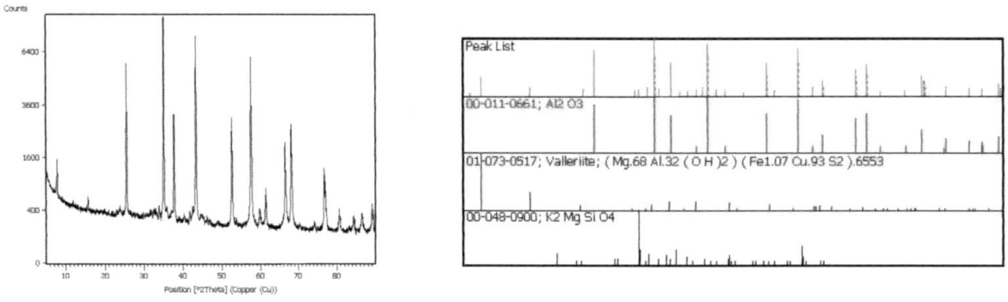

Figure 2. XRD analysis of Al_2O_3.

The main crystallographic phase in the sample is Al_2O_3. The sample contains small amounts of valerite and K_2MgSiO_4.

- WC

Table 3. XRF analysis of WC.

No.	Component	Result	Unit	Det. Limit	El. line	Intensity	w/o normal
1	**W**	86.3	%	0.02687	W-LA	1583.9812	88.4235
2	**Co**	6.16	%	0.00611	Co-KA	225.3081	6.3119
3	**Ti**	6.10	%	0.01478	Ti-KA	19.9021	6.2456
4	**Cr**	0.292	%	0.00834	Cr-KA	2.8675	0.2990

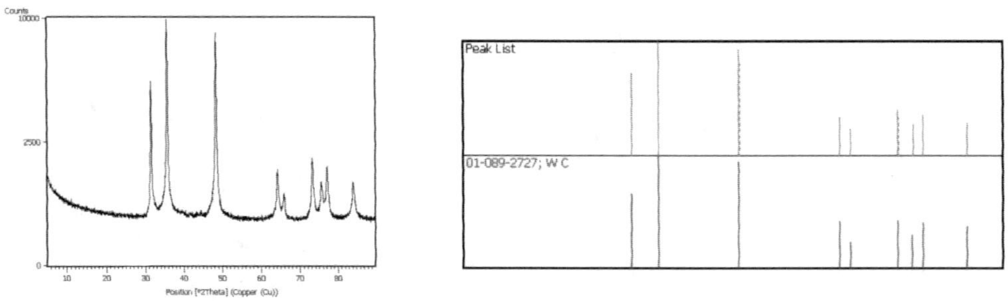

Figure 3. XRD analysis of WC.

The sample contains only WC. XRF analysis does not directly determine the content of light elements (eg, C, H, N) in the compound. The content of the light elements is then calculated by software. For this sample, however, the program was unable to calculate the carbon content (probably because of the high content of other heavy elements) and therefore is not represented in table 3. The tungsten carbide is shown further from the XRD analysis of figure 3.

- W

Table 4. XRF analysis of W.

No.	Component	Result	Unit	Det. Limit	El. line	Intensity	w/o normal
1	**W**	98.1	%	0.03829	W-LA	729.1112	86.9326
2	**Cu**	0.767	%	0.00872	Cu-KA	16.8511	0.6793
3	**Fe**	0.295	%	0.00764	Fe-KA	3.0183	0.2612

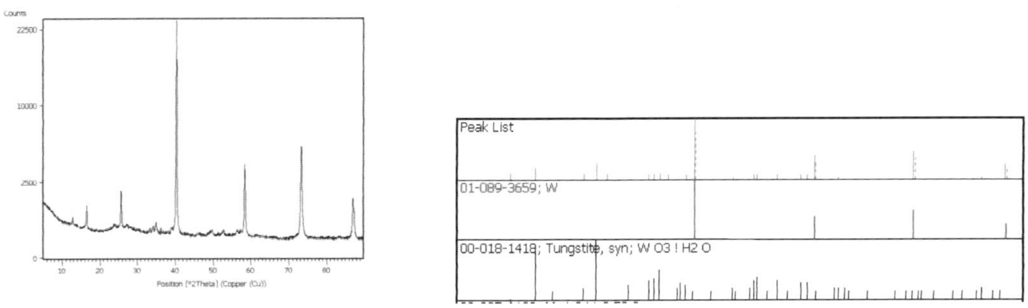

Figure 4. XRD analysis of W.

The main crystallographic phase in the sample is W. The sample further contains a small amount of tungstite.

3. Particle morphology

The samples were prepared (ground) in a planetary ball mill. They were then screened for size fractions

- < 20 µm
- 20 – 40 µm
- 40 – 80 µm
- > 80 µm

and subjected to SEM analysis.

The morphology of the particles milled in a ball mill is mostly sharp-edged, as confirmed by the SEM analysis in Figures 5 – 8 for the samples tested. The particle morphology also affects the properties of the coating. The sharp-edged particles can better hold the composite matrix and will not tear (for example, abrasion) than round or spherical.

- TiO_2

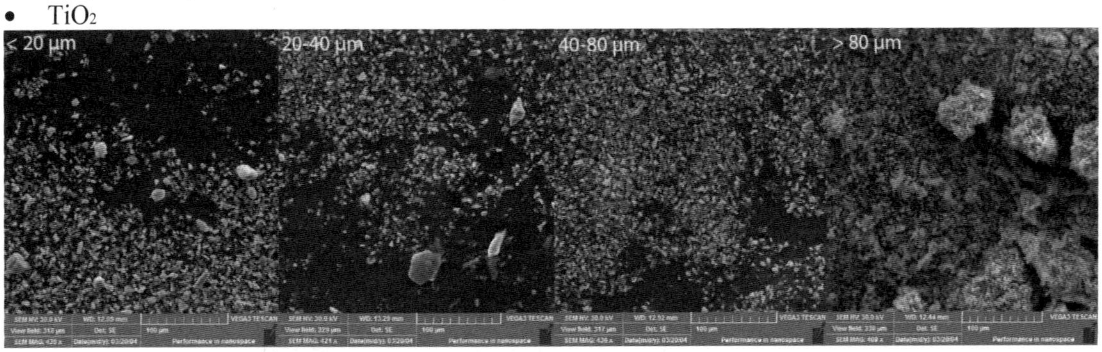

Figure 5. Morphology of TiO_2 particles.

Figure 5 shows the morphology of the individual TiO_2 fractions. The particles are irregular, sharp-edged and aggregates can be seen at a fraction > 80 µm.

- Al_2O_3

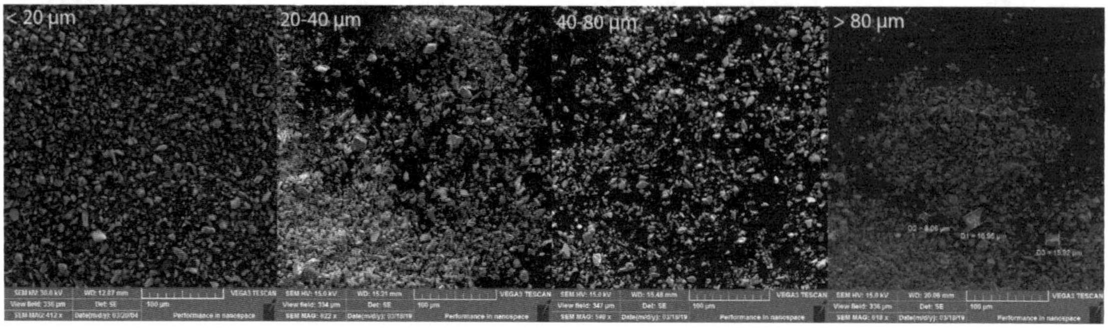

Figure 6. Morphology of Al_2O_3 particles.

Figure 6 shows the morphology of the individual Al_2O_3 fractions. The particles are irregular, sharp-edged. Aggregates are not present.

- WC

Figure 7. Morphology of WC particles.

Figure 7 shows the morphology of individual WC fractions. As an input sample, a powder was used, which was produced by the spraying method, which produced spherical particles that could be observed for a fraction > 80 µm. For the other fractions, which were already formed by mechanical milling in a ball mill (from the original sample with spherical particles), we observe again an irregular shape, but not as sharp as in the previous samples.

- W

Figure 8. Morphology of W particles.

Figure 8 shows the morphology of the individual W fractions. The particles are irregular, acute and flaky.

4. Conclusions

Composite coatings can extend the life of a component several times and significantly improve the surface properties, thus saving costs and time. In the following research, we will look at the development of coatings that will improve the durability, abrasion resistance and heat resistance of a component. For this reason, TiO_2, Al_2O_3, WC and W materials have been selected as fillers that meet these requirements. The tested PTFE emulsion, PMMA and geopolymer materials will be used as matrix.

We determined the exact composition and crystallographic phases of selected fillers by XRD and XRF analysis. Next, we ground the samples in a planetary ball mill and sorted according to fractions. Selected fractions were subjected to SEM analysis to determine the morphology of the particles, which is irregular and sharp-edged, except for WC input material, where the particles are spherical due to the production method used.

Acknowledgements

Supported by the OP VVV Project Development of new nano and micro coatings on the surface of selected metallic materials - NANOTECH ITI II., Reg. No CZ.02.1.01/0.0/0.0/18_069/0010045.

EUROPEAN UNION
European Structural and Investment Funds
Operational Programme Research,
Development and Education

MINISTRY OF EDUCATION,
YOUTH AND SPORTS

References

[1] Lysoňková I, Novotný J, Cais J and Michna S (2017). Extended of Durability Molds for Production Tires with New Types of Coatings. Proceedings of the International Conference on Science & Engineering for Sustainable Development (2017), 358-363. doi: http://dx.doi.org/10.21013/jte.ICSESD201734

[2] Müller, M. Effects of Aluminium Microparticles and Surface Treatment of $AlCu_4Mg$ on Mechanical Properties of Adhesive Bond Strength, In Manufacturing Technology, Vol. 17, No. 1, FVTM UJEP, Ústí nad Labem, pp. 66-71. ISSN 1213-2489

[3] Novotný J and Jaskevič M Nanocomposites and their use in technical practice, In Engineering for Rural Development, 2018 Vol. 16, Latvia, pp. 1270 - 1276, 2017, ISSN 1691-5976

[4] Novotný J, Lysoňková I and Jaskevič M Mesocomposites based on the PMMA matrix. Manufacturing Technology, Vol. XY No. Z, Ústí nad Labem: FSI UJEP, 2018, pp. XX. ISSN 1213-2489

[5] Novotný J, Lysoňková I and Sapieta M Nanocomposite coatings in polymeric matrices and their effect on friction coefficient. ITEP 2018 Bojnice, SR

[6] Novotný J, Jaskevič M and Cais J (2017) Morfologie tvaru a struktury vyrobených částic TiO_2, sborník příspěvků konference ERIN 2017, s. 150 – 156, ISBN 978-80-7561-056-0

[7] Novotný, J., MICHNA, S, JANOVEC, J.. Vybrané kapitoly z fyziky kovů a fraktografie, UJEP, 250 stran. ISBN 978-80-756114-6-8

[8] Novotný, J., Náprstková, N. Evaluation of possibilities of selected mechanically alloyed metal powders, in Toyotarity. Innovations. Improvement, 2017 Centrum Szkoleń Personalnych Monika Otrąbek, Poland, pp 63 -85. ISBN 978-83-948512-0-0

[9] Smetana, O. Hodnocení sypkých a konsolidačních vlastností v technické praxi, *Bakalářská práce,* FSI, UJEP, Ústí nad Labem, 2019

[10] Stančeková, D., Šajgalík, M., Petru, J., Náprstková, N., Balaz, P.: Implementation of coating for failure elimination of dial gauges. In METAL 2015 Conference Procedings of the 24Rd International Conference on Metallurgy and Materials, Czech Republic, Brno June 3rd - 5th 2015.

IOP Conf. Series: Materials Science and Engineering **723** (2020) 012024 doi:10.1088/1757-899X/723/1/012024

Methodology for observation of maraging tool steel after 3D printing using FIB and STEM mode

K Opatová, L Kučerová and I Zetková

University of West Bohemia, Regional Technological Institute, Univerzitní 8, Pilsen 30100

opatovak@rti.zcu.cz

Abstract. The components of the maraging tool steel produced by 3D printing are further heat-treated after printing. It includes a stress relief annealing and precipitation hardening. During this process precipitates of Ni_3Ti and Fe_2Mo or Fe_7Mo are formed, which significantly contribute to an increase in ultimate tensile strength, hardness, and toughness. To understand the precipitation processes, it is necessary to describe the particles in detail. However, they are very small in size and cannot be examined properly by a light microscope (LM) or by scanning electron microscope (SEM). A possible way to investigate and describe these precipitates is to produce a thin lamella using a focused ion beam (FIB) in the electron microscope chamber, where scanning transmission electron microscopy (STEM) is subsequently used for observation. A lamella was prepared from the DMLS (Direct Metal Laser Sintering) printed part by an ion beam for observation in STEM mode. The experiment took place at the Zeiss AURIGA scanning electron microscope, equipped with an ion gun and also with STEM capabilities for thin samples. The lamella preparation methodology was gradually optimized to achieve sufficient resolution for observation of these very fine microstructures produced by 3D printing.

Key words: FIB, STEM, maraging steel, 3D printing

1 Introduction

Due to their properties, maraging steels are suitable materials for use in many industrial applications. These are high strength steels with low carbon content, which are mainly alloyed with nickel, cobalt, molybdenum, and titanium. Maraging steels have a very fine martensitic microstructure whose mechanical properties are achieved by suitable heat treatment, in particular by precipitation hardening [1, 2]. A detailed microscopic analysis is necessary to observe the microstructure development after heat treatment and to investigate precipitates, which are developed during precipitation hardening. A light microscope will not provide sufficiently detailed information to describe such a microstructure. Using a scanning electron microscope, it is possible to describe the structure in more detail, but for a precise description of precipitates in the microstructure, it is necessary to use a transmission electron microscope or a scanning electron microscope with the possibility of transmission mode. For this purpose, it is necessary to prepare thin samples – lamellas. A scanning electron microscope equipped with two guns (dual beam microscope) was used for this experiment – dual beam microscope. It combines two independent microscope columns, a focused ion beam (FIB) and electron beam in the same machine [3, 4].

Although the preparation of thin samples by ion beam is already an understood phenomenon, it is necessary to find suitable parameters for each material, as well as for different conditions of its processing. In this case, it was a sample of maraging steel after 3D printing, which was annealed for stress relieving and subsequently it underwent the precipitation hardening treatment [5].

2 Experimental program

The experiment was performed on EOS MS1 maraging steel, whose chemical composition corresponds to US grade 18% Ni Maraging 300, the European 1.2709 and the German X3NiCoMoTi 18-9-5 grade steels (table 1). After printing, the material was annealed for stress relieving at 820°C and subsequently precipitation hardened at the temperature of 490°C for 6 hours. To prepare a thin sample for observation in STEM mode, a small piece was taken from the material using a precision metallographic saw, which was then ground and polished. It was then attached to a stub for an electron microscope using carbon tape for subsequent processing in the SEM. Different parameters of thin specimen preparation for STEM mode using ion gun were tested on the sample, which was printed using DMLS technology. This was done on a Zeiss Crossbeam 340 dual-beam scanning electron microscope equipped with a field emission gun of the Schottky type with an electron beam resolution of 1 nm. The system also featured a Focused Ion Beam (FIB) gun, detectors of secondary and back-scattered electrons (SE, BSE), an energy-dispersive X-ray spectroscopy (EDS) detector, electron backscatter diffraction (EBSD) detector, and also provided scanning transmission electron microscopy (STEM) capabilities for thin samples which features gas injection system (GIS) equipped with platinum.

Table 1. The chemical composition of the experimental material, EOS Maraging Steel MS1 [1].

wt. [%]	C	Si	Mn	P	S	Cr	Mo	Ni	Co	Ti	Cu	Al	Fe
MS1	≤ 0.03	≤ 0.1	≤ 0.1	≤ 0.01	≤ 0.01	≤ 0.5	4.5 - 5.2	17.0 – 19.0	8.5 – 9.5	0.6 – 0.8	≤ 0.5	0.05 – 0.15	bal.

2.1 Lamella preparation

The thin sample – lamella was prepared in several steps using FIB, GIS and nanomanipulator. The liquid metal ion source of gallium (Ga^{+}) is used as an ion source for FIB. Gallium ions are extracted from a liquid metal ion source. The ions are accelerated by the acceleration voltage up to 30 keV. The ion emission is regulated by the extractor and the suppressor. The GIS system, in combination with FIB, allows ion-beam induced deposition of platinum on the specimen surface. After selecting a suitable location for the observation that represented all investigated microstructural constituents, the sample was tilted at 54° to begin to remove material with an ion beam. First, a platinum layer was applied to the region of interest using GIS, which serves to protect the surface layer of the sample from deterioration by both ion and electron beams. Subsequently, two trapezoids were roughly excised using a probe of 30 kV to 30 nA (figure 1a). To achieve sufficient depth, a dose factor of 15 with a dose of 400 mC / cm^2 was chosen, which resulted in approximate depth of 14 μm, which was sufficient for subsequent handling of the lamella and its fine polishing and observation. The lamella dimensions required for observation of sufficient amount of microstructural constituents was about 3x4 μm. The scan mode during coarse cutting was bidirectional with cycle mode cross-section to ensure uniform cutting of the material over the entire surface of the prepared lamella.

The next step was to reduce the lamella from the original 2.5 μm to 500 nm in several steps using a lower probe, namely 30 kV: 15 nA and then 30 kV: 200 pA (figure 1b). During this process, the lamella was rotated by ± 0.5° to obtain a perpendicular cut. In case the rotation was not performed, the lamella would expand towards the base material and would not have good material removal at its base. Subsequently, the lamella had to be separated from the base material. Then a special stub with a Cu holder could be inserted into the chamber, to which the lamella was then attached with

Ga. Lamella was first undercut into a U-shape, which served as a preparatory work to move it. The nanomanipulator was used to remove the lamella and was also attached to it using GIS and Ga. Consequently, the lamella on the nanomanipulator could be transferred to the chamber space. After inserting the stub with Cu-holder, it was moved from the nanomanipulator to Cu-holder. There the lamella was reattached with Ga and the nanomanipulator was cut off by an ion beam (figure 1 d, e, f). The release of the nanomanipulator from the lamella is useful for re-sharpening the nanomanipulator with an ion beam (figure 2).

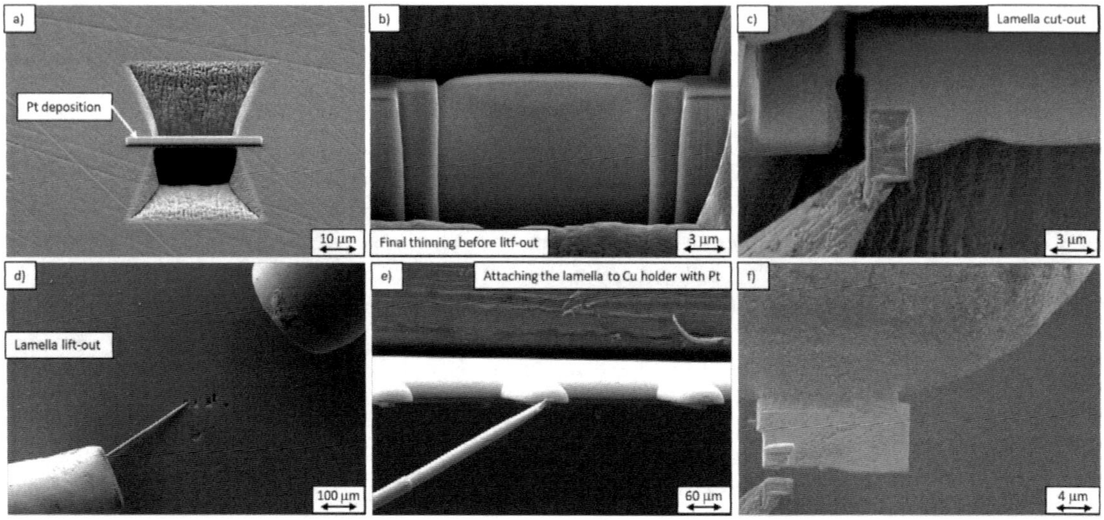

Figure 1. The scheme of lamella preparation: a) Pt deposition for surface layer protection and coarse trapezoides in the desired position on the sample, b) thinning before lift-out, c) Pt deposition to attach sample to the nanomanipulator and cut-out the lamella, d) litf-out, e) attaching lamella to the Cu holder by Pt using GIS, f) cut of the nanomanipulator.

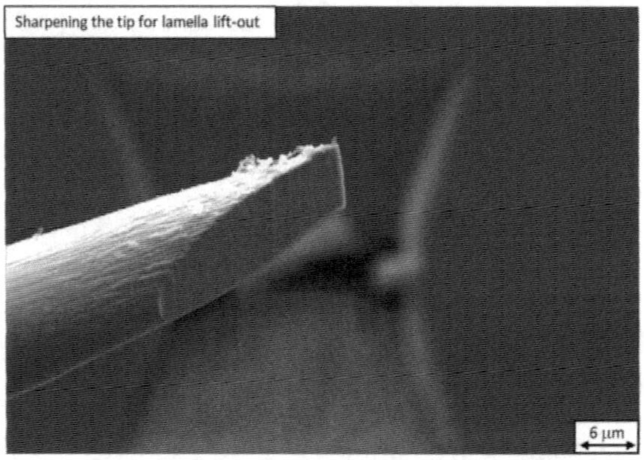

Figure 2. Sharpening of the nanomanipulator after lamella lift-out.

After shifting the lamella to the Cu-holder, further thinning of the individual layers of the material by ion beam is necessary. Lower operating parameters are used: 5 kV: 200 pA. The lamella is uniformly thinned from the previous 500 nm until a thickness of less than 100 nm is reached. The need for further thinning depends on the sample material and the acceleration voltage used for observation. For EOS MS1 maraging steel, the final thickness for observation in STEM mode was about 75 nm (figure 3).

The final step before it is possible to analyse the lamella in STEM mode is to transfer the Cu-holder with thin lamella to the special STEM holder, where the Cu-holder is placed in horizontal position in order to get its surface between electron gun and the detector of transmitted electrons.

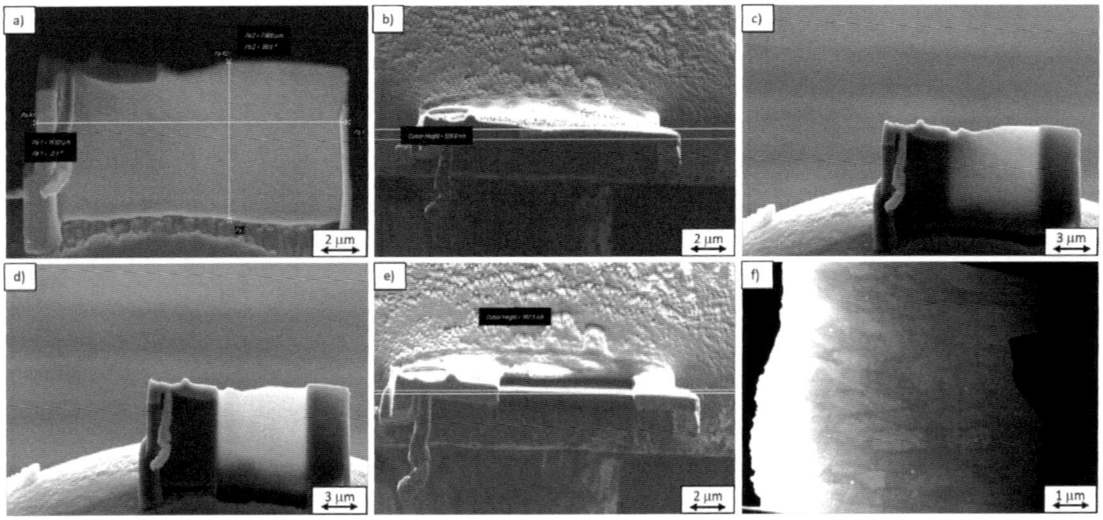

Figure 3. Continuous steps for obtaining the final dimensions of the lamella attached to the Cu-holder.

2.2 STEM observation of the lamella

The Zeiss electron microscope was equipped with 4-Channel aSTEM detector which detecting transmitted electrons in an FESEM with the annular STEM detector. The aSTEM detector collects electrons transmitted through a thin sample and it can separate unscattered and scattered electrons depending on the scattering angle. The detector constitutes of special annular arrangement of five diode segments (figure 4). The bright field (BF) segment placed in the centre of the detector detects the unscattered electrons. Around BF segment are diagonally placed two dark field (DF) segments and two segments for oriented dark field (ODF) imagining, one annular dark field (ADF) ring and the high angle annular dark field (HAADF) ring on the outer diameter to detect electrons scattered under higher angles (figure 4).

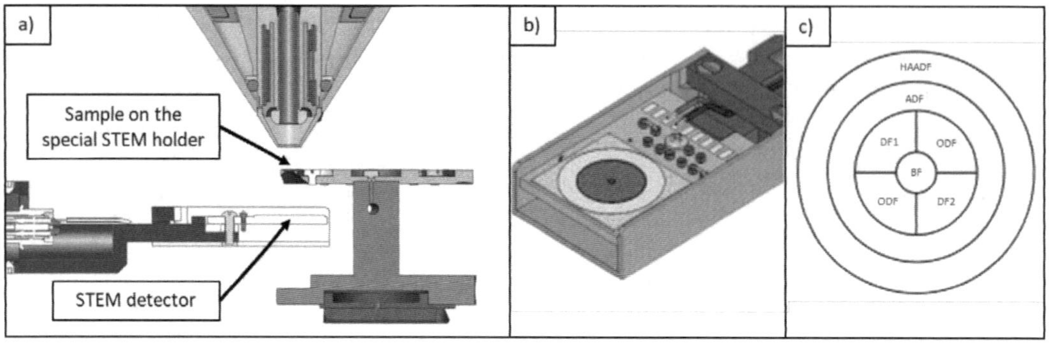

Figure 4. Arrangement for samples observation with STEM detector: (a) the sample is placed on a copper holder between the electron gun and the STEM detector, (b) head of the annular STEM detector, (c) Scheme of the semiconductor-based detector with its BF area in the centre, the four DF diodes, the ADF ring, and the HAADF outer ring [7].

With a given segments layout of the aSTEM detector, it is possible to combine signals from different segments to achieve the desired resolution and highlight precipitation or other areas of interest. The possibility of combining individual detector segments is the main advantage of a scanning electron microscope equipped with a STEM detector. By collecting and displaying multiple signals, it is possible to obtain as much information as possible from each scan. Each detector is used to highlight different points of interest in the structure of the material (figure 5). The BF detector includes a transmitted beam and is, therefore, suitable for highlighting holes that appear bright in this display. In contrast, the DF detector eliminates the transmitted beam, and the holes in this regard appear dark. An ADF detector that has the form of a ring forms an image by collecting scattered electrons similar to a DF detector in TEM. But ADF in STEM does not use an aperture to separate the scattered electrons from the main beam. The HAADF detector is, as the ADF detector, a ring, but the hole in its center is significantly larger than that of the ADF detector. Thanks to this, it detects electrons that are incoherently scattered to large angles. The result is an image called Z-contrast [8, 9]. It is obvious from the microstructure images that the basic matrix of the experimental material is formed by martensitic needles, among which spherical-shaped precipitates occur. Due to the shape and size of the precipitates and heat treatment of the material, most of these precipitates are Ni_3Ti or $Ni_3(Ti, Mo)$. The larger diameter particle that is visible in the lower-left corner of the HAADF image (figure 5d) is probably a Fe_7Mo6 precipitate that typically reaches a size between 10 - 30 µm. Precise determination of observed precipitates is possible only by using TEM and diffraction pattern, or by measuring chemical composition and stochiometry. The bright areas in figure 5e are retained austenite [10].

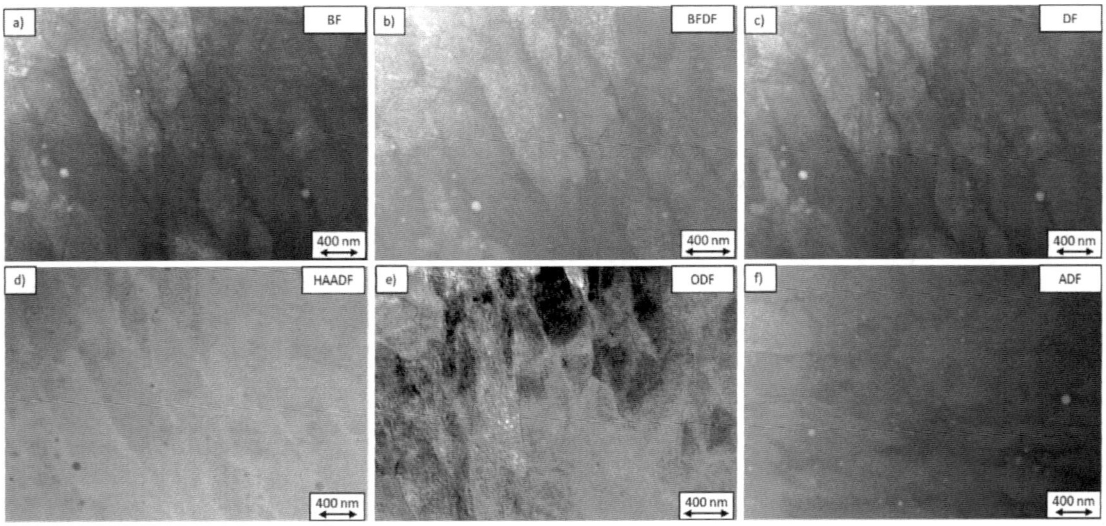

Figure 5. Different modes for sample observation.

3 Conclusion

Using an ion beam scanning electron microscope, a gas injection system and a detector for field emission scanning aSTEM, it was possible to prepare and observe a thin lamella from a maraging steel sample printed by the DMLS method. The lamella was prepared in several steps using various probes of the ion beam. Since the structure of the material is unconventional and very fine after 3D printing, it was necessary to continuously optimize the parameters to obtain sufficiently high-quality images in the transmission mode, which makes it possible to describe all the microstructural elements adequately. It has been found that to achieve lamella with dimensions of approximately 4 x 3 μm it is suitable to use probe 30 kV: 30 nA for coarse milling up to a thickness of 2.5 μm, followed by 30 kV: 15 nA and 30 kV: 200 pA for finer milling to a thickness of approximately 500 nm. After the lamella is moved to the Cu-holder, a 5 kV: 200 pA probe was applied for fine polishing to less than 100 nm thickness. In this state, the specimen on the Cu-holder were transferred to a special STEM sample holder, which was located between the electron gun and the aSTEM detector in the microscope chamber, allowing the transmission mode to be applied.

Acknowledgements

This research was funded by the Technology Agency of the Czech Republic under the project TJ01000161 'Systematic applied research of material properties of martensitic steel W-Nr. 1.2709 produced by 3D printing using DMLS technology with the application of research results in practice'.

References

[1] EOS GmbH 2011 Material Data Sheet - 18Ni300

[2] Mun S, Watanabe M, Li X, Oh K Y U H, Williams D B, and Lee H 2002 Precipitation of austenite particles at grain boundaries during aging of Fe-Mn-Ni steel *Metall. Mater. Tran.* A **33** pp 1057–67

[3] Kučerová L, Zetková I, Jandová A, and Bystrianský M 2019 Microstructural characterisation and in-situ straining of additive-manufactured X3NiCoMoTi 18-9-5 maraging steel *Mater. Sci. Eng.* A **750** pp 70–80

[4] Schaffer M, Schaffer B and Ramasse Q 2012 Sample preparation for atomic-resolution STEM at low voltages by FIB *Ultramicroscopy* **114** pp 62–71

[5] Tomus D and Ng H P 2013 In situ lift-out dedicated techniques using FIB – SEM system for TEM specimen preparation *Micron* **44** pp 115–9

[6] Kohl-Roscher U 2015 4-Channel-aSTEM Detector Detecting Transmitted Electrons in an FESEM with the new Annular STEM Detector *Carl Zeiss Microscopy GmbH* [Online]
[7] Pennycook S J et al. 2007 Scanning Microscopy for Nanotechnology: Techniques and Applications ed W Zhou, Z L Wang (Springer) chapter 6 152-91
[8] Petrova R V 2006 Quantitative high-angle annular dark field scanning transmission electron microscopy for materials science (University of Central Florida Orlando, Florida)
[9] Oveisi E, Spadaro M C, Rotunno E, Grillo V and Hébert C 2019 Insights into image contrast from dislocations in ADF-STEM *Ultramicroscopy* **200** pp 139–48
[10] Bodziak S *et al.* 2019 Precipitation in 300 grade maraging steel built by selective laser melting: Aging at 510 °C for 2 h *Mater. Charact* **151** pp 73–83

Induction hardening of steels with use of the device for incremental forming of round bars HDQT-R 30-12

M Peković, J Vrtáček, T Janda, J Volkmannová

Faculty of Mechanical Engineering – Regional Technological Institute, University of West Bohemia, Univerzitní 8, 306 14 Pilsen, Czech Republic

E-mail: pekovicm@rti.zcu.cz, vrtacekj@rti.zcu.cz, jandat@rti.zcu.cz, volkmann@rti.zcu.cz

Abstract: Induction hardening is mainly used for treating parts in which high hardness, wear-resistance, and ductile core are required, such as spur gears, wind turbine rings, and bearings. Innovative induction hardening processes are also developed for complex automotive components, such as camshafts and crankshafts. In the process, steel is heated by passing high-frequency alternating current through a coil which is equivalent to the primary winding of a transformer. The material to be heat-treated is equivalent to a short-circuited secondary winding. The HDQT-R 30-12 device for incremental forming of round bars includes modules which can be used for heat-treating bars after rolling or even without rolling. Heating is provided by five induction coils at a faster rate than in a chamber furnace. In addition, there is less oxidation of the material surface. This paper presents findings related to microstructures and hardnesses in 42CrMo4, 13CrMo4, and S235JR steel grades after heating at different inductor coil power settings with subsequent hardening in a water bath.

Keywords: induction hardening, quenching, AHSS, hardness

1 Introduction

The increasingly stringent requirements for mechanical properties of steels and the pressure on prices of final products lead to improvements in production processes. Induction hardening belongs to treatment methods which are widely used today. Electromagnetic induction was discovered by Michael Faraday in mid-1800s. The first use of induction heating for melting metals dates back to 1916. Induction heat treatment began to be used in the 1930s, mostly for induction hardening of pegs and crankshafts [1]. Today, induction hardening is important to the manufacture of gears, bearings, camshafts and crankshafts. In these applications, high hardness and wear resistance are sought, along with a good toughness in the core of the part [2]. Improved surface properties lead to longer life of treated products of this kind. Generally-recognized benefits of induction heat treatment include fast heating rates, high energy efficiency, rapid initiation and deactivation, environmental friendliness and a negligible extent decarburization of the part's surface. Induction hardening consists of two steps: induction heating and quenching with an appropriate quenchant [3].

Quenching belongs to heat treatment processes which are accompanied by phase transformations in steel. Typical quenching temperatures are higher than A_{c3}, which means quenching from the fully-austenitic region. Generally, the purpose of quenching operations is to produce martensitic structures. Formation of martensite depends on several factors. The first one is the critical cooling rate which must be attained if martensite is to form. The amount of carbon is another factor. With increasing carbon content, hardness of quenched steel increases [4].

Induction heating finds use in experimental forming processes where rapid heating to forming temperatures is needed. For the same purpose, this investigation was carried out on equipment for developing incremental forming processes which has induction heating capability.

2 Experimental programme

The HDQT – R 30 – 12 equipment for developing incremental forming processes (**H**igh **D**eformation **Q**uenching and **T**empering) is primarily intended for reducing the diameter of round bars. Its modular design makes it suitable for post-rolling heat treatment and heat treatment both with and without reducing the diameter. Available operations include induction hardening with quenching in water bath. The equipment features five induction coils in the front part of the rolling mill, upstream of the rolling stands, and a quenching bath in the rear part of the rolling mill (Figure 1) [5].

Figure 1. Device for incremental forming HDQT-R 30-12.

Induction coils are split between two circuits, where each has its own power supply. The first circuit comprises four heating coils with a power of 200 kW. The other includes a single 50 kW coil which reheats the bar to the desired forming or heat treatment temperature [6].

This experimental programme was concerned with induction hardening. The purpose was to find and test an appropriate power configuration for the induction coils to reach the desired optimum temperature for full austenitization of the materials under investigation and observe the effects of chemical composition of steel on its ability to be heated by induction coils. Three steel grades of different compositions were chosen for the experiments (Table 1).

Table 1. Chemical compositions of steels under investigation.

Material	C	Si	Mn	Cr	Mo	S	P	A_{c1} [°C]	A_{c3} [°C]
42CrMo4	0.44	0.3	0.8	1.2	0.2	0.04	0.01	744	764
13CrMo4	0.16	0.3	0.7	1,1	0,5	0.04	0.01	747	845
S235JRC	0.17	0.2	0.8	x	x	0.02	0.01	714	833

The data on their compositions given in their data sheets were input in JMatPro software to compute their CCT diagrams, A_{c1} and A_{c3} temperatures (Figure 2). The primary difference between the 42CrMo4 and 13CrMo4 grades was the level of carbon. Unlike S235JRC steel, 13CrMo4 grade is alloyed with chromium and molybdenum [8]. This similar chemical composition between pairs of steels differing either in carbon content or in alloying element content was the main reason for their choice.

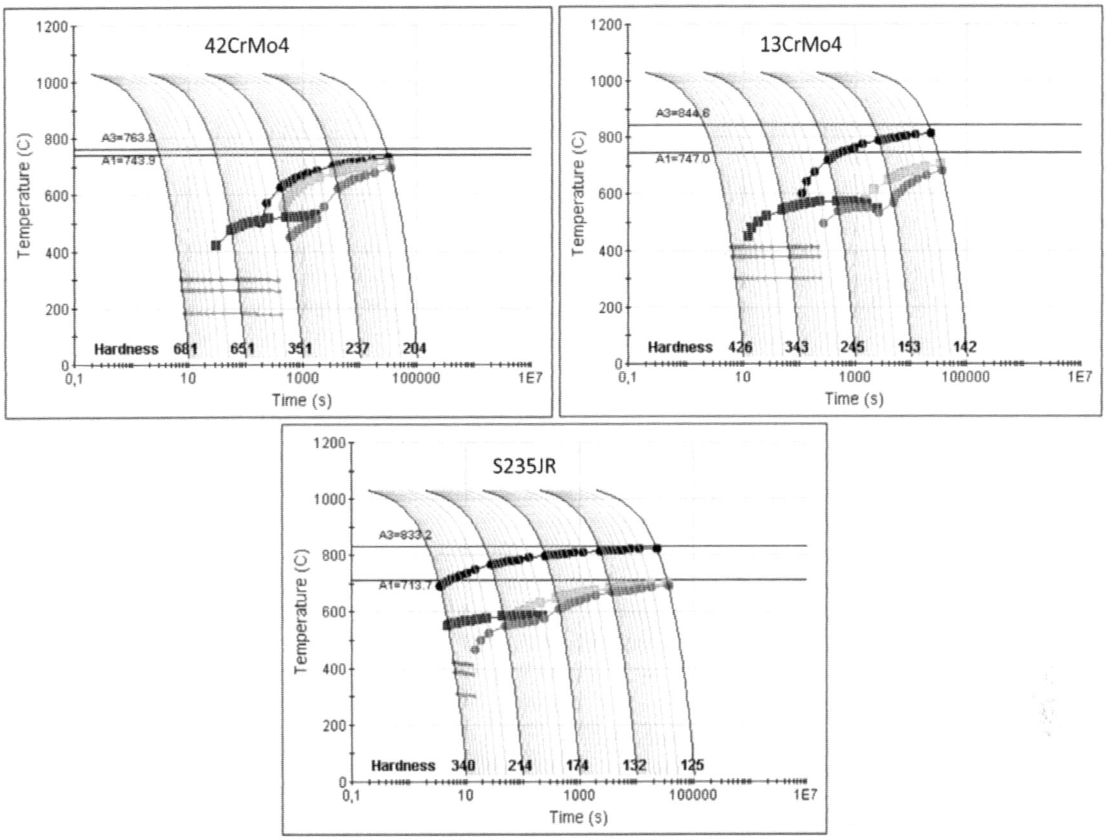

Figure 2. CCT diagrapms of experimental steels.

The as-received 42CrMo4 was in the quenched and tempered condition. Its microstructure consisted of tempered martensite (Figure 3a). 13CrMo4 contained a ferritic-pearlitic microstructure. Pearlite was globular (Figure 3b). The as-received condition was the condition after annealing. The as-received microstructure of S235JR consisted of ferrite and pearlite. In this case, pearlite was lamellar (Figure 3c).

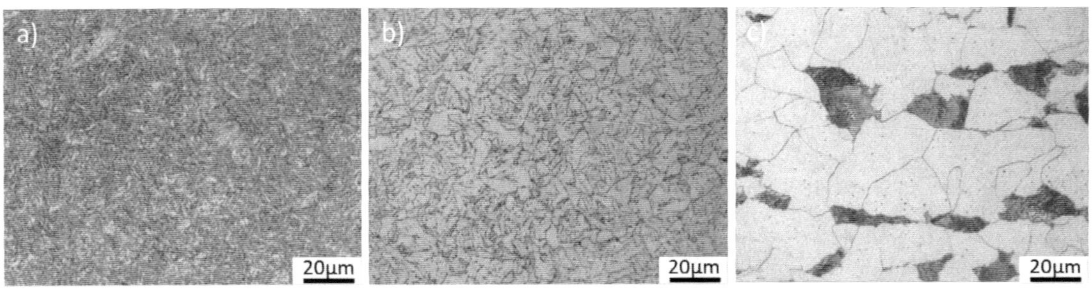

Figure 3. Micrographs of the materials in the initial condition: a) 42CrMo4, b) 13CrMo4, c) S235JR.

The material for experimental processing was bars 1000 mm in length and 30 mm in diameter. Three different configurations of induction coil power settings were proposed. In the first two configurations, both induction heating coil circuits were active. In the last configuration, only the first circuit was active. The other was set to the minimum power (Table 2). The bars were moving along the roller track at a speed of 135 mm/s, as the rollers were revolving at 20 rev./min. Temperature was measured using a pyrometer at the exit from the induction heating module. This means that the surface temperature of the bars was measured.

Table 2. Coils power settings.

Sequence	Coils 1 – 4 [%]	Coil 5 [%]
IH1	95	90
IH2	75	70
IH3	75	5

3 Results and discussion

Microstructures of the heat-treated specimens were examined using a light microscope. Their hardness was determined and reported using HV10 Vickers hardness number.

In configuration IH1, where the coil power setting was the highest, the surface temperature of bars of 42CrMo4 steel at the exit from induction heating was 850 °C. The resulting microstructure was martensitic with ferrite (Figure 4a). Hardness at the surface of this bar was 610 HV10 (Table 3). In agreement with the operation of induction heating, which heats material from the surface towards the core, the amount of martensite was decreasing from the surface to the centre of the bar. Accordingly, hardness dropped to around 510 HV10 (Table 3). In the second configuration, IH2, in which the power of induction heating coils was reduced, the surface temperature was no more than 730°C (at the same speed of feeding through the coils). The A_{c1} temperature, which is 744°C, was not exceeded. The microstructure consisted of martensite, a small amount of ferrite, and globular cementite (Figure 4b). The surface hardness was 375 HV10. Hardness values decreased towards the core, where the value was 255 HV10, which was due to a reduced amount of martensite (Table 3). In the last heating coil configuration, IH3, the power setting of the first four coils was 75%, whereas the last coil was set to 5%. The purpose was to ascertain the impact of heating by the last induction heating coil on the final temperature of the stock. At the exit from the coils, the surface temperature of the bar was a mere 620°C. The treatment in this configuration led to tempering of martensite. The final hardness was 260 HV10 (Figure 4c, Table 3).

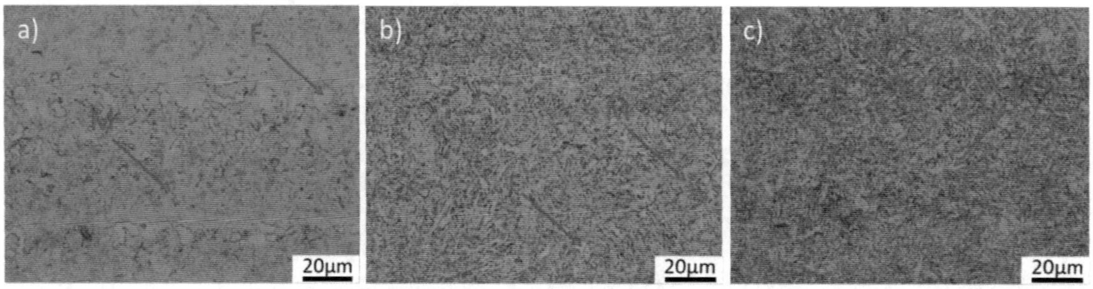

Figure 4. Microstructures at the centre of 42CrMo4 steel specimens after heat treatment: a) IH1, b) IH2, c) IH3.

13CrMo4 grade contains less carbon than 42CrMo4. The treatment with heating coil settings according to IH1 configuration led to a surface temperature of 870°C. With this configuration, a ferritic-martensitic microstructure was obtained (Figure 5a). Hardness was 215 HV10 (Table 3). Using the second heating coil configuration, IH2, which involved a reduced power settings, the steel was heated to 780°C. The resulting microstructure consisted predominantly of ferrite, with a fine

IOP Conf. Series: Materials Science and Engineering **723** (2020) 012025 doi:10.1088/1757-899X/723/1/012025

dispersion of globular cementite, and a very small amount of martensite (Figure 5b). Hardness was approximately 175 HV10 (Table 3). The last configuration, IH3, involved heating by the first heating coils only. It led to no significant changes in the initial microstructure, producing ferrite and pearlite. Pearlite was globular (Figure 5c). Hardness was approximately 165 HV10 (Table 3).

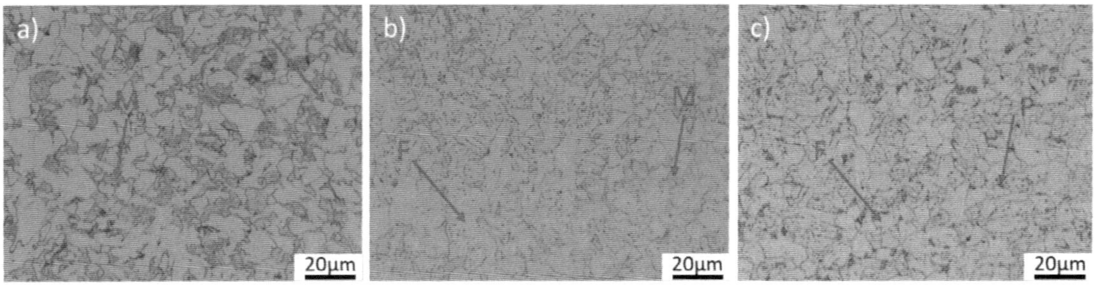

Figure 5. Microstructures at the centre of 13CrMo4 steel specimens after heat treatment: a) IH1, b) IH2, c) IH3.

S235JR grade was not alloyed with chromium. The heating configuration IH1 led to a temperature of 875°C. After subsequent quenching, the material contained ferrite and small portions of pearlite and martensite. Towards the core of the bar, the amount of martensite decreased, whereas that of pearlite increased. The hardness in the core was 156 HV10 and that at the surface of the bar was 193 HV10 (Figure 6a, Table 3). Heating configuration IH2 led to a surface temperature of 780°C. The microstructure was predominantly ferritic-pearlitic with a small fraction of martensite. Hardness levels were between 166 and 173 HV10 (Figure 6b, Table 3). Configuration IH3 led to a surface temperature of 670°C and to a ferritic-pearlitic microstructure. Its hardness was approximately 158 HV10 (Figure 6c, Table 3).

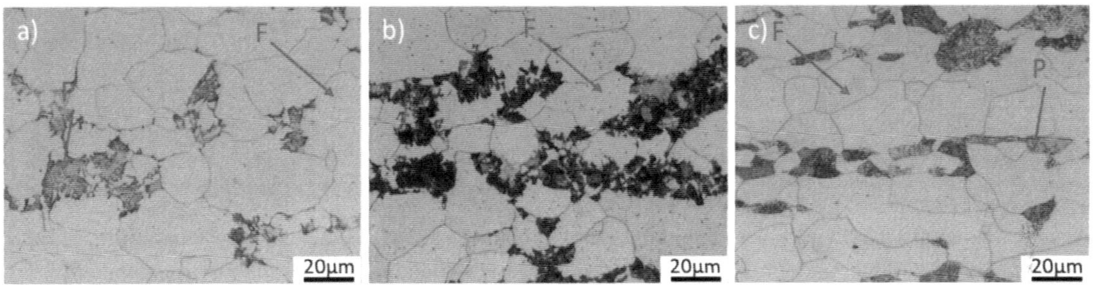

Figure 6. Microstructures at the centre of heat-treated S235JR specimens: a) IH1, b) IH2, c) IH3.

The resulting temperatures, microstructures and hardness levels suggest that chemical composition of the stock has a major impact on the outcome of heating in induction coils (Table 3).

Table 3. Results of the experimental programme.

Material	IH1 T [°C]	IH1 HV10 Core	IH1 HV10 Surface	IH2 T [°C]	IH2 HV10 Core	IH2 HV10 Surface	IH3 T [°C]	IH3 HV10 Core	IH3 HV10 Surface
42CrMo4	850	613	510	730	385	252	620	251	266
13CrMo4	870	230	202	780	196	173	660	171	156
S235JR	875	193	156	780	173	166	670	159	157

42CrMo4 steel contains higher amounts of carbon and chromium and was heated to the lowest temperatures. One can therefore presume that with increasing levels of carbon and other alloying elements, either higher induction heating power must be used to heat up steel sufficiently or the speed of stock passing through the induction heating equipment must be reduced. As expected, induction hardening has not produced hardening microstructures in the low-carbon steels 13CrMo4 and S235JR. The main reason was the low carbon level. Hardening steels generally contain more than 0.2% carbon.

4 Conclusions

This experiment revealed that in order to perform effective induction hardening in the HDQT-R 30-12 equipment, its induction heating coil settings must reflect not only the input diameter but also its chemical composition. 42CrMo4 grade steel is suitable for induction hardening. When heat treated in this equipment, it can develop hardening microstructures with a hardness exceeding 500 HV10 in the core of bars and more than 600 HV10 at the surface. In order to achieve full austenitization, the induction coil power should be increased or the speed of feeding the bars through the coils. 13CrMo4 and S235JR steels are not suitable for induction hardening, mainly due to their low carbon content, which reduces their ability to develop hardening microstructure. They were chosen for these experiments in order to ascertain the effect of the level of carbon and other alloying elements on the ability of material to be heated by induction heating coils. The findings show that the same inductor setting lead to higher temperatures in steels with lower levels of carbon and alloying elements. Further investigation will be devoted to the HDQT-R 30-12 equipment, primarily to the selection of materials and equipment settings leading to desired microstructures.

Acknowledgement

The present contribution has been prepared under project LO1502 'Development of the Regional Technological Institute' under the auspices of the National Sustainability Programme I of the Ministry of Education of the Czech Republic aimed to support research, experimental development, and innovation.

References

[1] Haimbaugh R 2001 *Practical Induction Heat Treating* (Ohio: ASM International) p 1-4
[2] Otero R, Albano L, Totten G, Meekisho L and Canale L 2015 *Encyclopedia of Iron, Steel, and Their Alloys vol 4* ed R Colás and G Totten (Boca Raton: Taylor & Francis) p 2706 – 2822
[3] Rudnev V 2015 *Encyclopedia of Iron, Steel, and Their Alloys vol 3* ed R Colás and G Totten (Boca Raton: Taylor & Francis) p 1807 – 1924
[4] Tensi H, Stich A, Totten G 2007 *Steel Heat Treatment Metallurgy and Technologies* ed G Totten (Boca Raton: Taylor & Francis) p 539 - 605
[5] Peković M, Jirková H, Vrtáček J, Opatová K, Janda T and Rubešová K 2018 Thermomechanical treatment of steels using device for incremental forming of rods HDQTR 30-12 *J. Kovárenství* (Brno: Svaz kováren ČR z.s.) **65** p 26-29
[6] GMT mbH 2015 Provozní návod zařízení HDQT – R 30 – 12 pro termomechanické zpracování

tyčové oceli (Berlin: Gesellshaft für metallurgishe technologie und softwareentwicklung)
[7] SENTE SOFTWARE LTD 2017 JMatPRo the Materials Property Simulation Package (Guildford: Sente Software LTD)
[8] Jirkova H, Kučerová L and Mašek B 2015 The effect of chromium on microstructure development during Q-P procesess. *J. Materials Today: Proceedings* (Amsterdam: Elsevier Ltd.) **2** p 627-630

IOP Conf. Series: Materials Science and Engineering **723** (2020) 012026 doi:10.1088/1757-899X/723/1/012026

Influence of temperature and speed of the laser head on the final structure surface hardened steel ČSN 12050 (EN 10083-2 steel 1.1191)

P. Peukert[1*] and M. Švec [1]

VÚTS, a.s., Svárovská 619, Liberec, Czech Republic

*pavel.peukert@vuts.cz

Abstract. Laser technologies are listed as unconventional methods of machining and heat treatment, which allows many advantages. In the area of heat treatment the laser technology allows a wide range of surface hardening. It provides a very fast and effective approach where the region of influence of the material is quite narrow. The method is also possible for hardening parts and materials which were not possible to process with the conventional processes. To get the required depth and quality of the hardened layer it is necessary to set up properly the laser process parameters (temperature in the process area, feed rate of the laser head, etc.), which varies for different materials. The paper presents a test for optimal process parameters for surface hardening of the material ČSN 12050 (EN 10083-2 steel 1.1191). The surface structure for temperatures of 900, 1100, 1200 and 1300°C and laser head feed rates of 0.18, 0.3 and 0.42 m/min were tested. Best results were obtained for a temperature of 1300°C and a feed rates of 0.18 m/min.

1 Introduction

The surface quality is one of the most important factors for the machine part life time, because near all fatigue cracks were initialized at the surface [1]. Therefore is taken high care to increase resistance of materials surface layers. It could be achieved by chemical-thermal treatment or hardening. Actual rise the trend for use of laser technology for surface hardening. It represents one of the top approaches of surface heat treatment. Despite to some technology limitations, the sophisticated laser hardening approach allows a lot of options, which were not possible with common processes. [2] The main advantages of laser hardening are high heat up rates (the heat up rate is around 1000°C/s), high effectivity, a thin heat effected layer and the option to process complex shaped parts. An advantage from the ecological and economical point the absence of the additional hardening medium. This laser technology uses the self-hardening process of the material. Another benefit is that the deformations caused by the laser-temperature treatment were significantly reduced. Further the laser spot size could be easily modified by changing the required lens. [3,4] Due to the common use of industrial robots and a high parameter level control, the repeatability of the hardening process is quite high. [5]

To obtain precise and homogenous hardened regions of required depth it is necessary to set up carefully laser parameters of temperature (eq. power), the feed rate of the laser head etc. [6,7,8] These parameters are different for various material. The goal of this article is a determination of optimal process parameters for different surface hardening of steel CSN 12050 (EN 10083-2 steel 1.1191).

2 Experiment of the parameter setup for laser hardening

Steel defined by CSN 12050 (EN 10083-2 steel 1.1191 – see Table 1) was used for the experiment. To perform the experiment four temperatures and three laser beam feed rates were chosen (see Table 2). The used laser was a Nd:YAG laser Laserline LDF 4000 with a maximum power of 4 kW and a wave length of 1020 – 1060 nm. Size of the laser spot was 4 by 16 mm. The sample surface temperature was permanently measured by the pyrometer LASCON (Mergenthaler GmbH) with a feedback to the laser power steering system to ensure constant temperature during the whole process. The feed of the laser head was provided by the industrial robot KUKA KR60 HA with an position accuracy of ±0.05 mm.

The determination of the depth of the hardened layer was determined from hardness gradient measurements. For this purpose was used the micro hardness tester MicroMet 6000, with an applied load of 1 kg (HV1) and a gradient step of 0.1 mm. For the evaluation for a fully hardened area was taken in account the ČSN EN 10328 that the hardness of 550 HV is an agreement as the lowest hardness for martensite. The hardness of 300 HV was then considered as the lowest level for bainitic structure for carbon steel. The depth of heat-treatment was determined by images taken by the optical microscope Nikon Epiphot 200 (LOM). Details of the microstructure of the samples was taken by the electron microscope Tescan Vega SB (SEM). The samples were prepared by a standard metallographic procedure and etched by Nital 3%.

Table 1. Nominal chemical composition of steel CSN 12050.

Chemical element		C	Mn	Si	P	S	Cr	Mo	Ni
Content	min	0.42	0.5						
(mass.%)	max	0.50	0.80	0.40	0.030	0.035	0.40	0.10	0.40

Table 2. Tested feed rates of the laser beam for selected hardening temperatures.

Velocity (m/min)	Temperature (°C)			
	900	1100	1200	1300
0.18	Sample V1	Sample V2	Sample V3	Sample V4
0.3	Sample V5	Sample V6	Sample V7	Sample V8
0.42	Sample V9	Sample V10	Sample V11	Sample V12

3 Optimization of process parameters for surface laser hardening of steel CSN 12050

3.1 Initial state of the material and the desired optimal state

The structure of the unprocessed steel CSN 12050 is shown in Figure 1. The structure is formed by ferritic and pearlitic grains. After the application of volumetric hardening (stay at a 820°C for 0,5 h and then cooled in water), a structure of martesite should appear in the whole volume like in Figure 2.

Figure 1. Image of the tested steel CSN 12050 surface (a, - LOM, ferrite – smooth light sides, pearlite – dark lamellar sides, magnification 500x; b, - SEM, ferrite – dark grey smoot sides, pearlite – grey white lamellar sides, magnification 2000x).

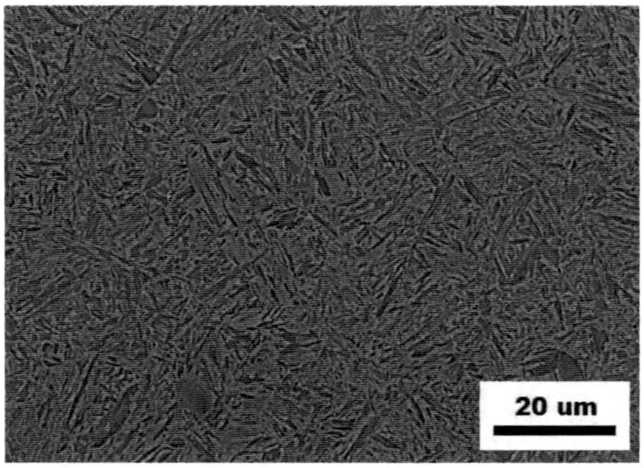

Figure 2. Detailed martensite structure of correctly hardened steel CSN 12050 (SEM, HV 20 kV, SE, magnification 2000x).

3.2 The material after laser surface hardening tests

The process of laser hardening formed in the heat treated side few layers. On the border a martensitic or bainitic structure was formed which changes with increased deepness in layers with decreasing hardness of lower bainitc and further non heat effected structures of the material. The thickness and presence of the layers is obvious from the used measurement method of the hardness gradient and also from the later image analysis done by LOM and SEM. The formation and quality of the layers is dependent on the experimentally tested parameters of surface temperature and feed rate.

For the test sample V1 with the temperature of 900°C and the lowest feed rate of 0.18 m/min the analysis get the inability of trough-hardening of the material layer to the martensitic structure. Figure 3 shows the hardness gradient measurement where the maximum reaches a value of 404 HV1. If it is considered that the lowest hardness for the bainitic structure is 300 HV, then the bainic layer reaches a depth about 400 µm. To the same conclusion leads the microscopic analysis shown in Figure 4a where clearly visible bainitic needles were recognizable.

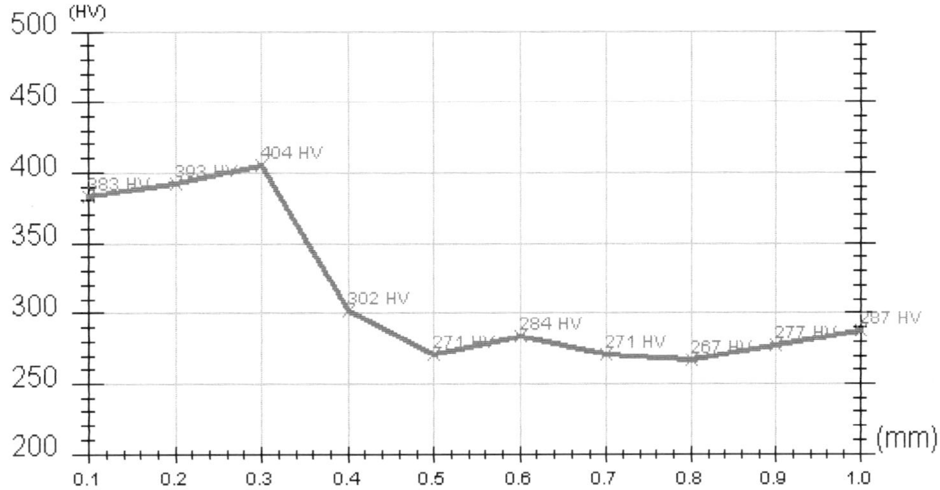

Figure 3. Record of the hardness progress through the material of Sample V1 (0.18 m/min, 900°C).

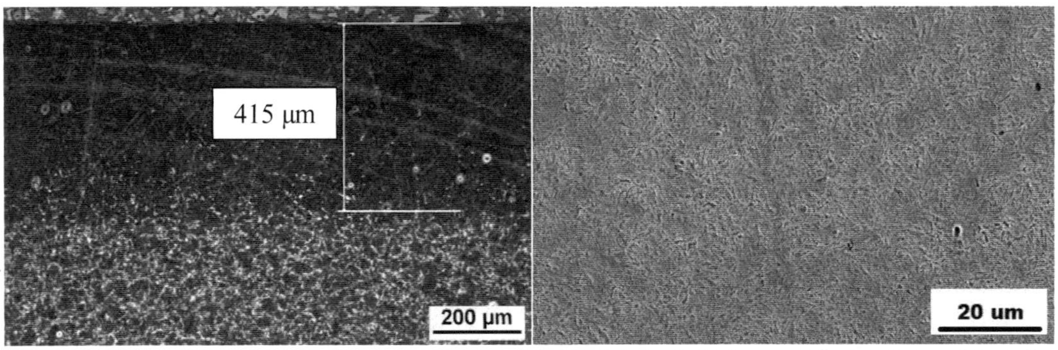

Figure 4. Depth of the heat effected region of Sample V1 (a, LOM, magnified 100x) and Detail of the bainitic hardened region of Sample V1 (b, SEM, HV 20 kV, magnified 2000x).

For the lowest feed rate of 0.18 m/min and the highest applied temperature of 1300°C the look of Figure 6 shows that a martensitic trough-hardening of a depth of 780 μm was reached. Figure 7 from the microscope shows that a bainitic layer below the martensitic was formed in a total depth of 1147 μm. In the transition between the base and heat effected region were visible martensitic needles and tiny regions of residual austenite. The highest hardness of this martensitic structure was 666 HV1.

IOP Conf. Series: Materials Science and Engineering **723** (2020) 012026 doi:10.1088/1757-899X/723/1/012026

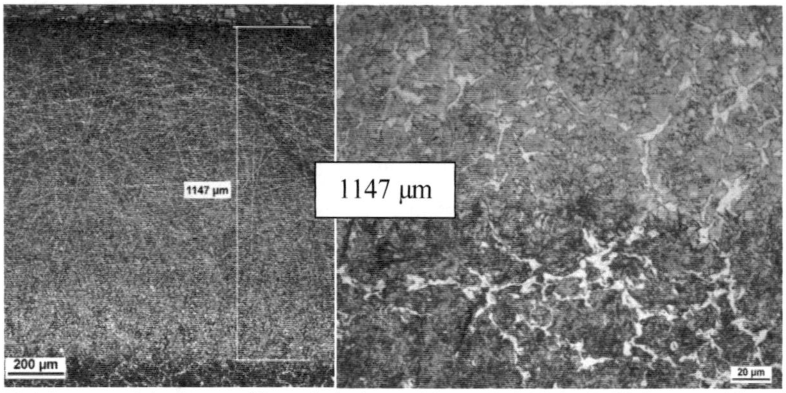

Figure 5. Thickness of the heat effected region of Sample V4 (a, LOM, magnification 100x) and Detail of the hardened region of Sample V4 (b, SEM, magnified 2000x).

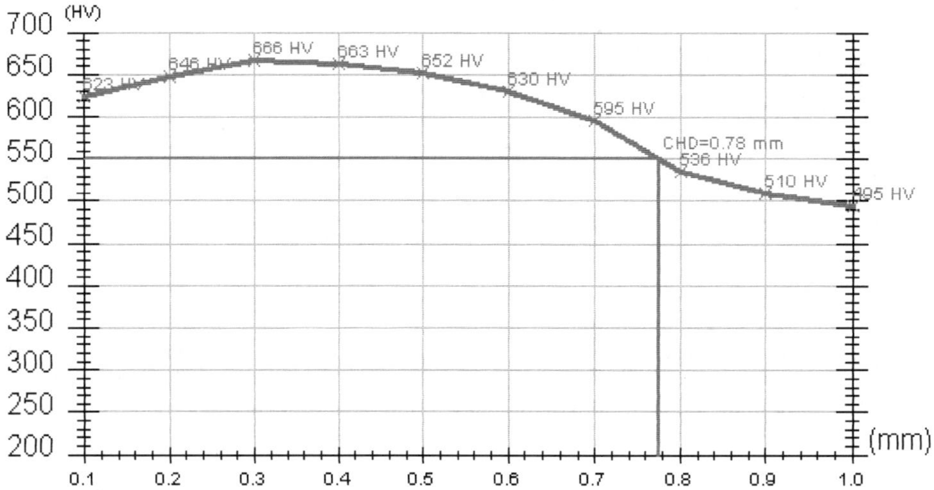

Figure 6. Record of the hardness progress through the material of Sample V4 (0.18 m/min, 1300°C).

For sample V9, the highest feed rate of 0.42 m/min and the lowest temperature of 900°C a depth of 240 µm for the martensitic layer was obtained (see Figure 7). The total hardened layer depth was 311 µm (see Figure 8). The highest value of hardness reached for sample V9 was 598 HV1. From Figure 8 is obvious that the austenitiyation process of the base material has not occurred in the heat effected region. Figure 8a shows a martensitic hardened layer with white pieces of the original ferrite. For some regions are noticeable location of lighter pieces between martensite, which represent remaining austenite. In the transition region between the hardened layer and the base bacterial the insufficient austenization is even more distinguishable.

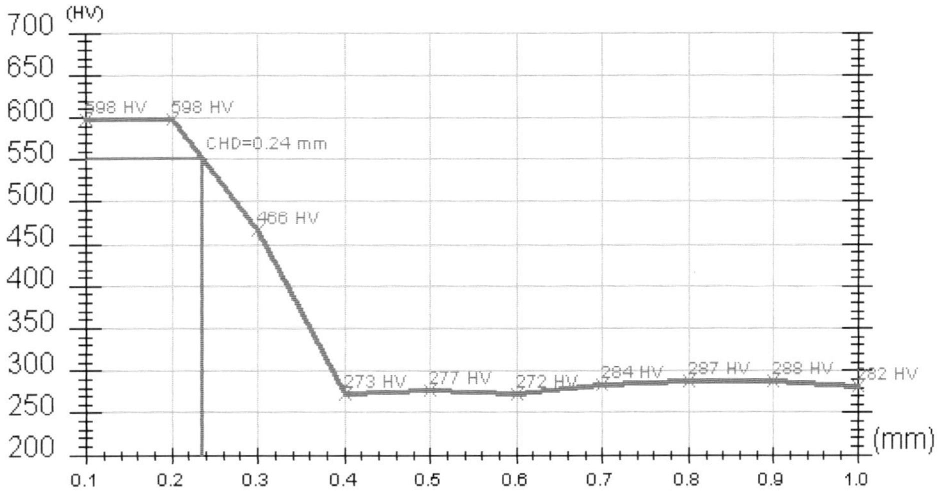

Figure 7. Record of the hardness progress through the material of Sample V9 (0.42 m/min, 900°C).

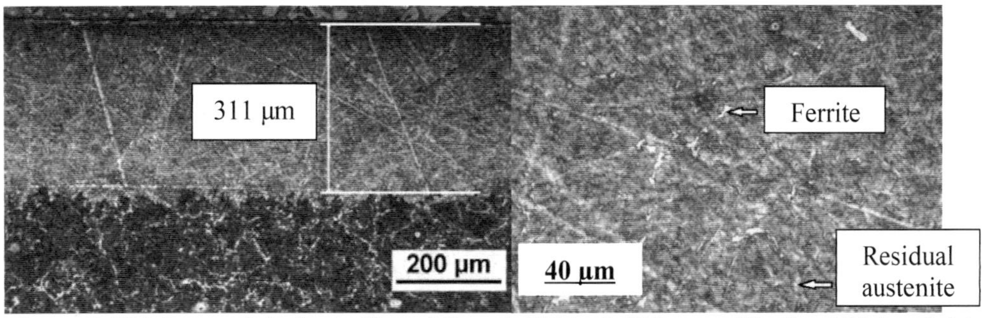

Figure 8. Thickness of the heat effected Sample V9 (a, LOM, magnification 100x) and detail of the martensitic structure of the hardened region of Sample V9 with recognizable ferritic pieces and regions with residual austenite (b, LOM, magnification 500x).

For Sample V12, the highest feed rate of 0.42 m/min and the highest applicable temperature of 1300°C, was the steel martensitic hardened to a depth of 1070 μm (see Figure 9). The martensitic layers passes to bainitic layers which reached a depth of 1202 μm in the hardened material (see Figure 10). The highest measured hardness reached 674 HV1.

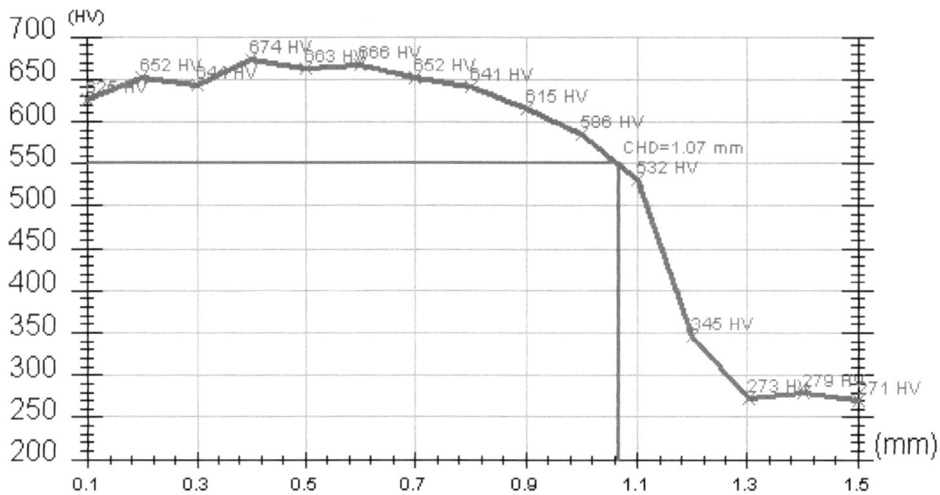

Figure 9. Record of the hardness progress through the material of Sample V12 (0.42 m/min, 1300°C).

Figure 10. Thickness of the heat effected region of Sample V12 (LOM, magnification 100x).

Final measured values from all tested samples were summarized in Table 3 for trough-hardened depth, in Table 4 for heat effected depth and Table 5 for maximum reached values of hardness.

The tables serve as a guideline for optimal choice of the laser technology parameters for resulting the quenching parameters, time consumption to a quality factor.

Table 3. Final depths of trough-hardening of tested samples for chosen laser technology parameters.

Depth of trough-hardening (µm)				
Velocity (m/min)	Temperature (°C)			
	900	1100	1200	1300
0.18	**-**	**630**	**710**	**780**
0.3	**210**	**600**	**710**	**690**
0.42	**240**	**730**	**870**	**1070**

Table 4. Final heat effected depths of tested samples for chosen laser technology parameters.

	Heat effected depth (μm)			
Velocity (m/min)	Temperature (°C)			
	900	1100	1200	1300
0.18	**415**	**824**	**966**	**1147**
0.3	**340**	**731**	**870**	**942**
0.42	**311**	**862**	**985**	**1202**

Table 5. Maximum reached hardness of the tested samples for chosen laser technology parameters.

	Maximum reached hardness (HV1)			
Velocity (m/min)	Temperature (°C)			
	900	1100	1200	1300
0.18	**404**	**652**	**660**	**666**
0.3	**570**	**610**	**646**	**577**
0.42	**598**	**660**	**671**	**674**

For comparison it is noted that the original material without any other processing was measured 277 HV1.

4 Summary

From tested laser technology parameters at tests samples of steel CSN 12050 was obtained that for lower surface temperatures original ferrite grain remnants had remained. Their presence is given due to insufficient temperature or time of austenitization of the matrix.

By the gradient-hardness measurement was found that the martensitic layer had the highest depth for the feed rate 0.42 m/min and applied temperature of 1300°C eq. sample V12. For these laser parameters was obtained the highest hardness of the processed area.

The lowest decrease of hardness with increasing distance from the hardened surface was detected for the Samples V4 and V8, the lower feed rate of 0.18 and 0.3 m/min and the highest temperature of 1300°C. For Sample V4 was the hardness only slightly lower than for Sample V12. The depth of the martensitic region was for Sample V4 only 780 μm, but the sum of the martensitic and bainitic layers was for Sample V4 and V12 similar.

From the evaluation of measurements and further analysis the optimal laser-hardening technology parameters for steel CSN 1205 were a feed rate of the laser head/beam of 0.18 m/min and an applied surface temperature of 1300°C – sample V4. This sample had not the highest hardness of all tested samples, but it is still quite high and the decrease of hardness was detected as relative evenly declining.

Acknowledgements

The work was supported within the frame of the TIP Project under the FR-TI1/604 grant. Our acknowledgment belongs to all donors that helped us to reach the results of the work.

Authors also thank to Laser Application Center of VUTS a.s. for preparing of laser quenched samples.

References

[1] Veles P 1989 Mechanické vlastnosti a skúšanie kovov (Bratislava: Alfa Publishing)

[2] Guan Y H, et al. 1997 The prediction of the Mechanical Properties of Metal during laser quenching. *J. Journal of Materials Processing Technology 63* p 614
[3] Zhaoyun Ch, et al. 2012 Microstructure and hardness investigation of 17-4PH stainless steel by laser quenching. *J. Materials Science and engineering A* p 536
[4] Lee J H, et al. 2009 Laser surface hardening of AISI H13 tool steel. *J. Trans. Nonferrous Met. Soc. China 19* p 309
[5] Martínez S, et al. 2012 Laser hardening process with 2D scanning optics. *J. Physics Procedia 39* p 309
[6] Qingbin L, et al. 1999 Experimental study of the laser quenching of 40CrNiMoA steel. *J. Journal of Materials Processing Technology 88* p 77
[7] Yao Ch, et al. 2010 Study of the softening in overlapping zone by laser-overlapping scanning surface hardening for carbon and alloyed steel. *J. Optics and Lasers in Engineering* p 20
[8] Jiang J 2011 Discrete lase spot transformation hardening of AISI 01 tool steel using pulsed Nd:YAG laser. *J. Surface and coatings technology* p 5156
Kirschner V, Švec M, Bruthansová R, Dvořák L 2014 The effect of the laser spots overlap on the structure during square parts quenching. (5th International Conference Recent Trends in Structural Materials) *IOP Conf. Series: Matterials Science and Engineering 461 (2019)*

IOP Conf. Series: Materials Science and Engineering **723** (2020) 012027 doi:10.1088/1757-899X/723/1/012027

Characterization of Low Cycle Fatigue Parameters of Rotor Steel using Sub-sized Specimens

R Procházka[1, 2] and P Konopík [1]

[1] COMTES FHT a.s., Průmyslová 995, 334 41 Dobřany, Czech Republic

[2] University of West Bohemia, Faculty of Mechanical Engineering, Department of Material Science and Technology, Univerzitní 8, Pilsen, Czech Republic

E-mail: rprochazka@comtesfht.cz

Abstract. The paper is dealing with strain controlled cyclic testing method employing a novel strain-control technique based on digital image correlation (DIC) in low-cycle fatigue (LCF) region. The cyclic behaviour of 22CrMoNiWV 8-8 rotor steel was investigated on sub-sized round specimens with a diameter of 2 mm in gage length and total length of 20 mm. These results were compared with results obtained using conventional specimens designed in accordance with the ASTM E606 standard. The attention was paid to confirm the suitability of the proposed sub-sized geometry, testing set up and procedure. The test procedure and results obtained enabled to record hysteresis loops, construct Manson-Coffin curves and obtain cyclic material properties in LCF region.

1 Introduction

There is still growing demand for cyclic properties of in service products. Many testing methods applied in this field are non-destructive (NDT) nature. But base on NDT results, which brings an advantage in some ways and disadvantages in others. The main issue is the accuracy in the mechanical properties evaluation which decreases with the increasing number of correlations and also there is a need for a large database of various materials. The problem becomes larger in case of cyclic loading and lifetime prediction. But there is no easy way to do that in case of low cycle fatigue test (LCF) where cyclic parameters are needed. To avoid using correlation relationships between two variables, the miniaturized testing technique using sub-sized specimens was developed [1] for the direct mechanical properties evaluation of LCF parameters. The idea was to use a size miniaturization geometry which is related to ASTM standard [2] and use the same loading conditions. That brings some simplification in evaluation processes and helps to extend the applicability of this method in case of acute shortage of experimental material. Very good results obtained with the sub-sized tensile specimens were confirmed in [3-6] and was already used for a number of applications like characterization of ECAP processes [7] or high strength chromium steel with hardness above 800 HV [8]. Furthermore, some studies were performed also with the use of small scale of LCF specimens [9-10]. But still here is an influence of failure probability caused by a presence of defects in material which depends on experimental volume of the material [11]. This factor is still an open question. A non-contact extensometer was employed for the strain controlled cyclic deformation. The Digital Correlation System (DIC) Mercury RT [12] provides high measuring accuracy of measuring volumes from some square milimeters to square meters depending on the camera settings.

The aim of the present work was to analyze the impact of size effect on cyclic material properties in LCF region obtained employing conventional and sub-sized specimens.

2 Materials and Methods

The rotor steel 22CrMoNiWV 8-8 was used as the experimental material. Standard quasi-static tensile tests and strain controlled low cycle fatigue (LCF) tests were carried out to obtain its LCF parameters employing the conventional and the sub-sized specimen geometry. All test were conducted at room temperature.

2.1 Material description

As the experimental material, a disc of steel 22CrMoNiWV 8-8 has been taken under the investigation. The disc was made by cutting a 50 mm slice from a 605 mm diameter rod. The microstructure of the rotor steel is composed of tempered bainite with an average grain size of 70 μm measured by the mean linear intercept method in accordance with ASTM E112 standard. The microstructure of investigated steel is depicted in **Figure 1** and chemical composition is summarized in **Table 1**.

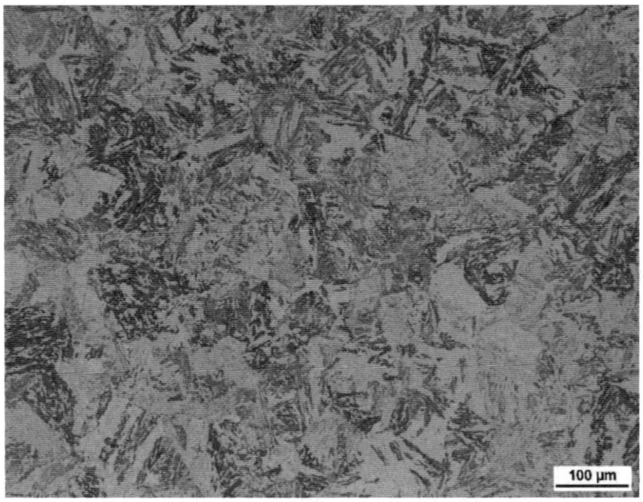

Figure 1. Microstructure of rotor steel 22CrMoNiWV 8-8, 100x.

Table 1. Chemical composition by weight % of 22CrMoNiWV 8-8 steel.

C	Si	*Mn*	Ni	*P*	S	*Cr*	*Mo*
0.23	0.038	0.67	0.68	0.004	0.001	2.07	0.85

2.2 Experimental methods

At the beginning of the experiment work, the differences in mechanical properties of the large steel disc were investigated by means of conventional tensile tests in accordance with European standard. The extraction of all tensile specimens was done from two sections close to the surface and the center of the disc in the tangential direction. Then, the conventional and sub-sized LCF specimens were extracted out in the same orientation as in the case of tensile tests. The sample preparation consists of three major steps in machining such as turning, longitudinal grinding and polishing in order to achieve a specific surface roughness. The strain-controlled LCF tests were conducted on conventional and sub-sized specimens. Based on LCF results, cyclic parameters were obtained.

2.2.1 Tensile tests. Round test bars with initial gage length of 50 mm and gage diameter of 10mm were prepared. Tests were conducted on electro-mechanical testing system Zwick Roell with a load capacity of 250 kN. Samples were pulled at a constant strain rate of 0.0004 s^{-1}. The tests were performed according to ČSN EN ISO 6892: Metallic materials – Tensile testing – Part 1: Method of test at room temperature. A longitudinal deformation of a loaded specimen was measured by means of mechanical extensometer with initial gauge length of 25 mm. Dimensions of the samples were measured prior to testing and after the test for determination of stress and plastic behavior quantities determination (Young´s modulus - E, Proof stress - YS, Ultimate Tensile Strength - UTS, Elongation – A and Cross Section Reduction - CSR).

2.2.2 Strain-controlled LCF tests. Tests were conducted under variable strain controlled loading conditions with the cyclic asymmetry coefficient R = - 1 and the sinusoidal shape. The aim was to cover the total strain amplitudes from until 1.5% if possible. The lowest strain level is given by the number of cycles which specimen not survived before reaching the maximum count of 100 000 cycles. As the reference values, the conventional specimen 8 mm in diameter was used in accordance with ASTM standard E606. The specimen geometry is illustrated in ***Figure 2***.

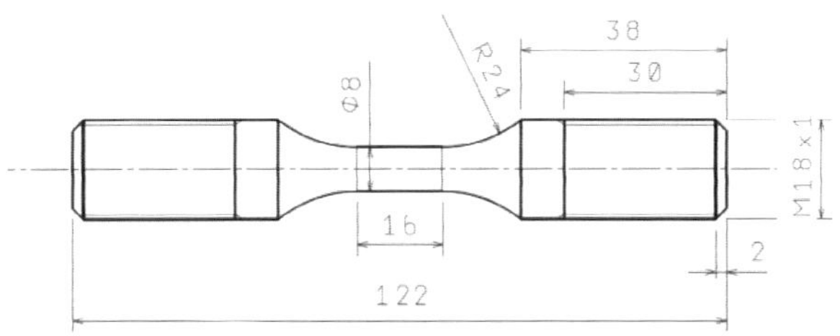

Figure 2. Conventional specimen geometry in accordance with ASTM standard E606, D8.

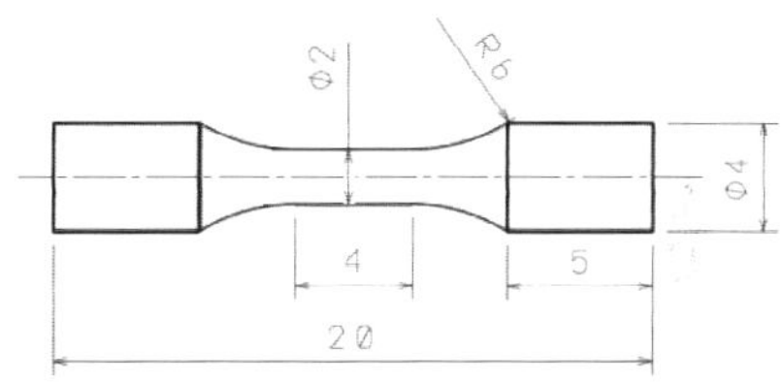

Figure 3. The draft of sub-sized specimen geometry, D2.

The sub-size specimen was designed to meet some recommendations from above mentioned standard as much as possible. The attention was also paid to design the reduced section which would fit in case of on-site sampling where the extracted sample has dimesions 4x20x20 mm. The draft of the sub-sized specimen geometry with 2 mm diameter is depicted in **Figure 3**.
In case of the sub-sized specimen, the initial gage length *L1* of the optical extensometer was set to 3 mm. Prior to testing, the speckle pattern was spreyed on specimen surface in order to ensure deformation measurements using 2D version of DIC system, see **Figure 4**.

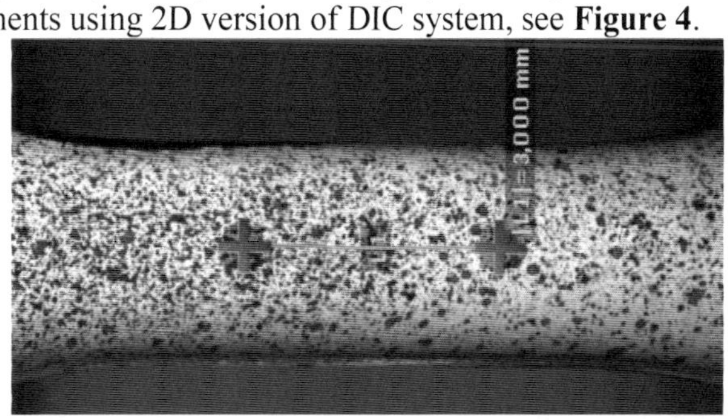

Figure 4. Sub-sized specimen with 3mm initial gage length of the optical extensometer.

Further evaluation was done based on Basquin's law (1) where stress amplitude σ_a depends on number of cycles to failure N_f and fatigue strength exponent b and fatigue strength coefficient σ_f'.

$$\sigma_a = \sigma_f'(2N_f)^b \tag{1}$$

Mansson-Coffin relation in plastic deformation ε_{ap} is described by equation (2) where ε'_f is coefficient of ductility in fatigue and ***c*** is ductility exponent.

$$\varepsilon_{ap} = \varepsilon_f'(2N_f)^c \tag{2}$$

The total amplitude of the deformation ε_{at} can be expressed as

$$\varepsilon_{at} = \frac{\sigma'}{\bar{E}} \cdot (2N_f)^b + \varepsilon' \cdot (2N_f)^c \tag{3}$$

where $\bar{E}$ is the Young modulus of elasticity. The elastic modulus was determined from two unloading parts of a saturated hysteresis loop in both tension E_T and compression E_C sides. E_T and E_C were determined in accordance with ČSN ISO 12106 [13]. Young modulus was used in order to refine the determination of the plastic deformation when the total strain is decomposed into elastic and plastic components. The cyclic deformation curve can be estimated based on equation (4) where ***K'*** is cyclic hardening coefficient and ***n'*** is cyclic hardening exponent.

$$\sigma_a = K' \cdot \varepsilon_{ap}{}^{n'} \tag{4}$$

The material cyclic properties were estimated in accordance with geometrical meaning of formula (1), (2) and (4). The parameters were determined based on power law regression analyses of variables in a fully-logarithmic scale. Moreover the transition life N_T is calculated based on relationship (5).

$$N_T = \frac{1}{2} \cdot \left(\frac{\sigma'_f}{\bar{E} \cdot \varepsilon'_f}\right)^{\frac{1}{c-b}} \tag{5}$$

3 Results and Discussion

Tensile test results did not show a differences in tensile properties over the large steel bar cross section area. The values in **Table 2** are represented by mean and standard deviation σ_{SD} of three repetitions.

Table 2. Tensile test results of 22CrMoNiWV 8-8 steel.

	E	σ_{SD}	YS	σ_{SD}	UTS	σ_{SD}	A	σ_{SD}	CSR	σ_{SD}
	GPa	GPa	MPa	MPa	MPa	MPa	%	%	%	%
Surface	209.3	1.3	679.0	2.9	785.2	0.8	20.7	0.1	70.5	0.1
Center	210.3	0.8	676.7	3.1	782.0	0.7	20.9	0.1	70.5	0.1

The evaluation of $\bar{E}$ was done on saturated hysteresis loops depicted in **Figure 5**. Each strain level corresponds to one loop of each specimen type. In this picture, 7 points can be seen which are represented by total strain amplitudes ε_{at} in percentage 0.25, 0.4, 0.6, 0.8, 1.0, 1.2 and 1.5. Relationships among variables have been found and the material cyclic properties of conventional and sub-sized specimens were estimated based on formula (1), (2), (3), (4) and (5). For both specimen geometries the results are shown in **Table 3**.

Table 3. Cyclic material properties of 22CrMoNiWV 8-8 steel.

	$\bar{E}$	σ'f	b	ε'f	c	K'	n'	N_T
	GPa	MPa	-	-	-	MPa	-	-
D2	196.2	787.0	-0.0419	1.13	-0.7324	767.5	0.0541	1764
D8	193.1	797.6	-0.0437	0.90	-0.7146	711.9	0.0584	1525

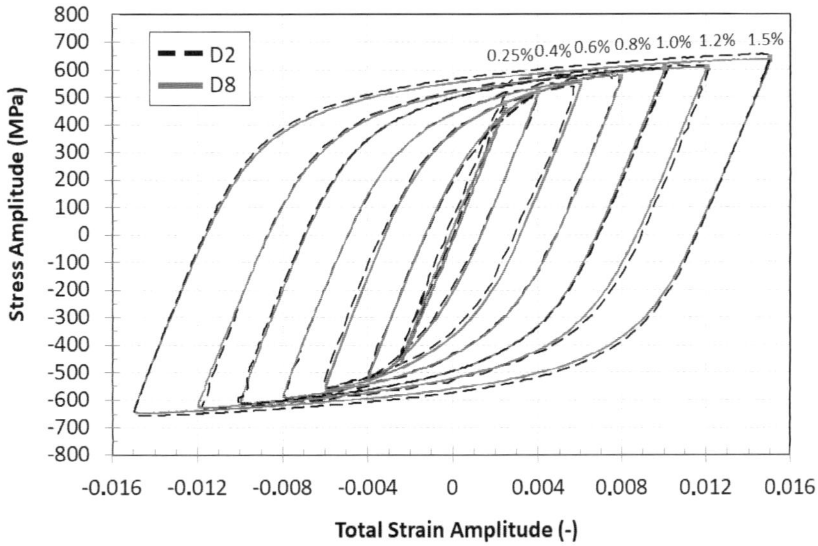

Figure 5. Hysteresis loops at saturation.

Using Manson-Coffin-Basquin equation (4), the fatigue life behaviour was estimated and the fully-logarithmic plot of the strain amplitude versus the number of cycles to failure was done in **Figure 6**.

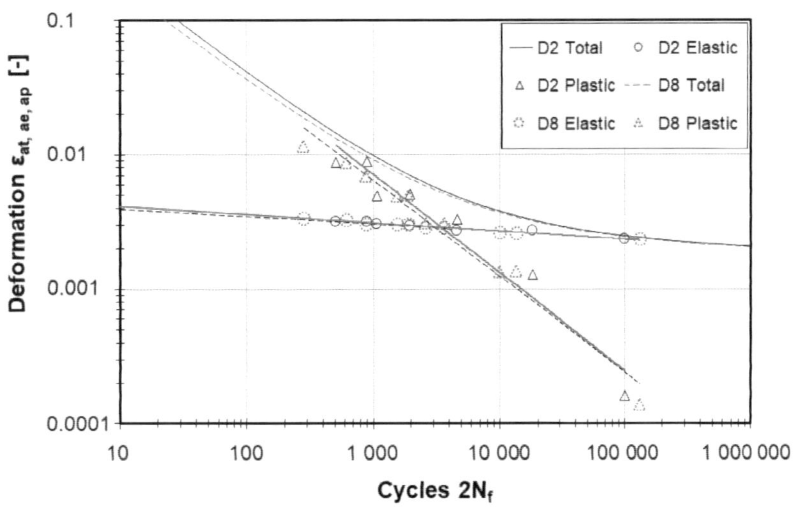

Figure 6. Manson-Coffin curves.

4 Conclusion

It was shown that the tensile properties of experimental material are homogeneous despite of large diameter of steel disk, see tensile test results in **Table 2**. Based on these results, the negligible impact of the sampling location on LCF test results can be noted. For decomposition of total strain the $\overline{\boldsymbol{E}}$ modulus was used (3) due to the fact that its value is decreasing along with the rising total strain amplitude. $\overline{\boldsymbol{E}}$ was determined from loading and unloading part of a saturated hysteresis loops, **Figure 5**, where the range for this data set is from 185 to 208 GPa. Loops obtained using both specimen types look very similar throughout the measured strain range. The average modulus of conventional ($\overline{\boldsymbol{E}}$ = 193.1 GPa) and sub-sized ($\overline{\boldsymbol{E}}$ = 196.2 GPa) specimens was found. Slight differences can be observed between both results in MC curves, **Figure 6**, and cyclic material properties, **Table 3**. The differences in these properties are not significant but they are observable in elastic region (E 1.5%; σ'_f 1.3%; b

4.5%). In plastic region the differences are greater (ε'_f 25.0%; c 2.5%; K' 7.8%; n' 7.3%). The shift in transition life N_T (5) is 15.6%. Increasing disparities between both obtained results appearing at higher total strains above ε_{at} 1% which may be associated with different buckling response of both test setups and specimen geometries to increasing applied stress.

The possibility of using the sub-sized specimens to obtain cyclic material properties by LCF tests was presented in this paper. The results obtained with sub-sized specimens appear to be sufficient for material description in fatigue life behavior equal to conventional ones without any correlation. The results of cyclic material properties once more confirmed that the sub-sized geometries are comparable to the conventional ones in LCF region.

Acknowledgement
This study was created by project Development of West-Bohemian Centre of Materials and Metallurgy No.: LO1412, financed by the MEYS of the Czech Republic.

References

[1] R. Prochazka, J. Dzugan: Strain controlled cyclic tests on miniaturized specimens, *IOP : Material Science and Engineering, 179, 2017, DOI: 10.1088/1757-899X/179/1/012060*
[2] ASM International E606-04 (2004): Standard Practice for Strain-Controlled Fatigue Testing
[3] Kumar K, Madhusoodanan K and Rupani B 2006 Miniature Specimen Technique as an NDT Tool for Estimation of Service Life of Operating Presurre Equipment *International Conference &Exhibition on Pressure Vessel and Piping*
[4] Konopík P, Dzugan J and Prochazka R 2013 *Mat. Sci. and Tech. Conf. and Exh.* Evaluation of local mechanical properties of steel weld by miniature testing technique, pp 2404-11
[5] Dzugan J, Konopik P., Rund M and Prochazka R 2015 *ASME Pressure vessels and piping conf. - 2015* Determination of local tensile and fatigue properties with the use of sub-sized specimens, Volume 1A: Codes and Standards (New York: Amer. Soc. Mech. Eng.)
[6] Cyclic Stress-Strain Curve for Low Cycle Fatigue Design and. Nishimura, A., Nogami, E. and Wakai, S. Japan : *National Inst. for Fusion Science, 2012. NIFS--1055.*
[7] Trojanova Z, Halmesova K, Dzugan J, et al. Micro-Tensile Behavior of Mg-Al-Zn Alloy Processed by Equal Channel Angular Pressing (ECAP), MATERIALS, 2018, Volume: 11 Issue: 9 Article Number: 1644
[8] Jirková J, Rubešová K, Konopík P and Opatová K 2018 Effect of the Parameters of Semi-Solid Processing on the Elimination of Sharp-Edged Primary Chromium Carbides from Tool Steel, *Metals - Open Access Metallurgy Journal* 8(9):713
[9] Dzugan J, Konopik P, Podany P, Prochazka R, Rund M; Lewandowski J. J.; Seifi M.: Effects of thickness and orientation on the small scale fracture behaviour of additively manufactured Ti-6Al-4V. *Materials Characterization, Vol. 143, 2018, https://doi.org/10.1016/j.matchar.2018.04.003*
[10] Nogami, S. and Nishimura, A. Development of fatigue life evaluation method using small specimen. *Journal of Nuclear Materials. 2013, Vol. 441, ISSN: 0022-3115*
[11] Tomaszewski T, Sempruch J and Piątkowski T: Verification of selected models of the size effect based on high-cycle fatigue testing on mini specimens made ofen AW-6063 aluminium alloy. *Journal of Theoretical and Applied Mechanics. 2014, Vol. 52.*
[12] Stamborska M, Lapin J, Bajana O and Losertova M 2015 Tensile Deformation Bahaviour of Ferritic-Pearlitic Steel Studied by Digital Image Cerrelation Method *Kovove materialy*
[13] ČSN ISO 12106 (2018): Metallic materials – Fatigue testing – Axial-strain-controlled method

Effects of Deformation on the Behaviour of Chromium Carbides in Tool Steel Studied by Use of Semi-Solid Forming

K Rubesova, M Pekovic, H Jirkova, M Bystriansky

Faculty of Mechanical Engineering – Regional Technological Institute, University of West Bohemia, Univerzitní 8, 306 14 Pilsen, Czech Republic

E-mail: krubesov@rti.zcu.cz, pekovicm@rti.zcu.cz, hstankov@rti.zcu.cz, mbyst@rti.zcu.cz

Abstract: Convetional technology is mainly used for processing parts where high hardness, although conventional treatment of tool steels is ordinarily used in industrial practice, engineers continue to seek new procedures to rid tool steels of objectionable primary sharp-edged chromium carbides, which impair toughness. Fortunately, research into metal forming yielded new methods of modifying the microstructure of hypereutectoid steels. Using these methods, mechanical properties can be improved by virtue of eliminating objectionable sharp-edged carbides. These carbides resist dissolution and their size and shape make them undesirable microstructural constituents. Although they do improve wear resistance of the matrix, they also impair toughness and may act as stress concentrators. The microstructures produced by a sequence involving semi-solid processing and subsequent forming operations were different from conventional semi-solid-processed microstructures. In the former microstructures, the prior carbide network was broken up, dispersed, and became a strengthening constituent. Brittleness which plagues materials with prominent carbide networks was thus removed.
The experimental material used in this study was X210Cr12 tool steel. Two semi-solid processing temperatures were used: 1240°C and 1260°C. There were two holding times: 30 minutes and 60 minutes. Another variable was the number of reductions. The resulting microstructures were examined with respect to individual sequences and reductions applied. Detailed microstructure analysis was carried out using a scanning electron microscope (SEM). Chemical compositions of carbides were determined by means of EDS (Energy Dispersive X-ray Spectroscopy). Microhardness was measured in order to gather comprehensive materials data. The purpose of the study was to identify trends, if any, in microstructural property evolution in response to the above-described processing sequence.

Keywords: semi-solid processing, primary chromium carbides, tool steel, carbide refinement

1 Introduction

Unconventional treatment of material by heating to the semi-solid state has been known since the 1970s. At that time, it was primarily used for low-melting materials. Most often, they were aluminium and magnesium alloys [1, 3]. Years later, the process was adopted for higher-melting materials, typically steels [2-4]. For the desired microstructure to be obtained, a liquid fraction interval of 10%-60% must be maintained. Microstructures produced by semi-solid processing typically consist of polyhedral austenite grains embedded in a ledeburite network. In terms of the effect of the network on mechanical properties, this configuration is undesirable as the network is very brittle. It is therefore necessary to crush, refine and disperse its fragments uniformly across the part. The microstructure obtained in this manner will be fine-grained and homogeneous, containing a fine dispersion of carbides and showing improved mechanical properties.

Semi-solid processing involves heating to high temperatures. This was one of the reasons for using this process for eliminating large sharp-edged chromium carbides. These carbides greatly improve wear and creep resistance but reduce toughness of the material due to their large size. Typically, they are found in high-chromium tool steels. Tool steels produced by a conventional metallurgical route contain mostly primary carbides of the M_7C_3 type. These are very difficult to eliminate by conventional processes. Consequently, some steels must be made by powder metallurgy [5-7]. To overcome these difficulties, a sequence was developed which combines semi-solid processing and subsequent forming. It can eliminate sharp-edged chromium carbides by converting them into a very fine dispersion within a martensitic matrix. This sequence was tried on X210Cr12 tool steel. It was first applied to small specimens 30 mm in diameter and 55 mm in height [8-10]. Following the verification of all key parameters, the sequence was carried out again on stock which was almost three times larger. Specimens of this size were expected to provide enough material for manufacturing real-world products

2 Experimental program

The material which was studied in these experiments was X210Cr12 tool steel, a typical representative of a group of steels with large sharp-edged chromium carbides. Given its chemical composition and a relatively wide solidification range, this steel is suitable for semi-solid processing (Table 1). It was obtained in the form of annealed bars. The as-received microstructure contained primary sharp-edged chromium carbides and very fine cementite in a ferritic matrix (Figure 1). In order to be able to specify the values of key variables for this sequence, one has to identify the relevant processing window, i.e. the solidification range and liquid fractions at particular temperatures [1, 3]. Another variable was the temperature at which chromium carbides dissolve. As the rate of cooling from a forming temperature controls the decomposition of metastable austenite, it becomes another variable of importance. The heating temperatures and other parameters of the experimental sequence were identified using JMatPro software [11].

Table 1. Chemical composition of experimental steels.

X210Cr12	C	Cr	Mn	Si	Ni	P	S
Content [%]	1.8	12	0.3	0.35	max 0.5	max 0.03	max 0.035

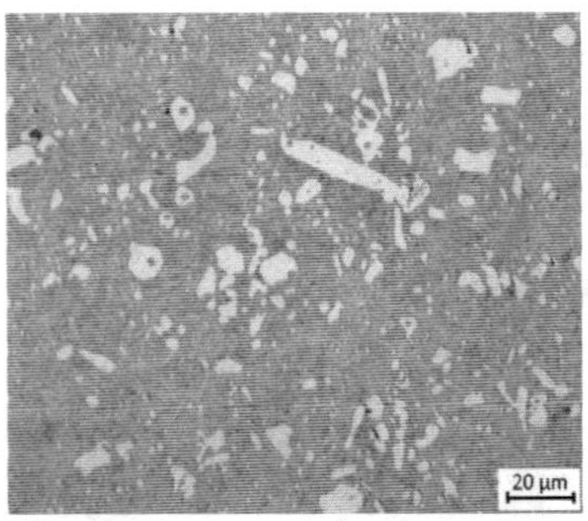

Figure 1. As-received condition of X210Cr12 steel, confocal micrograph.

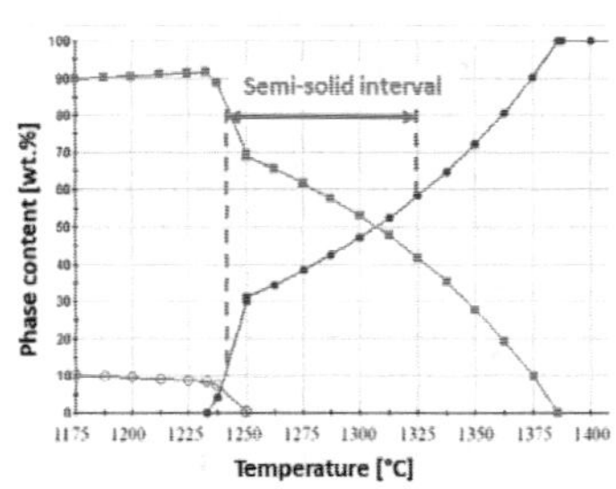

Figure 2. The calculated solidification range (semi-solid interval) for heating the stoc.k

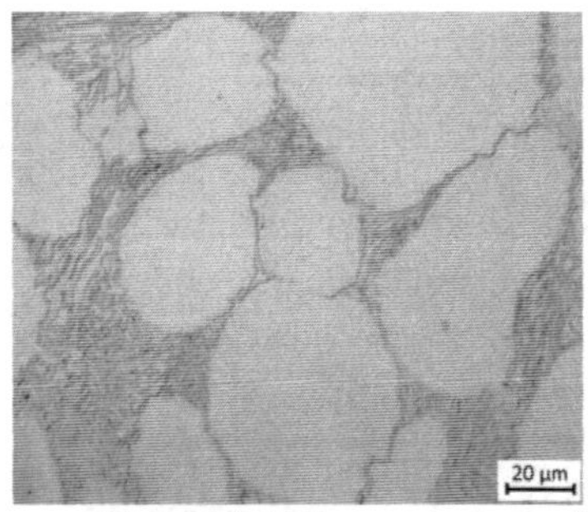

Figure 3. Microstructure of material which has been heated to semi-solid condition, no deformation was applied, light micrograph.

Calculations performed using JMatPro revealed that up to 758°C, the material retains a ferritic-cementitic microstructure. Incipient melting takes place at 1225°C (Figure 2). Melting is finished at 1373°C. Chromium carbides should dissolve completely at 1255°C.

In earlier experiments, the sequence was tried and fine-tuned on smaller specimens 30 mm in diameter [9, 12]. The aim of the present experimental programme was to transfer the procedure to larger stock in order to obtain a larger amount of uniform material. It was essential to bring the material into the semi-solid state in order to obtain an optimal liquid fraction. After cooling to forming temperature, appropriate deformation was to disintegrate and uniformly disperse ledeburite network (Figure 3), (which forms upon semi-solid processing) and initiate dynamic recrystallization of austenite [9, 12].

In this experimental programme, several heating temperatures and amounts of reduction were used. Specimens were heated in an air furnace with no protective atmosphere and formed in CKW 1000 hydraulic press.

The resulting microstructures were examined using optical (OM) and scanning electron microscopy (SEM). Mechanical properties were determined by means of hardness and microhardness testing.

2.1 Semi-solid processing and subsequent thermomechanical treatment

Experimental specimens were 75 mm in diameter and 150 mm in length. For ease of handling of partially-melted stock, the specimens had to be inserted into containers of SJ355 steel, which has a higher melting point.

A bar of X210Cr12 tool steel 65 mm in diameter was inserted into a low-carbon steel tube which was sealed with lids on both ends and welded closed. Four experimental sequences were designed, involving two different temperatures and several other variables (Table 2).

The heating temperature for the first two sequences was 1240°C and the time at temperature was 30 minutes. Soaking was followed by water quenching for 20 seconds, after which the stock was placed into a furnace at 1080°C and held for 10 minutes. Temperature of 1080°C is the recommended forging temperature, as set forth in the material data sheet for X210Cr12. After holding in a furnace, the stock was formed between flat dies in a hydraulic press. In sequence 1, three reductions were used. The first was a reduction of the diameter by 20 mm along the entire length of the workpiece. The second reduction was by 20 mm, but only along 2/3 of the workpiece length. The last reduction, by another 20 mm, only spanned one third of the workpiece length at its end. This produced three different regions for comparing the impact of deformation within sections reduced to different thicknesses on refinement of carbides and disintegration of the carbide network (Table 2).

In sequence 2, only two reductions were applied. The first one, to half thickness, i.e. to 37.5 mm, covered the entire length of the workpiece. The second one, to 18 mm, only spanned one half of the length.

Sequence 3 involved a longer time at the semi-solid process temperature 1240°C: instead of 30 minutes, it was 60 min. In addition, the time at 1080 °C, the forging temperature, was longer than in previous sequences. Deformation was imparted in a manner identical to sequence 1. In the last sequence, no. 4, the semi-solid process temperature was increased above the previous level, to 1260°C, and the time at temperature was extended to 60 minutes. The time at 1080 °C was extended as well: from 10 minutes to 30 minutes. Deformation was imparted in a manner identical to sequence 1 (Table 2).

IOP Conf. Series: Materials Science and Engineering **723** (2020) 012028 doi:10.1088/1757-899X/723/1/012028

Table 2. Heating temperatures, times, and deformation parameters.

Sequence	Semi-solid process temperature [°C]/ time [min]	Cooling method	Forming temperature [°C] /Time at temperature [min]	Number of reductions [-]	Mode of reduction	Cooling method
1	1240/30	Quenched in water/20s	1080/10	3	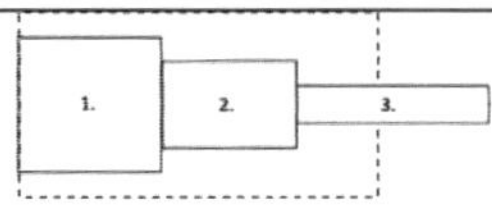	Quenched in oil to RT
2				2	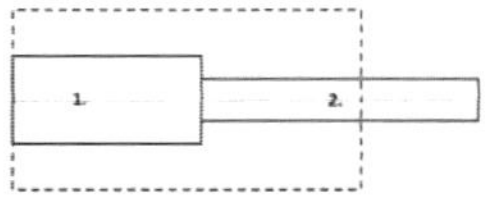	
3	1240/60		1080/30	3	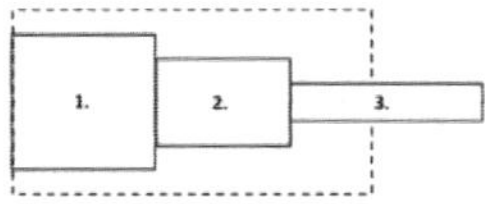	
4	1260/60			3	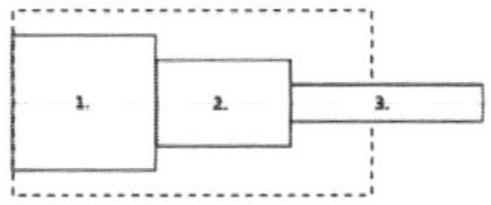	

3 Discussion and Results

Microstructures of the workpieces were examined on longitudinal metallographic sections, with reference to their portions which experienced different amounts of reduction.

Sequence 1 involved a temperature of 1240°C. The resulting specimens contained austenite and a fine dispersion of chromium carbides and cementite particles. It was thus clear that the temperature was too low and failed to cause dissolution of all primary chromium carbides. If melting occurred, it was probably limited to grain boundaries. In the single-reduction portion of the workpiece, the carbide network was already fragmented, with some fragments dispersed through the austenitic matrix (Figure 4). With increasing magnitude of deformation, austenite grains became finer. In the portion subjected to two reductions, the grain size was approximately 20 μm (Figure 5). Where the largest deformation was applied, the austenite grains were smaller than 10 μm (Figure 6). Larger strains dispersed chromium carbides and cementite more effectively.

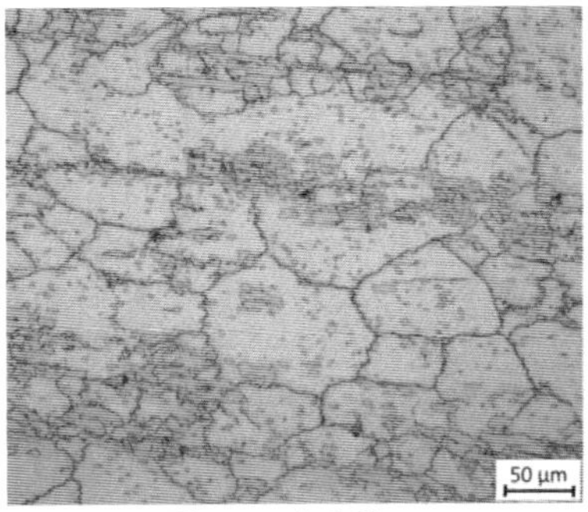

Figure 4. Microstructure produced by heating to semi-solid state and subsequent deformation - sequence 1 - reduction 1.

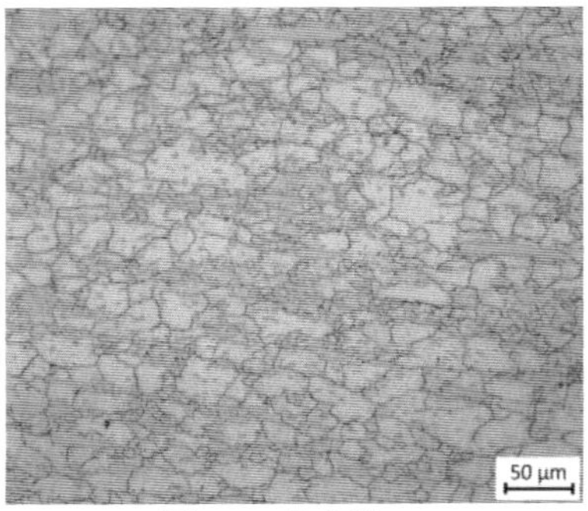

Figure 5. Microstructure produced by heating to semi-solid state and subsequent deformation - sequence 1 - reduction 2.

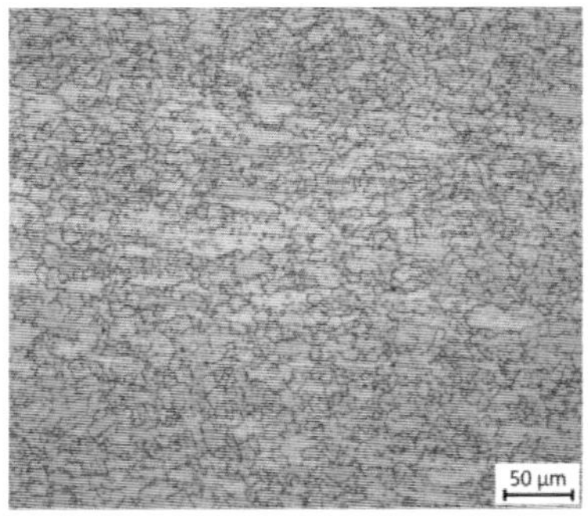

Figure 6. Microstructure produced by heating to semi-solid state and subsequent deformation - sequence 1 - reduction 3.

The same results were obtained with sequence 2 which involved two reductions. The portion of the workpiece which had been subjected to a single reduction from 75 mm to 37.5 mm contained an austenitic structure with dispersed chromium carbides (Figure 7). The portion which also received the second reduction by 19.5 mm to the final thickness of 18 mm contained a very fine microstructure thanks to dynamic recrystallization of austenite. The grain size was less than 10 µm (Figure 8).

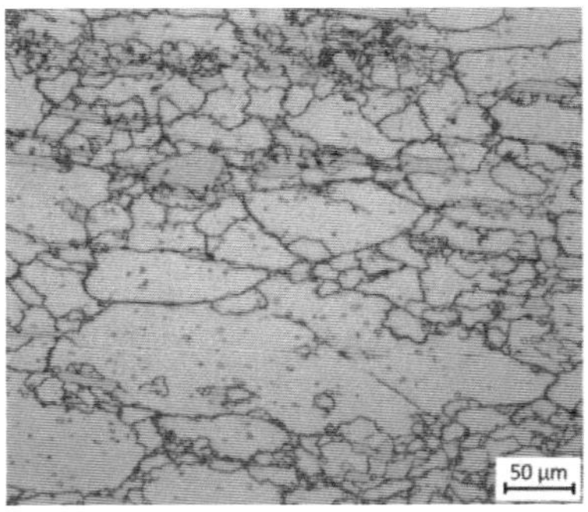

Figure 7. Microstructure produced by heating to semi-solid state and subsequent deformation - sequence 2 - reduction 1.

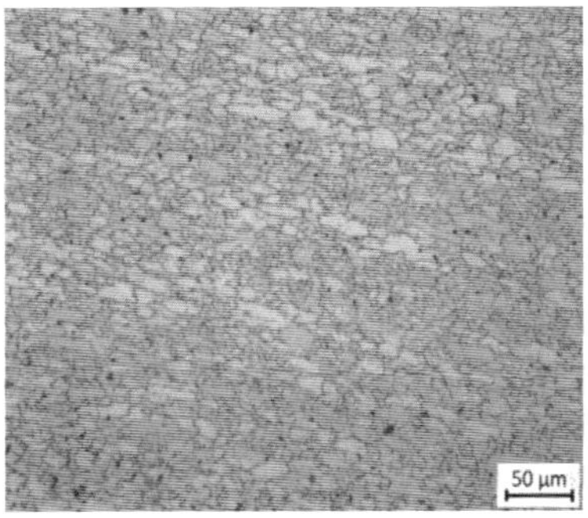

Figure 8. Microstructure produced by heating to semi-solid state and subsequent deformation - sequence 2 - reduction 2.

Holding time was extended in sequence 3 from 30 minutes to 60 minutes. Microstructural evolution initiated by this sequence was similar to those in the previous sequences. In addition, the time at the forging temperature of 1080°C was also increased from 30 min to 60 min. The workpiece portion which had been reduced from 75 mm to 55 mm contained a rather coarse microstructure. It was obvious that the long time at 1240°C caused partial dissolution of primary chromium carbides. The resulting carbide particles are smaller and uniformly distributed throughout the material (Figure 9). The second reduction from 55 mm to 37 mm reduced the austenite grain size. In the relevant portion of the specimen, the grain size was around 20 µm (Figure 10). Where additional strain was introduced, it continued to refine the grain. The third reduction, from 37 mm to 17 mm, produced a grain size of less than 10 µm (Figure 11).

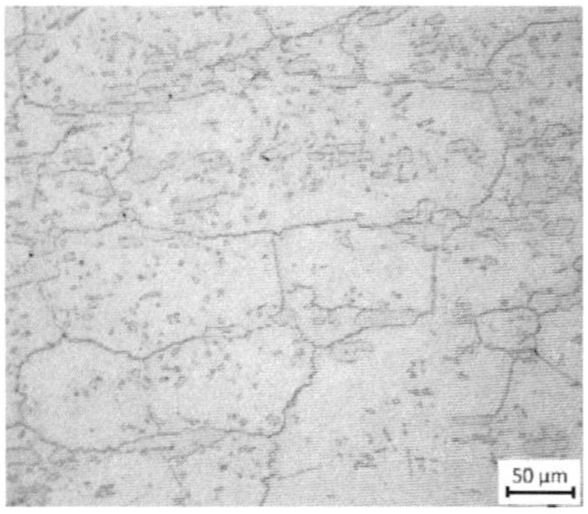

Figure 9. Microstructure produced by heating to semi-solid state and subsequent deformation - sequence 3 - reduction 1.

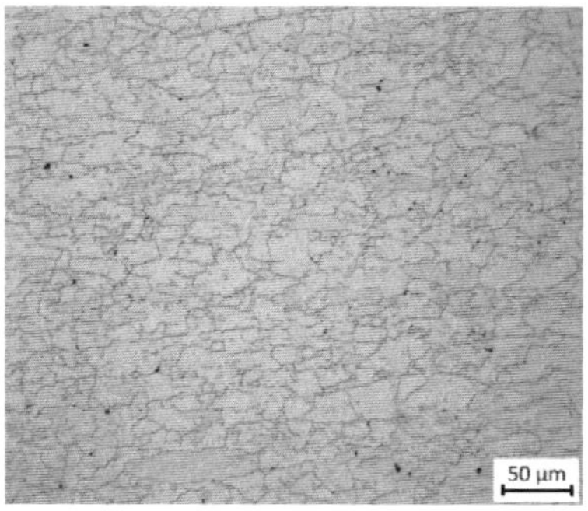

Figure 10. Microstructure produced by heating to semi-solid state and subsequent deformation - sequence 3 - reduction 2.

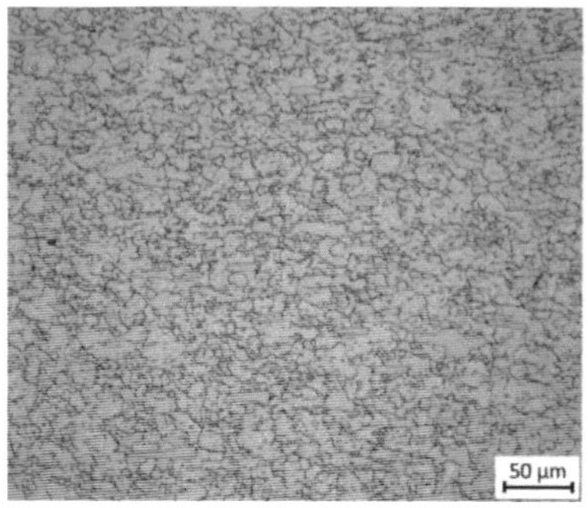

Figure 11. Microstructure produced by heating to semi-solid state and subsequent deformation - sequence 3 - reduction 3.

In the last sequence, the semi-solid process temperature was increased to 1260°C, and the time at temperature was extended to 60 minutes. This higher temperature caused partial melting in the structure, and dissolved fine cementite and primary chromium carbide particles. The network which formed upon cooling was very fine and consisted of austenite and carbides. The first reduction led to no substantial distortion of austenite grains (Figure 12). After the second reduction, the strain imparted was manifested in elongated austenite grains and in stronger dynamic recrystallization in some regions. The mean grain size was around 20 µm (Figure 13). The last reduction resulted in notable elongation of austenite grains and alignment of the ledeburite network with the deformation direction. Recrystallization led to a number of very fine austenite grains. Along some austenite grain boundaries, austenite transformed to martensite. This occurred primarily in regions with a large proportion of carbide network. In these regions, carbon was bound in the carbide network, rendering austenite unstable during cooling to room temperature (Figure 14).

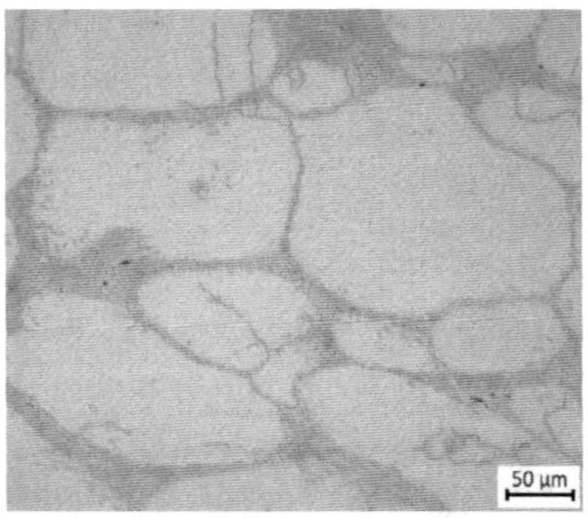

Figure 12. Microstructure produced by heating to semi-solid state and subsequent deformation - sequence 4 - reduction 1.

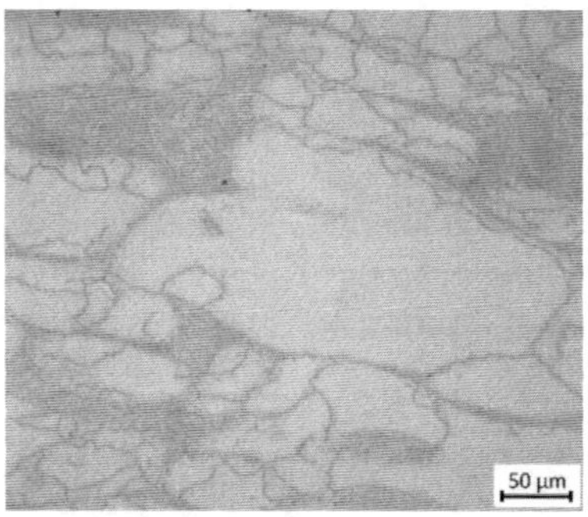

Figure 13. Microstructure produced by heating to semi-solid state and subsequent deformation - sequence 4 - reduction 2.

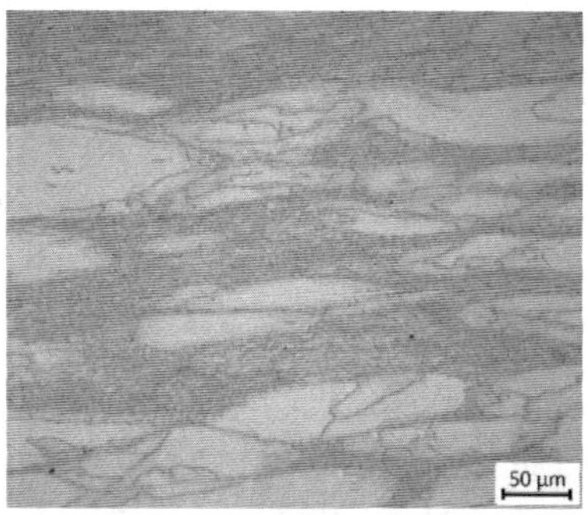

Figure 14. Microstructure produced by heating to semi-solid state and subsequent deformation - sequence 4 - reduction 3.

Detailed examination of the microstructure was conducted with the aid of scanning electron microscopy. Using high magnifications, the presence of martensite was even confirmed upon sequences involving the lower temperature, 1240°C. Martensite needles were found not only in the vicinity of reprecipitated fine carbides but also around the original chromium carbides (Figure 15). Close examination of ledeburite network in a specimen upon sequence 4, the sequence involving the higher temperature, 1260°C, revealed martensite even in those austenite grains within the network, in addition to new austenite grains. Very fine secondary chromium carbides sized approximately 2 μm were found as well (Figure 16).

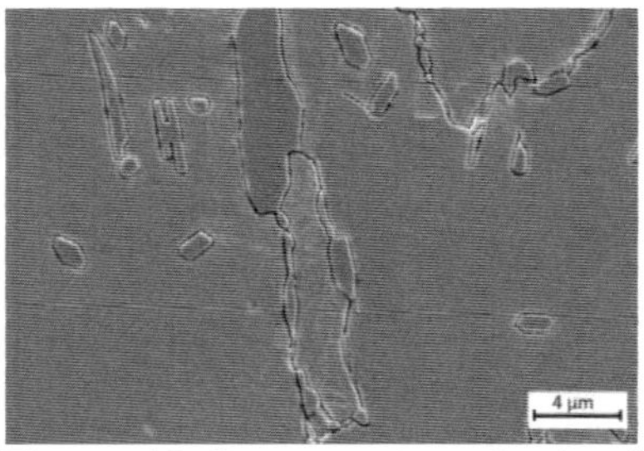

Figure 15. Sequence 4 - reduction 3 - strings of secondary carbides along grain boundaries, martensite needles within austenite, scanning electron micrograph.

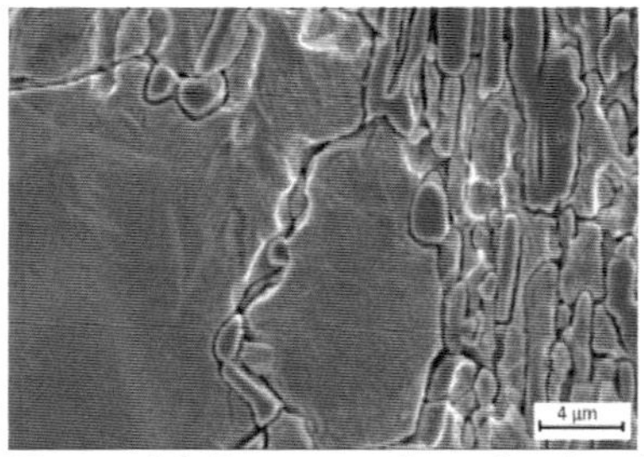

Figure 16. Sequence 3 - reduction 1 - undissolved primary Cr carbides, scanning electron micrograph.

EDS was employed for chemical analysis of carbides (Figure 17). It was found that both primary chromium carbides and newly-formed strings of secondary carbides were rich in chromium, whereas the matrix was depleted of this element.

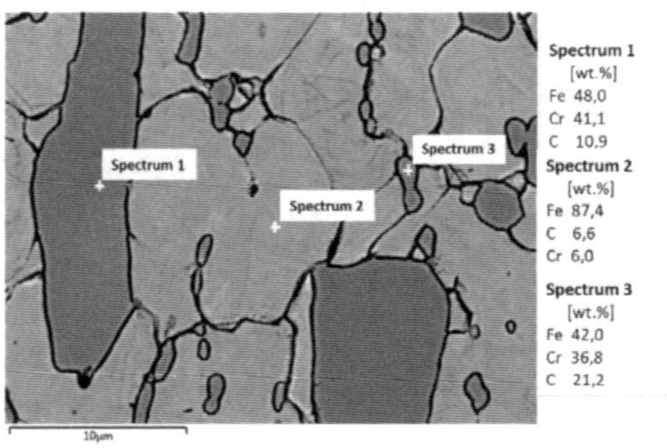

Figure 17. EDS analysis of primary carbides and the matrix – sequence 3 – reduction 3.

Measurement of Microhardness

Microhardness HV0.1 was measured in individual microstructure constituents: the matrix (Figure 18) and chromium carbides (Figure 19). These measurements were made for all sequences and for all combinations of reductions. In each region, three readings 3 were taken and their arithmetic mean was calculated.

In the matrix, hardness values reflected the number of reductions performed. In regions subjected to a single reduction, most values were within the range 360-415 HV0.1. Three reductions led to higher hardness levels: 580–950 HV0.1. Another parameter of importance was temperature. Sequence 4 which involved a temperature of 1260°C, led to higher hardness in the matrix, up to 600 HV0.1 in the region subjected to a single reduction. Hardness levels of carbides were within 1200-1400 HV0.1.

Table 3. Mean values of microhardness (HV0.1) regardless of the number of reductions.

Sequence	Location	HV0.1
Sequence 1	Matrix	724
	Cr carbide	1272
Sequence 2	Matrix	373
	Cr carbide	1052
Sequence 3	Matrix	474
	Cr carbide	1085
Sequence 4	Matrix	640
	Cr carbide	1039

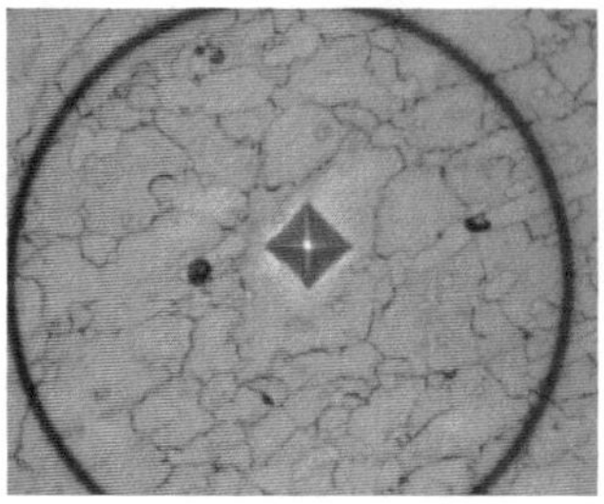

Figure 18. Sequence 3 - reduction 2 – an impression in the matrix – 415 HV0.1.

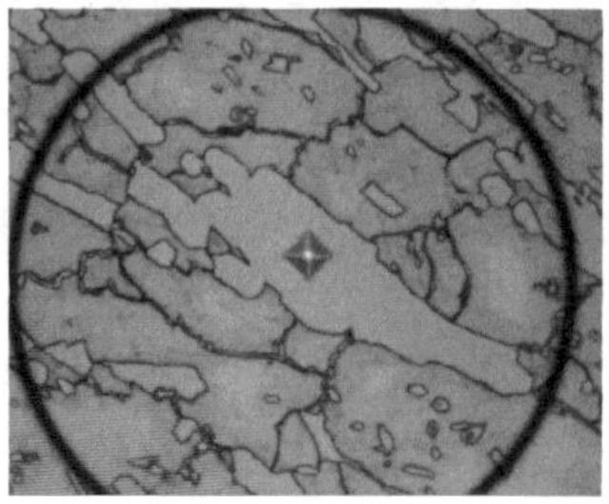

Fig. 19. Sequence 3 - reduction 1 – an impression in a primary chromium carbide – 1267 HV0.1.

4 Conclusion

Experimental semi-solid processing with subsequent forging in a press was carried out on X210Cr12 tool steel. The purpose was to identify the effects of temperatures of 1240°C and 1260°C and times at these temperatures on dissolution of primary chromium carbides and microstructural evolution. Where the temperature of 1240°C was maintained for 30 minutes, the specimens developed microstructures consisting of austenite with a fine dispersion of chromium carbides and cementite. In locations which experienced a single reduction, the carbide network became fragmented, with some fragments dispersed through the austenitic matrix. Higher strains led to further austenite grain refinement. Larger strains also dispersed chromium carbides and fine cementite particles more effectively. Sequence 3 involved holding at 1240°C for an extended period of 60 minutes but failed to produce any substantial changes in the microstructure. In the single-reduction region, the microstructure was rather coarse. It became clear that the longer time at 1240°C and added reductions had produced finer grains. Each reduction reduced the grain size by approximately 10 μm. After the third reduction, the grain size was approximately 10 μm. In the last sequence the semi-solid process temperature was increased to 1260°C, and the time at temperature was extended to 60 minutes. This higher temperature caused partial melting in the structure, and dissolved fine cementite and primary chromium carbide particles. The network which formed upon cooling was very fine and consisted of austenite and carbides. The last reduction resulted in notable elongation of austenite grains and alignment of the ledeburite network with the deformation direction. Recrystallization led to a number of very fine austenite grains. Along some austenite grain boundaries, austenite decomposed into martensite.

Distribution of individual phases in the microstructure was mapped using EDS. It was found that both primary chromium carbides and newly-formed strings of secondary carbides were rich in chromium, whereas the matrix was depleted of this element. Microhardness increased with the amount of strain, and finally ranged from 580 to 950 HV0.1. Hardness of carbides remained virtually constant for all sequences: 1200-1400 HV0.1.

Acknowledgment

This research was funded by the Technology Agency of the Czech Republic under the project TJ02000182 „Improvement of wear resistance of tool steels by a combination of semi-solid treatment with subsequent hot forming and deep cryogenic treatment.

References

[1] Fleming M C, Behavior of Metal Alloys in the Semisolid State 1991 Metall Trans A 22 957-981
[2] Hirt G. And Kopp R. Thixoforming: Semi-solid Metal Processing WILEY-VCH Verlag GmbH & Co. KGaA Weinheim Copyright 2009 pp 1 – 27 ISBN: 978-3-527-3220 4-6
[3] Rassili A., Atkinson H. V. A review on steel thixoforning Trans Nonferrous Met. Soc. China,2010. 20 1048-54
[4] Puttgen, W. Hallstedt, B. Bleck, W. Uggowitzer, P.J. On the microstructure formation in chromium steels rapidly cooled from the semi-solid state Acta Mater. 2007 55 1033–1042, doi:0.1016/j.actamat.2006.09.021
[5] T. Večko Pirtovšek G. Kugler M. Terčlelj Materials Characterization, Vol. 83 2013 p. 97-108, http://dx.doi.org/10.1016/j.matchar.2013.06.008
[6] ČSN 41 9436. Ocel 19436 chromová Praha: Český normalizační institut, 1994
[7] Jirková H. Aišman D. Mašek B. Unconventional structure of X210Cr12 steel obtained by thixoforming. Journal of Alloys and Compounds, 2010. 504 S500–S3
[8] Rubešová, K. M. Pekovic and H. Jirková. Innovative process to eliminate ledeburite network in tool steel. Manufacturing Technology. 2017 17(6) ISSN 12132489
[9] Jirková, H. K. Rubešová K. Opatová M. Peković and B. Mašek. Effect of heating temperature during semi-solid processing on structure of X210CR12 steel Acta Metallurgica Slovaca [online]. 2017, 23(3). ISSN 13381156 Dostupné z: doi:10.12776/ams.v23i3.969
[10] Jirková H. Aišman, D. Rubešová K. Opatová K. Mašek B. Semi-solid processing of high-chromium tool steel to obtain microstructures without carbide network IOP Conference Series Materials Science and Engineering 2017
[11] JMatPro Release 9.0 Sente Software Ltd. 2016.
[12] Jirková H. Rubešová K. Konopík P Opatová K. Effect of the Parameters of Semi-Solid Processing on the Elimination of Sharp-Edged Primary Chromium Carbides from Tool Steel METALS 2018 roč. 8, č. 9, s. 1-15. ISSN: 2075-4701 Haimbaugh R 2001 Practical Induction Heat Treating (Ohio: ASM International) pp 1-4

IOP Conf. Series: Materials Science and Engineering **723** (2020) 012029 doi:10.1088/1757-899X/723/1/012029

The effect of SiO_2 nanoparticles addition on lubrication properties of 10W-40 engine oil.

Ondřej Seibert, Pavel Kejzlar, Totka Bakalová

Department of Material Science, Faculty of Mechanical Engineering, Technical University of Liberec, Studentska 1402/2, 461 17, Liberec, Czech Republic, EU.

ondra.seibert@centrum.cz

Abstract. Proper lubrication of mechanical parts in the car engines shall minimize the wear of friction pair and thus to ensure long lifetime. Another goal of the lubrication is to decrease coefficient of friction and thereby reduce energy losses. Many publications are devoted to nanoparticle addition into simple oil (e.g. paraffin) but there is only limited attention to more complex lubricants as synthetic oils. In addition, there is paid very little attention to the stability of resulting suspensions. Present paper deals with the effect of different ways of SiO_2 NPs addition into the advanced synthetic oil on the wear ratio of friction pair using ball-on-disc tribological test.

1. Introduction

Nowadays, a lot of people has a car, motor bike or any other vehicle. It is necessary to regularly change the motor oil to keep the engine going properly. The change is time and money consuming. These reasons are strong arguments for intensive research in the fields of oil nanoaditives.

Silica dioxide nanoparticles are widely used in various research, such as material engineering [1], medicine [2] etc. For our research, silica nanoparticles were chosen as appropriate nanoparticles due to their availability and low price on the market, very good potential of creation stabile suspension and homogenous size.

The shape of the chosen nanoparticles was also important. Globular shape silica nanoparticles may louver the friction and wear due to "ball bearing" principle [4].

For ensuring the suspension stability, part of authors are putting they effort to functionalization of nanoparticles [5].

The aim of presented study is to figure out the effective method for dispergation of nanoparticles in the real, easily available motor oil with solid properties, not only paraffin [6] or any other lubricants such as organic oils [7], which cannot be used in the real vehicle. The last studied step was to evaluate the effect of used SiO_2 nanoparticles to the tribological properties of synthetic motor oil using ball-on-disk method.

2. Experiment

2.1. *Tested oil.*

Castrol GTX Ultraclean was chosen as commonly used oil, with solid tribological properties. This motor oil, with standard specification 10W-40 perfectly withstand condition of central European weather and effectively protects the engine itself.

2.2. *Nanoparticles addition.*

As it was mentioned previously, globular shaped silica dioxide nanoparticles, with average diameter of 20 nm were chosen as potentially suitable nanoaditives.

Based on studied literature [3] and authors of previous experiences, the right amount of added nanosilica was chosen. The optimal amount of additive is an important part of whole experiment. The lack of additive cause less tribological properties differentiation. On the other side, the surplus of additive cause deterioration of tribological properties. With all of this on mind, 0.2 weight % was chosen as an ideal amount. This exact amount needs to be dispergated to establish stabile suspension.

2.3. *Nanoparticle dispergation.*

The common problem with addition of nanoparticles to the fluids is their clustering, which directly lowers the suspension stability. This is nature behaviour of nanoparticles. From the view of energy, it is very beneficial for the particles to create clusters, instead of staying separate and having bigger active surface.

In this paper, 3 different approaches in nanoparticle dispergation and are presented. To be concrete – usage of eccentric blender, ultrasonic waves and high energy ultrasonic pulses. All oil suspensions were treated for 20 minutes.

Firstly, both methods using ultrasonic waves were tested on pure oil to be sure that the dispergation method itself does not negatively affect the tribological properties.

Also, chemical functionalization of the surface of the silica nanoparticles was studied. Firstly, pH of base oil was measured for preventing inappropriate choice of chemical treatment. With slightly acidic pH, APTES ((3-aminopropyl) triethoxysilane) was chosen. Even functionalized nanoparticles were dispergated with eccentric blender.

After evaluating the tribological grooves, successful methods were used for testing the nanoaditive.

2.4. *Measuring of tribological properties – tribometer and confocal microscope.*

After homogenization and dispergation of the nanoaditive, next step naturally was to evaluate the effect of nanosilica on the tribological properties of the oil. For the measurement, ball-on-disc tribomemer was used. Al_2O_3 was chosen as ball and chrome-manganic steel ČSN 14 220 as a disc. In the measuring bath, 70 ml oil with nanoaditive was introduced and heated to the temperature of 90 °C, which was chosen as simulation of real temperature in the engine. On the steel disk, path of 500 m on diameter of 9 mm was measured. The whole measurement took 3 hours.

After cooling to the room temperature, steel disk was cleaned by ethylalcohol and introduced to the confocal microscope to collect profile data of the groove.

3. Results and discussion

3.1. *Measuring of tribological groove.*

Software Gwyddion was successfully used for the groove visualisation and measuring of desired width and depth. The groove 2D model is clearly visible on Fig. 1. The software is also capable of 3D visualisation (Fig. 2) for detailed study of the groove.

Groove width and depth were measured on the mentioned 2D model (Fig. 1) as critical points in evaluation of silica nanoparticles impact,.

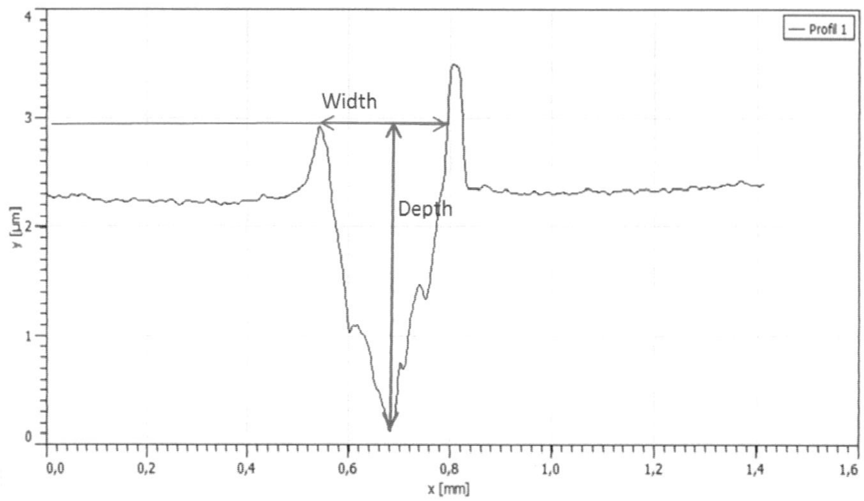

Figure 1. 2D profile of tribological groove.

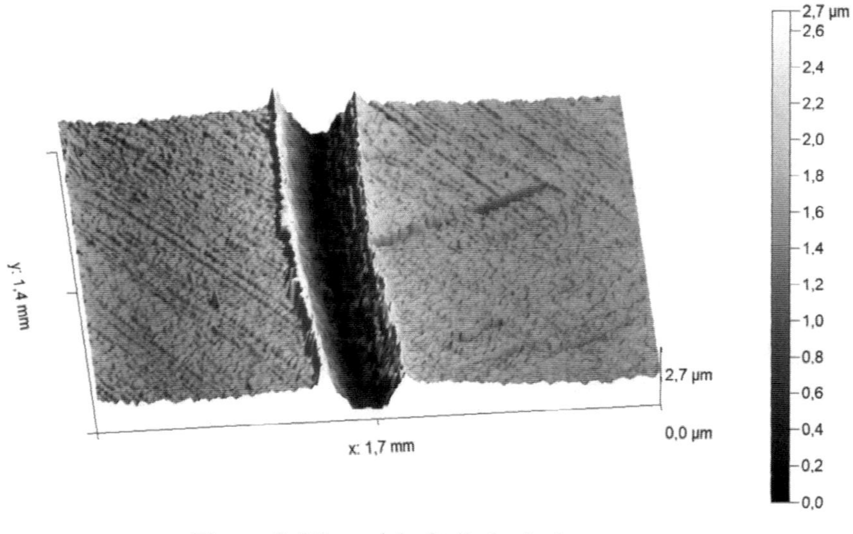

Figure 2. 3D model of tribological groove.

The summarization of tribological measurement is in Tab. 1. First measurement (**1** Pure oil) is reference measurement to determine tribological properties of pure oil. Second measurement (**2** Pure oil high energy ultrasonic) was evaluate as inconvenient. With a high probability, the serious degradation of base oil was caused by high energy ultrasound waves. According to this measurement, the high energy pulses were no longer used in the study. As last tested method on pure oil, ultrasonic from ultrasonic cleaner remain (**3** Pure oil ultrasonic). This measurement proved that no negative impact occurs after using standard ultrasonic. The impact on eccentric blender was not deeply studied due to the simplicity of mixing and no impact to the oil properties.

Addition of silica nanoparticles, which were dispergated using ultrasonic (**4** SiO_2 Ultrasonic) brought improvement in the width and depth of the groove. The improvement of 12,7 % seems to be promising for further research.

Measurement number 5 (**5** SiO_2 eccentric blender) was evaluated as most satisfying. The improvement of 13,8 % in groove width and 32,9 % in groove depth is more than promising result.

Long term ultrasonic treatment was tested to make sure even several hours of exposition does not cause the degradation of base oil. According to the results, also the long term exposition has nearly similar effect as the usage of ultrasonic high energy pulses.

The functionalized nanoparticles with APTES created whey more stable suspension and also brought slightly better results than usage of simple blender.

Table 1. Overview of tribological measurement.

	Groove width [μm]	Groove depth [μm]	Width differention [%]	Depth differention [%]
1 Pure oil	281	1,94	-	-
2 Pure oil high energy ultrasonic	290	2,03	-3,1	-4,3
3 Pure oil ultrasonic	265	1,83	6	6,0
4 SiO_2 Ultrasonic	264	1,72	6,4	12,7
5 SiO_2 Eccentric blender	247	1,46	13,8	32,9
6 SiO_2 Ultrasonic 3 hours	286	2,09	-1,7	-7,0
7 SiO_2 APTES	246	1,45	14,2	33,7

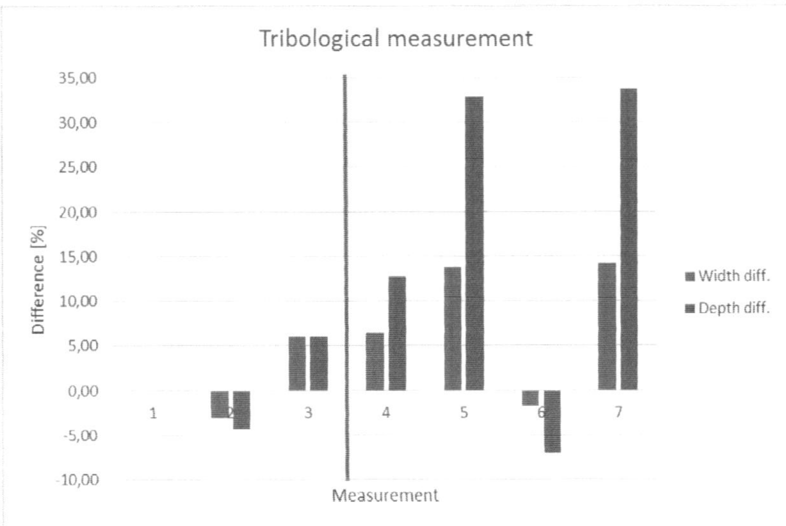

Figure 3. Graph of resulting tribological measurements.

3.2. *Discussion.*

With the grooving pressure for industry to be more and more ecological, the addition of nanoparticles to the motor oil to prolong the efficient time of usage and lower the wear of lubricated parts seems to be potentially good way to keep suitable industry development.

According to the results of our research, functionalization seems to be the most sufficient method for the creating stabile suspension and for improvement of tribological properties.

But, even the suspension is more stable, after several days, the nanoparticles are also sedimented and the price and time consumed of the chemical process are strong arguments to use the eccentric blender.

Eccentric blender has also convincing results to the tribological properties, but the price for the dispergation process is minimal comparing to the chemical functionalization. Also, thinking of potential industrial usage, the technology of eccentric blender is well known and for reasonable price. To sum up, the problems of sedimentation and clustering are still the most challenging problems in recent studies. It is necessary to put all effort to solving this problem and prepare stable suspension for wide common usage.
In this state, the oil with additive may be used in non-stopping processes, where the particles have no chance to sediment due to never ending flow of the oil, such as gearboxes in wind power plans or 24 hours production in the industry.

Acknowledgements:
This paper was supported by the project of Technical University of Liberec SGS "The study and evaluation of structures and properties of metallic and non-metallic materials"

References:
[1] Nazari A and Riahi S 2011 The effects of SiO_2 nanoparticles on physical and mechanical properties of high strength compacting concrete. Composites Part B: Engineering, **42**(3), 570–578.
[2] Mitran R A. et all. 2019 Mesoporous Silica as Carrier for Drug-Delivery Systems. Nanocarriers for Drug Delivery, 351–374.
[3] Peng D X, et all. 2009 Tribological properties of diamond and SiO2 nanoparticles added in paraffin. Tribology International . , **42**(6), 911-917
[4] Maharaja K et all 2016 Tribological Effect of Size, Shape and Structure of Nanoparticle in Lubricant Oil – A Review. In: International Conference on Systems, Science, Control, Communication, Engineering and Technology, 730 – 734.
[5] Peng et all 2009 Tribological properties of diamond and SiO2 nanoparticles added in paraffin. Tribology International ., **42**(6), 911-917
[6] Yang G et all 2012 Preparation and tribological properties of surface modified Cu nanoparticles. Transactions of Nonferrous Metals Society of China, **22**(2), 366-372,
[7] Khan M S et all. 2019 Measurement of tribological properties of Cu and Ag blended coconut oil nanofluids for metal cutting. Engineering Science and Technology, an International Journal

IOP Conf. Series: Materials Science and Engineering **723** (2020) 012030 doi:10.1088/1757-899X/723/1/012030

The effect of heat treatment on the structure and high – temperature strength of Fe_3Al – based iron aluminide doped by carbon and carbide-forming element

Martin Švec, Věra Vodičková, Pavel Hanus, Vojtěch Keller

Technical University of Liberec, Studentská 1402, Liberec, Czech Republic
Material Science Department

E-mail: martin.svec@tul.cz

Abstract. A structural study as well as the determination of $\sigma_{0.2}$ compression yield stress at high temperatures have been performed on the Fe_3Al-based iron aluminide doped by Nb (5 at. %) and C (1.5 at. %). The carbon occurrence leads to the formation of three-phase structure. Fe_3Al matrix, niobium carbides and (Fe, Al)$_2$Nb Laves phase were observed in the structure of investigated alloys. Precipitates were inhomogeneously distributed in as cast state alloy. The annealing at 1000 °C for 50 hrs. leads to more homogenously distribution of secondary phase particles and also to refinement of grain size. The strengthening mechanism of material at high temperatures was detected as a combination of strengthening by incoherent precipitates (NbC and Laves phase) and a solid solution hardening by Nb atoms. Effect of two reinforcing mechanisms together with homogeneously precipitate's distribution increase the values of $\sigma_{0.2}$ compression yield stress about 50 MPa in whole measured temperature range (600 – 800 °C) in comparison to similar intermetallic's system Fe-Al-Zr-C.
Phase composition of the alloy was studied by means of scanning electron microscopy (SEM) equipped by energy dispersive X-ray spectrometer (EDX). The electron backscattered diffraction (EBSD) was used for phase verification and grain size visualization.

1. Introduction

Alloyed iron aluminides have been recognized as candidate materials for high-temperature (it means temperature range 600 – 800 °C in case of Fe-Al-X alloy) structural application due to their very good corrosion resistance as well as their low density. The binary iron aluminides show reasonable strength and ductility at room temperature but only mediocre creep resistance at high temperatures [1-4].

The creep resistance improvement is possible by the secondary phase particles, which should form the stable and fine dispersion. These particles may be precipitates of carbides, borides or intermetallic phases [5, 6]. In last period the research works are focused on the improvement of high-temperature creep resistance by alloying of binary alloys by carbide-forming or boride-forming elements [7, 8]. Many recent studies have examined a variety of carbide or intermetallic precipitates such as Laves phase [9]. An alternative method of material reinforcement is incorporation of the secondary phase through directional solidification of eutectic alloys [10].

Recently the iron aluminides doped by carbon and other carbide – forming element(s) are investigated [11, 12]. The primary motivation is to use inexpensive raw materials as we are currently focusing on high performance materials but made from cheaper raw materials (and cheap raw iron usually used to be contaminated by carbon). In the present the focus is on the materials with the high utility materials properties but made from cheaper raw materials. At the same time, it is required the

IOP Conf. Series: Materials Science and Engineering **723** (2020) 012030 doi:10.1088/1757-899X/723/1/012030

restriction of use of strategic element such as Cr, Ni etc. The research of iron aluminides doped by carbon and carbide-forming elements is also motivated by options to replace expensive high alloyed (especially alloyed by high amount of strategic Cr) steels P-91 or P-92 type by these intermetallic alloys in high – temperature applications.

Fe-Al-Nb system is one of the Fe-Al-X systems which are of interest for development of Fe-Al – based alloys. Niobium has potential for both solid solution hardening and precipitation strengthening. The precipitation of intermetallic Laves phase $(Fe, Al)_2Nb$ in Fe-Al-Nb system can lead to improvement of high-temperature creep resistance of alloys. The knowledge of Laves phase $(Fe, Al)_2Nb$ formation processes and their stability against coarsening during high-temperature annealing is very important. It was shown that fine precipitates of Laves phase were able to produce good strengthening for temperatures up to ca. 750 °C [13]. Recently the effect of heat treatment on the composition of niobium doped iron aluminides was investigated [14, 15]. The aim of article is to show the effect of heat treatment on the structure of alloy and to compare the influence of different type of carbide forming elements on high-temperature strength.

2. Materials and experimental methods

The investigated alloys were prepared by vacuum induction melting and casting. The nominal chemical composition of the samples is given in table 1. Two states of alloy were investigated for selection of sample with suitable microstructure for the next high-temperature compressive testing. Initial state was as cast (marked as FA5Nb1C AC), annealed state (marked as FA5Nb1C HT) was heat treated at 1000 °C for 50 hrs. and air cooled. The microstructure was observed in five different cross-section places of cast in both alloys. No structure differences were observed between each places.

The samples for study of microstructure were oxide-polished by suspension OP-S in the final step. This leads to the enhance of visibility of carbide phase in topographical contrast [16]. The structure was investigated by scanning electron microscopy (SEM) Tescan Vega 3SBH equipped with Oxford X-Max20 detector for energy dispersive analysis (EDX) and by SEM Zeiss Ultra Plus equipped with Oxford X-Max20 EDX detector and Nordlys Nano EBSD detector used for phase verification and for determination of grain maps (scanning area was 500 x 700 μm in both cases, step was 2 μm).

The samples (prisms 6 x 6 x 10 mm) for high – temperature compression tests were prepared from annealed material (FA5Nb1C HT) by spark machining. The compression yield stress $\sigma_{0.2}$ was measured using INSTRON 1186 at temperature 600, 700 and 800 °C with the initial strain rate 1.2×10^{-4} s^{-1}. The values of yield stress were measured twice for each temperature. The obtained results were comparable.

The values of high-temperature strength of alloy FA5Nb1C was compared with values for 1181 alloy [17] for comparison of carbide – forming element effect. The quantitative chemical composition of 1181 alloy (Fe-26Al-4.9Zr-1.4C in at. %) was analogous to FA5Nb1C, structure was different only in type of carbide-forming element (Zr by 1181, Nb by FA5Nb1C). The samples for compression test had the same dimensions (prisms 6 x 6 x 10 mm) and surface finishes (polished). They were also made from annealed material (at 1000 °C for 50 hrs.), the applied initial strain rate was the same. The carbon occurrence in Fe-Al-Zr system shifts the phase stability and therefore no τ phase was observed in the case of 1181 alloy. $(Fe, Al)_2Zr$ Laves phase and ZrC were present in the structure only.

Table 1. The nominal chemical composition of investigated samples.

sample	chemical composition (at. %)			
	Fe	Al	Nb	C
FA5Nb1C	bal.	26.1	4.9	1.4

IOP Conf. Series: Materials Science and Engineering **723** (2020) 012030 doi:10.1088/1757-899X/723/1/012030

3. Results and discussion

3.1. Structure of FA5Nb1C AC

The structure of matrix was coarse-grained with average grains dimensions of micrometers of 75 ± 30 μm – see figure 1 (the dark areas enclosed by light grey net of precipitates) and figure 2 (EBSD grain size measurement). Two types of precipitates were observed in the structure. The niobium carbides (NbC) were situated inside of grains. Distribution of these precipitates was very inhomogeneous (figure 1) and their size was about 5 μm. Second type of particles was determined as Laves phase (Fe, Al)$_2$Nb. Laves phase started to nucleate along the grain boundaries and some particles were present also inside the grains (see detail in figure 1). Laves phase particles had elongated shape with dimensions about 1 μm or less. Small amount of added niobium was also dissolved in the matrix (see table 2). Preliminary phase identification was performed by EDX and verified by EBSD (see table 2 and figure 3).

No eutectic net formed in the structure of FA5Nb1C AC sample in comparison to the alloy with the same content of aluminum and niobium but without addition of carbon (Fe – 27.3 Al – 4.8 Nb alloy) [14]. In FA5Nb1C AC there appeared the precipitates NbC in the structure preferably due to high affinity of carbon to niobium. The niobium excess is sufficient for Laves phase (Fe, Al)$_2$Nb formation, but it is not sufficient for formation of Fe3Al matrix + (Fe, Al)2Nb eutectic net as in the case of Fe – 27.3 Al – 4.8 Nb alloy [14].

Table 2. The average chemical composition of phases in FA5Nb1C AC alloy.

phase	Fe	Al	Nb	C
Matrix (at. %)	balance	26.6 ± 0.2	1.0 ± 0.1	-
Laves phase (at. %)	balance	18.6 ± 0.8	27.3 ± 0.5	-
NbC (at. %)		verified by EBSD as NbC		

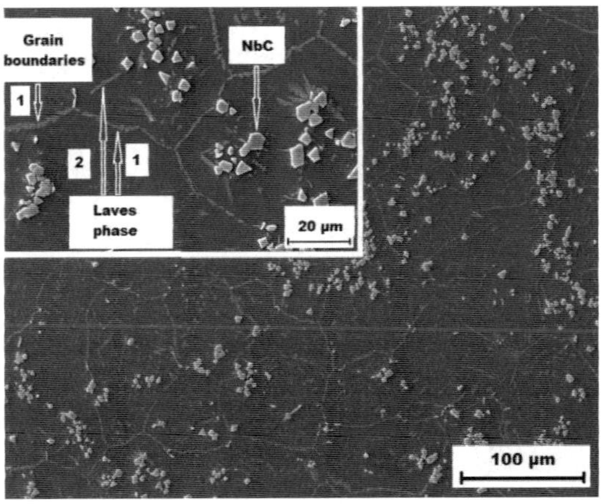

Figure 1. The structure of alloy FA5Nb1C AC (SE detector). Dark grey – Fe$_3$Al matrix, Light grey – (Fe, Al)$_2$Nb Laves phase (in detail: arrow 1 – Laves phase on grain boundaries, arrow 2 – Laves phase inside the grain,), White – NbC.

IOP Conf. Series: Materials Science and Engineering **723** (2020) 012030 doi:10.1088/1757-899X/723/1/012030

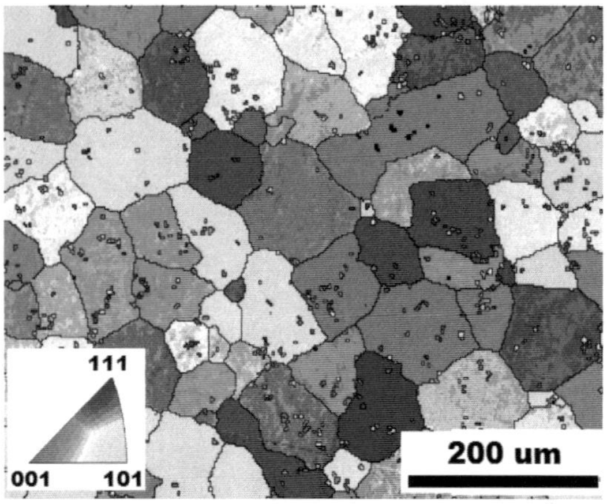

Figure 2. The EBSD map of grains of FA5Nb1C AC alloy.

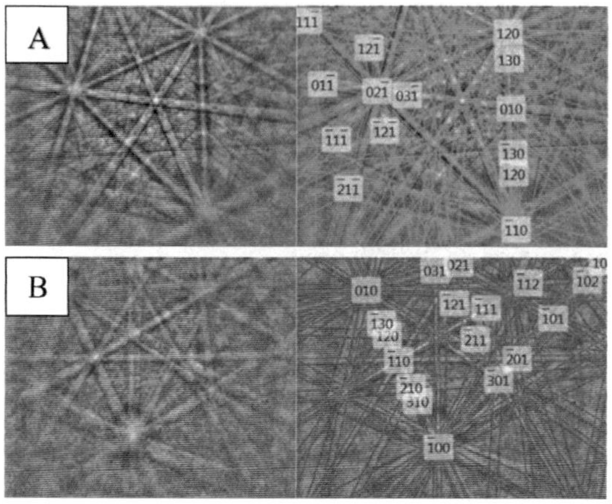

Figure 3. The Kikuchi bands of NbC (A) and Laves phase (B) in FA5Nb1C AC alloy.

3.2. Structure of FA5Nb1C HT

No changes were observed in the phase composition after heat-treatment. Annealing at 1000 °C for 50 hrs. had beneficial effect on the NbC distribution in the structure of investigated sample. Niobium carbides are presented inside the grains and their distribution is much more homogeneously (figure 4). It is clearly visible from figure 4 (in comparison to figure 1) that NbC were not only redistributed, but also they were partly refined (size in order of units of micrometers). On the other hand, precipitates of Laves phase got coarsen and individual small particles on grain boundaries (and partly also inside the grains) started to coagulate. Annealing at 1000 °C for 50 hrs. also significantly refined the grains of matrix. The average grains size was 19 ± 6 μm after heat treatment (see EBSD figure 5).

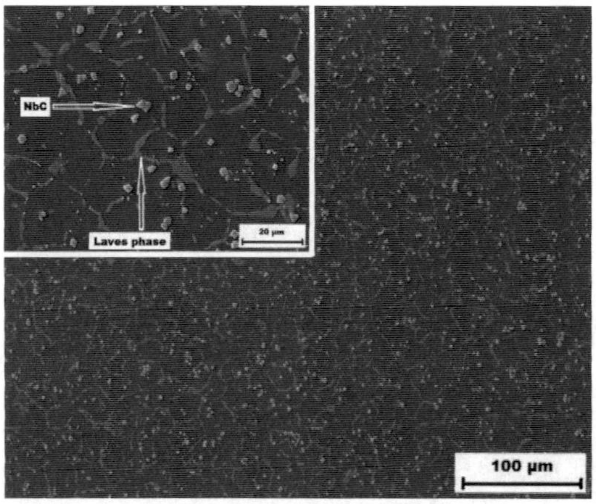

Figure 4. The structure of alloy FA5Nb1C HT (SE detector).

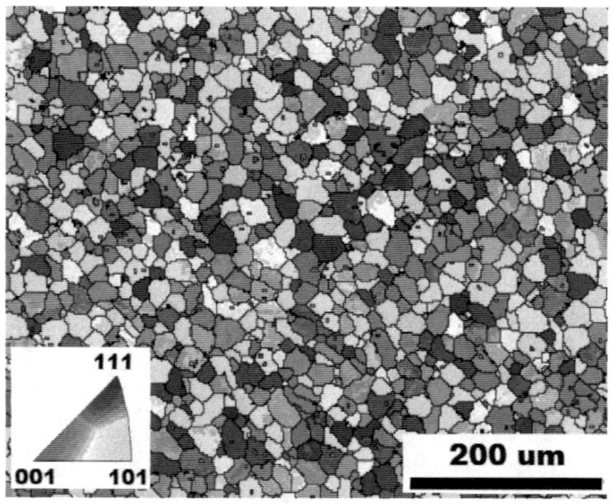

Figure 5. The EBSD map of grains of FA5Nb1C HT alloy.

3.3. The yield stress at high-temperatures

The structure of FA5Nb1C HT sample was more homogenous, therefore this material was chosen for all high-temperature compressive tests. As expected the values of yield strength $\sigma_{0.2}$ of FA5Nb1C were decreasing with the increasing temperature. However, they were sufficient for high-temperature applications up to 700 – 800 °C.

The values of high-temperature yield stress of investigated FA5Nb1C HT sample were compared with values for the alloy 1181 [17] – see figure 6. The carbides and the precipitates of Laves phase formed in both alloys. The $\sigma_{0.2}$ values of FA5Nb1C were about 30 – 50 MPa higher at all temperatures than that for 1181 [17]. There may be more reasons for the higher strength of FA5Nb1C at high temperatures:

The first, in FA5Nb1C there is a combination of strengthening mechanism by incoherent precipitates (NbC and Laves phase) and strengthening by solid solution hardening (Nb). Solid solubility of niobium is about 1 at. % in Fe_3Al matrix [18], however the solid solubility of Zr is only 0.1 at. % [19, 20] and therefore the effect of incoherent precipitates in 1181 takes place in strengthening only.

The second reason is the secondary phases form and distribution. No coarsening of precipitates during compression tests was noticed (see figure 7). The Laves phase precipitates and NbC are very homogeneously distributed in the structure of FA5Nb1C. The grain boundaries are also reinforced by fine precipitates of Laves phase. These precipitates could work as obstacles for dislocation movement and they also act similar effect as hardening phase in composites. No cracks appeared in the matrix or in precipitates after high-temperature compressive tests (see Fig. 7 – structure after deformation at 800 °C). Compared to that, the eutectic net of Fe_3Al matrix + $(Fe, Al)_2Zr$ Laves phase and inhomogeneously distributed clusters of ZrC were observed in the structure of 1181 [17]. These clusters of brittle ZrC could act as initial areas for cracks. Also, the Laves phase in 1188 is present only as part of eutectic net. There are the places completely without precipitates in the structure. No dislocation movement obstacles were present in these places between eutectic net.

It can be summarized, that both Nb and Zr have high affinity to carbon. However, the carbon affinity to Zr is slightly higher than that to Nb and it is the reason, why the clusters of ZrC are formed in 1181 [17] and why they are stable also after annealing [21]. The clusters of NbC formed to a certain extent also in FA5Nb1C AC, but they dispersed after annealing.

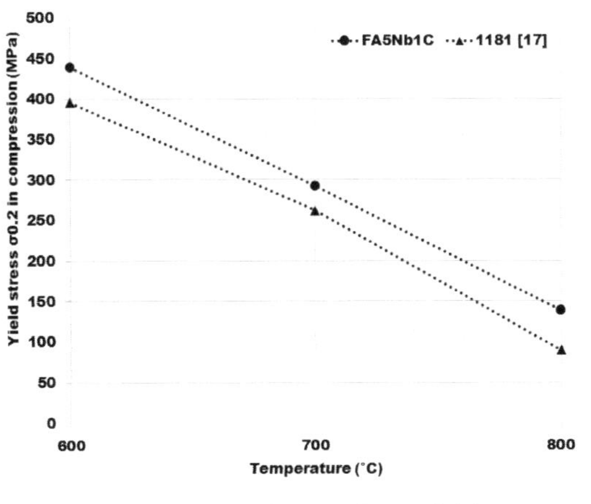

Figure 6. The dependence of $\sigma_{0.2}$ yield stress on the temperature for investigated FA5Nb1C and 1181 [17].

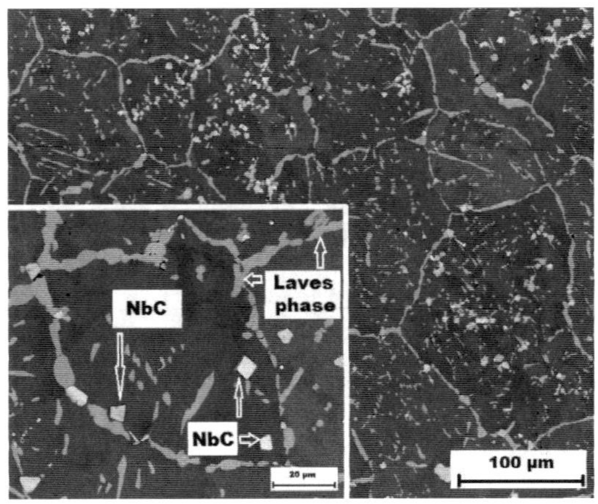

Figure 7. The structure of FA5Nb1C HT alloy after compression deformation at 800 °C (BSE detector).

4. Conclusions

The structures of investigated samples are three-phase – Fe_3Al matrix + NbC niobium carbides + $(Fe, Al)_2Nb$ Laves phase (phase composition was the same for both as cast and heat-treated states).

The heat-treatment at 1000 °C for 50 hrs. leads to achieve of the homogeneous structure and technically usable values of high-temperature strength.

The strengthening mechanism in FA5Nb1C is a combination of strengthening by incoherent precipitates (NbC and Laves phase) and a solid solution hardening by Nb atoms. However, the precipitate's strengthening holds the main effect.

The addition of Nb + C into Fe_3Al iron aluminides increases the high – temperature strength more effectively than alloying by Zr + C (the values of yield stress of FA5Nb1C are about 30 – 50 MPa higher at all tested temperatures than those for 1181 alloy [17]. This could be caused by more homogeneous distribution of precipitates in case of FA5Nb1C HT alloy as well as by additional effect of solid solution hardening by Nb atoms.

Acknowledgements
This publication was written at the Technical University of Liberec as part of the project "The study and evaluation of the structure and properties of metallic and non-metallic materials" with the support of the Specific University Research Grant, as provided by the Ministry of Education, Youth and Sports of the Czech Republic in the year 2019 and as part of the project supported by the Ministry of Education, Youth and Sports of the Czech Republic and the European Union - European Structural and Investment Funds in the frames of Operational Programme Research, Development and Education - project Hybrid Materials for Hierarchical Structures (HyHi, Reg. No. CZ.02.1.01/0.0/0.0/16_019/0000843).

References

[1] Morris D G and Munoz - Morris M A 1999 The influence of microstructure on the ductility of iron aluminides *Intermetallics* **7** p 1121
[2] Mc Kamey C G 1994 Physical Metalurgy and processing of Intermetallic Compounds p 351
[3] Stoloff N S 1998 Iron aluminides: present status and future prospects *Mater. Sci. Eng. A* **258** p 1
[4] Deevi S C and Sikka V K 1996 Nickel and iron aluminides: an overview on properties, processing and applications *Intermetallics* **4** p 357
[5] Palm M, Schneider A, Stein F and Sauthoff G 2005 Strengthening of iron aluminide alloys for high-temperature applications *Mater. Res. Soc. Symp. Proc.* **842** p S1.7.1
[6] Palm M 2005 Concepts derived from phase diagram studies for the strengthening of Fe – Al based alloys *Intermetallics* **13** p 1286
[7] Mc Kamey C G and Maziasz PJ 1998 Heat treatment effects for improvement creep-rupture resistance of a Fe_3Al-based alloy *Intermetallics* **6** p 303
[8] Dobeš F, Kratochvíl P, Pešička J and Vodičková V 2015 Microstructure and creep behavior of Fe-27Al-1Nb alloys with added carbon *Metall. Mater. Trans.* **46A** p 1580
[9] Stein F, Palm M and Sauthoff G 2005 Mechanical properties and oxidation behaviour of two-phase iron aluminium alloys with $Zr(Fe, Al)_2$ Laves phase or $Zr(Fe, Al)_{12}$ τ_1 phase *Intermetallics* **13** p 1275
[10] Milenkovic S and Palm M 2008 Microstructure and mechanical properties of directionally solidified Fe-Al-Nb eutectic *Intermetallics* **16** p 1212
[11] Schneider A, Falat L, Sauthoff G and Frommeyer G 2003 Constitution and microstructures of Fe-Al-M-C (M=Ti, V, Nb, Ta) alloys with carbides and Laves phase *Intermetallics* **11** p 443
[12] Falat L, Schneider A, Sauthoff G and Frommeyer G 2005 Mechanical properties of Fe-Al-M-C (M=Ti, V, Nb, Ta) alloys with strengthening carbides and Laves phase *Intermetallics* **13** p 1256
[13] Moriss D G, Munoz – Morris M A and Baudin C 2004 The high-temperature strength of some Fe_3Al alloys *Acta mater.* **52** p 2827
[14] Švec M and Vodičková V 2014 The effect of niobium addition and heat treatment on the phase structure of Fe_3Al – type intermetallic alloys *Manufacturing Technology* **14** p 456
[15] Morris D G, Munoz – Morris M A, Requejo L M and Baudin C 2006 Strengthening at high temperatures by precipitates in Fe – Al – Nb alloys *Intermetallics* **14** p 1204
[16] Kejzlar P, Švec M and Macajová E 2014 The usage of backscattered electrons in scanning electron microscopy *Manufacturing Technology* **14** p 333
[17] Kratochvíl P, Vodičková V, Král R and Švec M 2016 The effect of Laves phase $(Fe, Al)_2Zr$ on the high-temperature strength of carbon-alloyed Fe_3Al aluminide *Metall. Mater. Trans.* **47A** p 1128
[18] Prymak O and Stein F 2010 Solidification and high-temperature phase equilibria in the Fe-Al-rich part of the Fe-Al-Nb system *Intermetallics* **18** p 1322
[19] Raghavan V 2010 *J. Phase Equi. Diff.* **31** p 459

[20] Kratochvíl P, Kejzlar P, Král R and Vodičková V 2012 The effect of Zr addition on the structure and high temperature strength of Fe-30 at. % Al type alloys *Intermetallics* **20** p 39
[21] Palm M 2014 Max – Planck Institut für Eisenforschung GmbH, Düsseldorf, unpublished research

Static and dynamic testing of a bogie

J Tittel[1], M Kepka[1] and P Heller[1]

Regional Technological Institute, Research Centre of the Faculty of Mechanical Engineering, University of West Bohemia, Univerzitni 2732/8, Pilsen 30614, Czech Republic

E-mail: tittejan@rti.zcu.cz

Abstract. The development process of railway bogie frames consists of three stages. The first stage involves stress calculations, the second stage comprises static and fatigue tests and the third one is focused on validating fatigue life data by stress monitoring during test runs on an actual track. The aim of this research is static and dynamic testing of a functional bogie model by hydraulic cylinders and subsequent evaluation of measured values. The load is determined according to the requirements specified in CSN EN 13749, Railway applications – Wheelsets and bogies – Method of specifying the structural requirements of bogie frames. In the fatigue test, the loads are composed of the static load, the quasi-static load, and the dynamic load. The bogie frame is a weldment made of S355 steel sheets, which has two transoms and two side rails ending with suspension brackets. It should be noted that the fatigue strength of a welded frame under dynamic loading is limited by the resilience of the welded joints. Attention is therefore paid to analyzing these joints. The development experience derived from this new bogie can be used to optimize the production of rolling stock. In the present experiments, the frame was examined using NDT to detect possible defects.

1 Introduction

Regional Technological Institute, a research facility affiliated with the University of West Bohemia, has developed a new concept of a bogie which departs from traditional solutions in that it uses leaf springs for primary suspension and involves innovative wheelset guidance. There are two leaf springs arranged above each other, one on top and the other below the axle box housing, which provide suspension and guide the wheelset. Their spring action is applied in an unconventional manner. One loop of each leaf spring is attached to the frame, while the centre clip supports the frame and the other loop is fixed in the axle box. Rubber bushings are inserted in each loop. The secondary suspension is provided by the bolster and coil springs. Narrow-gauge bogies require wheelsets with outer bearings and a frame with outer longitudinal members. The bogie frame is an open structure, consisting of two transoms connected with short longitudinal members ended with brackets for springs. The wheelsets are powered by a traction motor via a spur-and-bevel gearbox, with a disk brake mounted on the other end of the engine shaft. The drive is longitudinal, mounted outside the frame on full suspension. The bogie concept is illustrated in Figure 1 [1, 2].

IOP Conf. Series: Materials Science and Engineering **723** (2020) 012031 doi:10.1088/1757-899X/723/1/012031

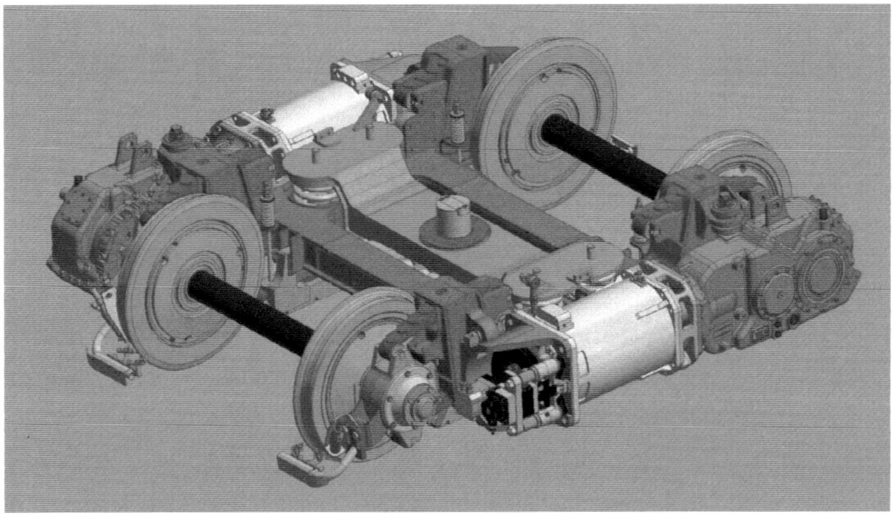

Figure 1. Traction pivoting bogie for a narrow-gauge tramway car.

The benefits of this configuration include a small wheelbase of 1700 mm, with radial steering of the wheelset in a curve, and therefore less wear on the wheelset and rails, less noise, and less weight of the bogie, approx. 3600 kg. The welded bogie frame consists of two transoms and two longitudinal members ending in brackets for attaching leaf springs. Both longitudinal members and transoms are weldments from two strips and two vertical webs. The sheets used were of standard S355NL steel. The weight of the bogie frame alone is 454 kg.

2 Basic specifications of the problem

Transport vehicles are an ordinary part of our daily lives. The class of rail vehicles includes tramways, train units and underground railway units. These vehicles experience static and dynamic loads in service. Their key assemblies include welded bogie frames and other structures to which individual components are attached. Development of railway bogie frames comprises three stages. The first stage involves stress calculations, the second one comprises static and fatigue testing, and the third stage is concerned with validation of fatigue life data by on-track testing. This research work focused on static and dynamic testing of a functional model of a bogie on a test rig fitted with hydraulic cylinders, and subsequent evaluation of measured data [3].

3 Loading

The structure was subjected to test loads in Inova testing equipment. The functional model shown in Figure 2 was supported on fixtures attached firmly to T-slots in the test bed for adequate stiffness. Static and dynamic forces were then exerted by hydraulic cylinders. Vertical loads were applied by two cylinders acting on the points of attachment of springs. Lateral force, i.e. the transverse load, was applied by a single cylinder. The magnitudes of loads were determined in line with CSN EN 13749, Railway applications - Wheelsets and bogies - Method of specifying the structural requirements of bogie frames [4]. In a fatigue test, the load comprises a static component, a quasi-static component and a dynamic load component.

Figure 2. Loading on bogie frame.

The deformation response of the frame was measured by means of fourteen strain gauges positioned according to Figure 3. The strain gauges were mounted onto those locations of the frame where the highest stresses were anticipated, based on a computational model and measurements on an earlier type of frame. The strain gauges were 1 – LY 11 6/120 type from Hottinger Baldwin Messtechnik GmbH. They were attached using a two-component adhesive. They were coated with a sealant to provide protection from air humidity. Data from the strain gauges were captured and recorded in QuantumX module and evaluated using the software Catman HBM.

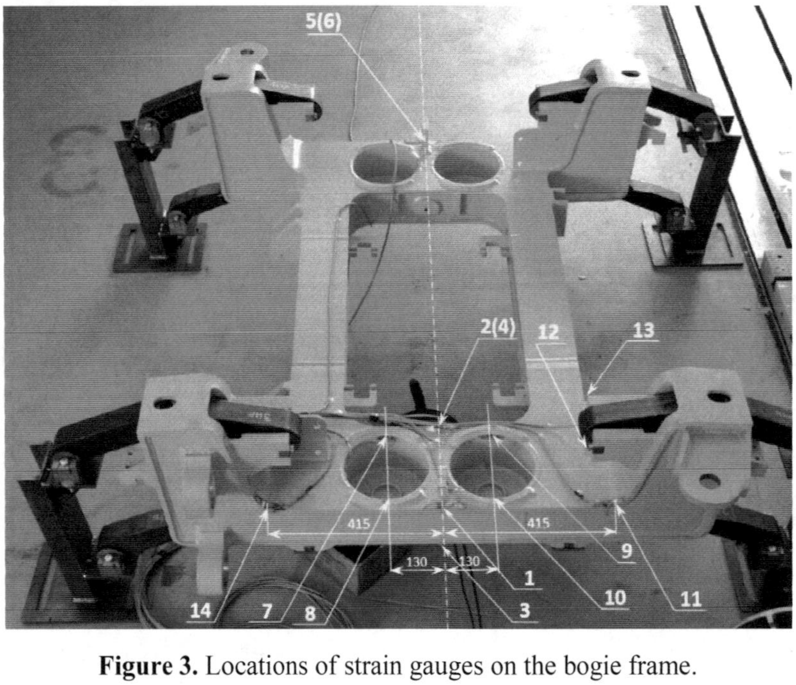

Figure 3. Locations of strain gauges on the bogie frame.

3.1 Static strength test

The static test involves applying what is referred to as "exceptional load" on the bogie. They are the highest forces which may occur in service under exceptional circumstances. Permanent structural deformation must not occur as a result of this test. The purpose of static testing of the bogie frame was to ensure that no residual deformation remains in the frame after removing the load. In the static test, stress was first measured without applied load. Then the bogie was subjected to a vertical force of 157 kN, a load exerted by a full tram car with a superimposed dynamic component. After releasing the load, the values from strain gauges were recorded. The strain gauges on the bogie frame yielded the values shown in Table 1. None of the values exceeded the minimum yield stress (355 MPa) of the material of the bogie frame.

Table 1. Values measured in static test.

Load	Strain gauge number Measured stress value [MPa]				
0 kN	**1**	**2**	**3**	**4**	**5**
	0.08	-0.15	0.05	-0.11	-0.23
	6	**7**	**8**	**9**	**10**
	0.08	-0.09	-0.08	0.36	0.10
	11	**12**	**13**	**14**	
	-0.02	0.01	-0.01	-0.07	
157 kN	**1**	**2**	**3**	**4**	**5**
	-143.16	-120.10	74.23	66.92	-141.07
	6	**7**	**8**	**9**	**10**
	76.60	-189.95	-224.81	-188.02	-223.76
	11	**12**	**13**	**14**	
	-98.77	-165.52	123.84	-91.07	
0 kN	**1**	**2**	**3**	**4**	**5**
	-0.45	-0.04	-0.08	-0.16	0.03
	6	**7**	**8**	**9**	**10**
	-0.07	0.15	0.36	0.56	0.45
	11	**12**	**13**	**14**	
	0.18	0.57	-0.92	0.13	

The values measured after removing the load on the bogie frame suggest that the residual stress is zero.

IOP Conf. Series: Materials Science and Engineering **723** (2020) 012031 doi:10.1088/1757-899X/723/1/012031

3.2 Fatigue tests

The goal of dynamic testing of the bogie frame was to verify its life according to CSN EN 13749. The standard prescribes the use of loading forces which are ordinarily encountered in service. The loading cycles (Figure 4) simulate vertical forces from the load carried by the vehicle, loads from curve negotiation and from riding over switches and associated structures. Three load levels were applied according to Figure 5. The first load level corresponds to the full load on the bogie frame over 6×10^6 cycles. The second involves 2×10^6 cycles, in which quasi-static and dynamic components have been raised by 20 %. The third level involves another 20 % higher quasi-static and dynamic loads [2].

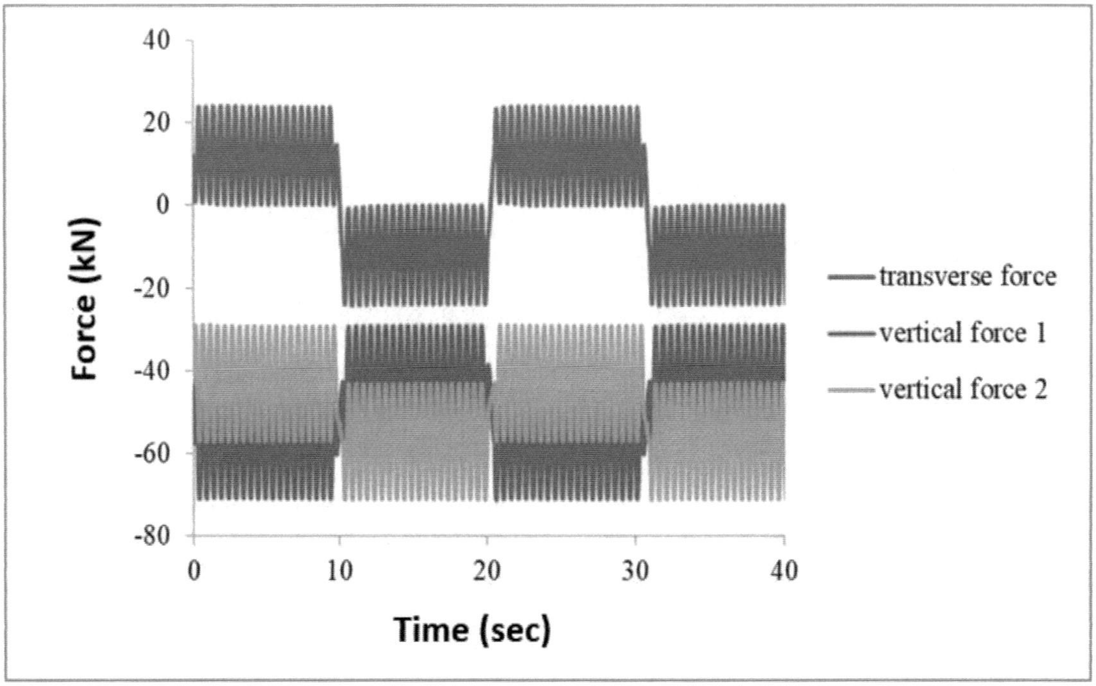

Figure 4. Example of loading during fatigue testing.

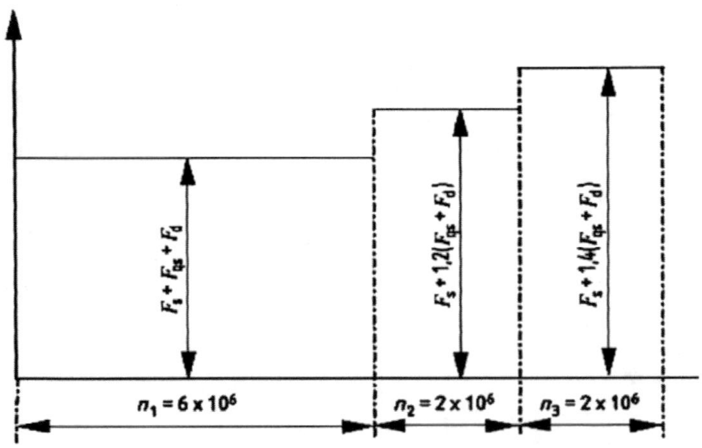

Figure 5. Distribution of dynamic load on the frame.

3.3 Non-destructive testing

Non-destructive magnetic testing was performed in order to find any defects which may have occurred. This method can detect flaws in the surface or just below the surface. The procedure for detecting flaws with a UV lamp which highlights fluorescent magnetic particles followed the standards CSN EN ISO 17638 and CSN EN ISO 9934 – 1. This method belongs to rapid, simple and inexpensive non-destructive techniques. The test was performed prior to load application and then after 6×10^6, 8×10^6 and 10^7 cycles.

4 Conclusions

A bogie frame was fatigue tested under laboratory conditions. Experimental static and dynamic tests of a bogie frame for a narrow gauge tramway according to CSN EN 13749 yielded data which are useful for further development and application of these components. Strain gauge measurements after static loading of 157 kN provided evidence of zero residual deformation in the bogie frame.

The fatigue test verified the life of the bogie frame after 10^7 cycles under loads at increasingly higher levels. Non-destructive inspections for defects were performed by the magnetic particle method before each increase in the load level. No defects were found by this method.

Acknowledgements

This article "Static and dynamic testing of bogie" was created with financial support of TA CR - project TH02010553.

References

[1] Kalina, T., Heller, P., Chval, Z., Sedlácek, F., Křížek, M., Barton, L. Numerical Simulation and Experimental Testing of Two-axle Chassis of Low-floor Trams. *Manufacturing Technology*, 2018, roč. 18, č. 2, s. 248-254. ISSN: 1213-2489

[2] Kepka, M., Kepka, M., Žlabek, P., Heller, P., Chvojan, J., Mentl, V. Fatigue Tests - Important Part of Development of New Vehicles. In MATEC Web of Conferences. New York: Curran Associates, Inc, 2018. s. 1-7. ISBN: neuvedeno , ISSN: 2261-236X

[3] Kraus V., Kepka M., Doubrava D., Chvojan J. (2019) Strength Analysis of Tramway Bogie Frame. In: Correia J., De Jesus A., Fernandes A., Calçada R. (eds) Mechanical Fatigue of Metals. Structural Integrity, vol 7. Springer, Cham

[4] CSN EN 13749, Železnicní aplikace - Dvojkolí a podvozky - Metoda specifikování konstrukcních požadavku na rámy podvozku (Railway applications – Wheelsets and bogies – Method of specifying the structural requirements of bogie frames)

Residual stress investigation in additively manufactured samples

Jaroslav Vavřík, Josef Hodek

COMTES FHT a.s., Průmyslová 995, Dobřany, Czech Republic

jvavrik@comtesfht.cz

Abstract. In the present study, additively manufactured (AM) samples of different sizes were analysed by the contour method (CM). Builds of three different volumes were examined to determine the influence of residual stress (RS) distribution along the z-axis on volumetric changes in the part. They were made using the direct energy deposition (DED) technique. Only a handful of reports cover the use of CM for finding the RS distribution in DED builds. This study aims to complement those with new findings.
The contour method is a destructive technique whose main advantages are the absence of sample size limits and the ability to predict residual stresses at greater depths than other methods. In this study, residual stresses were evaluated in three stainless steel samples (35×35×35 mm, 25×25×25 mm, 15×15×15 mm) prepared by the DED method. Stainless steel 316L powder was blown into a melt pool created by a high energy laser source. The melted powder solidifies as the laser head moves along a predefined path. In order to protect the material from oxidation during deposition, argon is used as a shielding gas-fed continuously to the melt pool.
316L austenitic stainless steel was chosen for this experiment because it undergoes no phase transformations during deposition. This reduces the complexity of the metallurgical processes involved. Evaluation of the sections produced by the contour method was carried out using an optical precision measuring machine (3D scanner) and FE software.

1 Introduction

The concept of additive manufacturing includes a widespread family of production techniques available for processing almost any material. Plastics are a common choice followed by relatively recently-adopted materials like metals, ceramics, concrete, etc. Additive manufacturing comprises SLM (selective laser melting), DED (direct energy deposition) and other techniques. SLM is a method where the powder is spread by a roller and then melted by a high energy source. The amount of porosity in SLM builds is usually somewhat greater than in DED builds. Powder particles are smaller in SLM. DED uses a continuous flow of powder, which is then completely melted by a high power laser source. Both SLM and DED require a protective atmosphere, which usually consists of argon gas. The main advantages of DED are the possibility of using multiple powders at once for deposition and the use of a multiaxial deposition. Both methods are relevant for special applications in the automotive, aerospace, energy and healthcare sectors [1, 2].

Deposition leads to substantial levels of various kinds of stresses (namely residual stress) in the structure of the build (depending on the type of material). Residual stress is caused by localized heating by high energy laser during deposition. This results in non-uniform cooling, thermal expansion and contraction of the material during cooling or heating [3, 4]. It is essential to know the magnitudes of these stresses because they can significantly influence the shape, properties and also the lifetime of the build.

Finding residual stress levels using FE software and simulation of thermomechanical processes is not an easy task. Temperature-dependent material data are required for this purpose. In addition, phase transformations data must be available. Excessive computational time may be required in some cases. Efforts to eliminate difficulties or reduce the costs involved lead to approximations and simplifications and may compromise the accuracy of the simulation of residual stress distribution. The contour method is an experimental-computational destructive test method for finding residual stress [5]. Although there are articles dealing with the evaluation of residual stress by the contour method in SLM parts [6, 7], reports on DED build analysis by the contour method are rare [8]. This method is based on the principles of solid mechanics. The sample contains some amount of initial stress. After dividing the sample by cutting, residual stresses are released which leads to the deformation of the cut plane. The most widely-used and also the best cutting method for CM measurement is the wire electric discharge machining (WEDM). It is a non-contact technique. The material is eroded from the work piece by repetitive spark discharges (between the cutting wire and the work piece) generated by a pulsed power supply source. The goal is to produce the best possible quality of the surface [5]. The distorted cut plane is scanned by a 3D scanner to collect surface displacement values caused by the RS redistribution. The output from the scanning procedure (the measured displacement) is processed by smoothing. The measured and processed displacement data are input into an FE simulation as a boundary condition. For this purpose, an FE model of the sample prior to cutting is created, meshed and characterized using mechanical properties, such as elastic modulus and Poisson´s ratio. The result of the simulation is a 2D map of RS distribution prior to cutting [5, 7, 8].

2 Experimental procedures

In this study, three 316L samples of different sizes were produced by additive manufacturing, specifically by the method of direct energy deposition using Insstek MX-600. The sizes of the samples were 35×35×35 mm, 25×25×25 mm and 15×15×15 mm. Each sample was built separately on a new platform (at room temperature), as illustrated in Figure 1. The platform dimensions were 95×95×10 mm. The platforms were heat-treated at 400 °C/4 hours to relieve stress.

The chemical composition of the platforms indicated in Table 1 was analyzed using the Bruker Q4 Tasman optical spectrometer. The 316L stainless steel powder was produced by Sandvik Osprey LTD. The size interval of the powder particles was 53–150 µm. Its chemical composition is listed in Table 1.

Table 1. Chemical compositions of the powder and the platform in wt. %.

	Fe	Cr	Ni	Mo	Mn	Si
Powder	Bal.	17.20	10.40	2.30	1.30	0.80
Platform	Bal.	16.24	10.49	2.14	1.12	0.44

Models of samples for the deposition were developed by slicing using Magic 3D software. The slicing method was CF ZZ, where C stands for "contour", F indicates "filling" and ZZ means "Zig Zag". ZZ is a method of filling. At the end of deposition of each layer, the filling path was rotated by 90°. Each layer was 0.25 mm in height and 0.8 mm in width; with a hatch spacing of 0.5 mm. The average power of the laser was about 400 W. In order to protect the build from oxidation during the deposition, argon gas was fed continuously as the shielding gas.

Figure 1. The build: a 25 mm-side cube on a platform.

After the AM process, the samples were cut off at 1 mm above the platform, as shown in Figure 2 using WEDM (the diameter of the wire was 0.25 mm) for analysis of RS along the z-axis. During cutting, the sample was submerged in a tank of deionized water. An optical precision measuring machine (OPMM), ATHOS Capsule from GOM, was used for scanning the surface and recording the displacements. The resolution of ATHOS Capsule is 2 μm. The cut surfaces had to be coated with a special mixture of TiO_2 with ethanol in order to reduce reflectivity, which would otherwise influence the quality of the scanning output, see Figure 3.

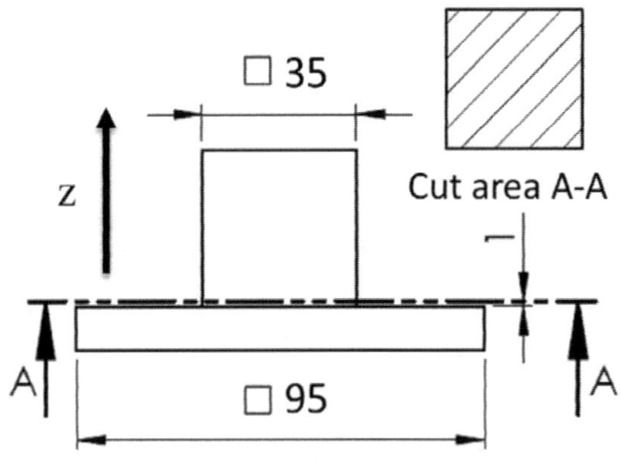

Figure 2. Cutting plan.

The values measured by ATHOS Capsule were processed by smoothing, averaging and inverting with user script Octave 4.4.1. The results were input into MSC Marc software as a boundary condition. Using the elastic analysis for both parts of the cut sample, normal stress on the cut plane prior to cutting was determined. This linear elastic analysis used Young's modulus E = 192.968 GPa and Poisson's ratio $\nu = 0.2989$. Type7 8-point isoperimetric elements with a characteristic element length of 1 mm were used.

Figure 3. A cut sample prepared for scanning.

3 Results

Residual stress perpendicular (σ_z) to the cut surface calculated using MSC Marc software is shown in Figures 4-5. Positive values of residual stress indicate tensile stresses, whereas negative values represent compressive residual stress. Compressive residual stresses can be seen in the middle of the cube and the platform. Tensile residual stresses are found along the edges (corresponding to average residual stress in Figure 6). The same pattern was observed by Rangaswamy et at [8] and by Beuth et al [10]. The possible reason is the non-uniform cooling of the sample [8, 10].

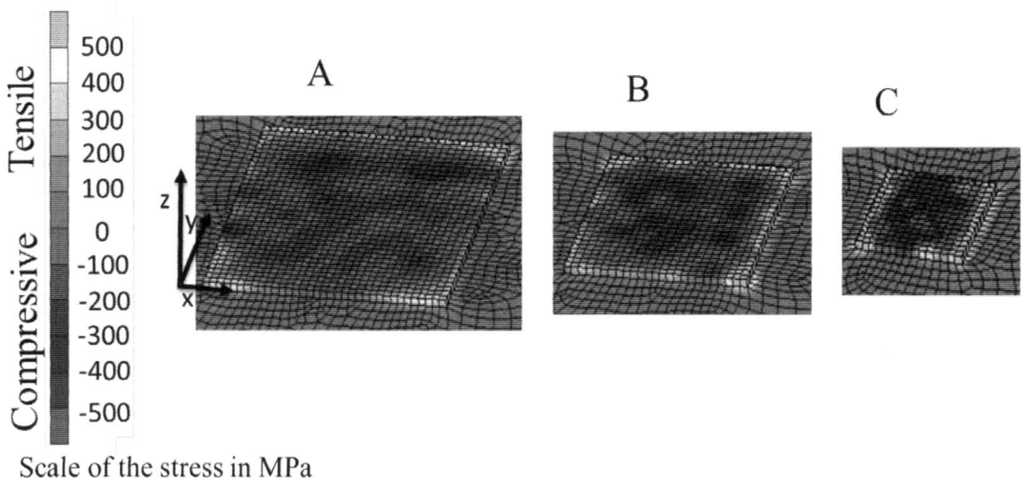

Figure 4. Residual stress in platforms in Z direction (σ_z) A) 35 mm cube, B) 25 mm cube, C) 15 mm cube.

The values of the residual stress in the z-direction (σ_z) (perpendicular to the cutting plane) for samples and platforms are summarized in Figure 6. These values are averages (both vertical and horizontal direction specified in Figure 6) for samples and platforms. Relative lengths were used for a good comparison among all samples. The highest compressive residual stress was present in the smallest sample. The larges sample had the smallest amount of compressive residual stress. The maximum of compressive residual stress was approximately in the middle of each sample. High tensile residual stress occurs near the edges but the value has to be taken with care due to the limitations of the contour method [5]. For further improvement of residual stress data, neutron diffraction should be used.

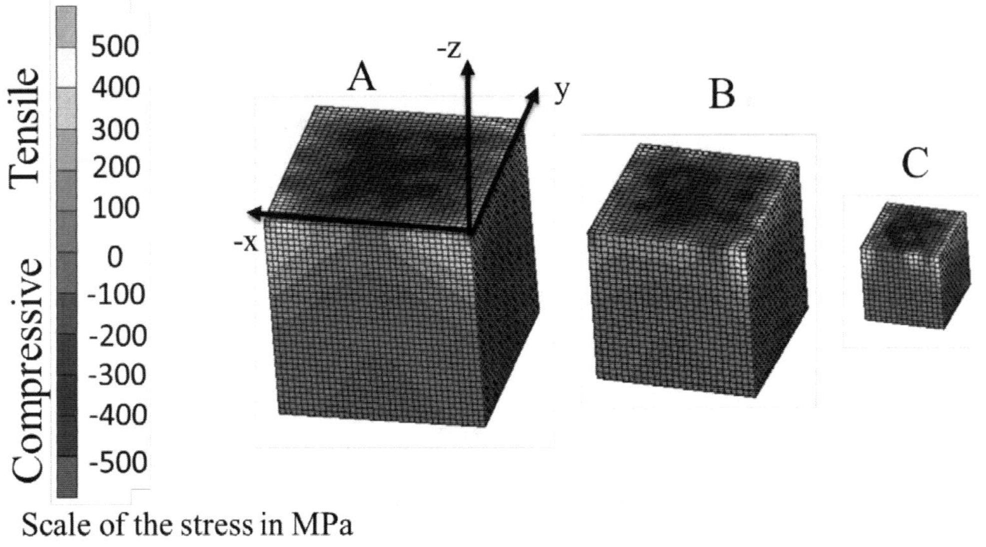

Figure 5. Residual stress in the builds along z-direction (σ_z) A) 35 mm cube, B) 25 mm cube, C) 15 mm cube.

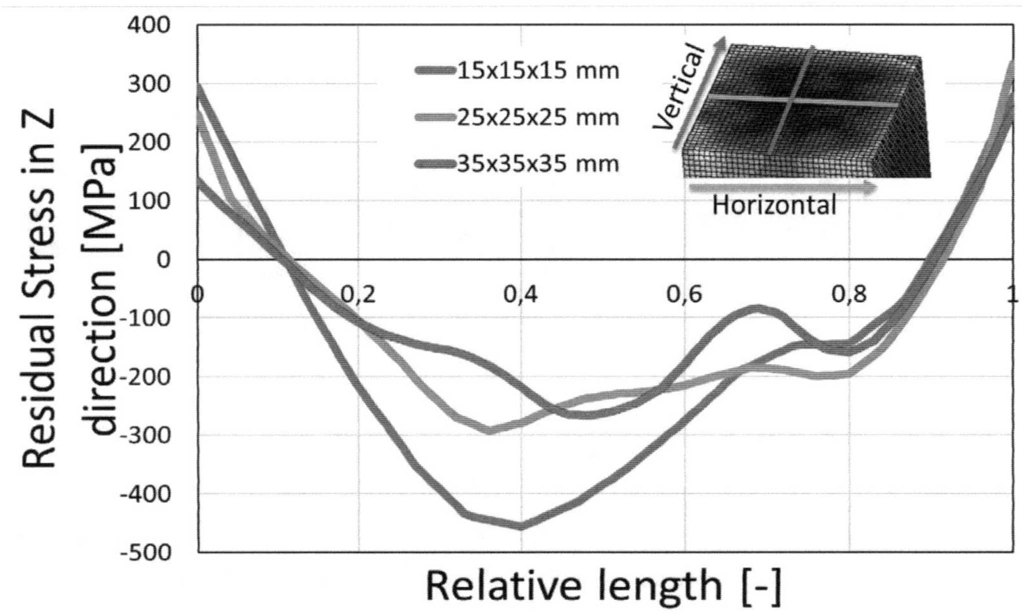

Figure 6. Average residual stress in the z direction (σz) in the cubes and platforms.

4 Conclusion

Three samples of different sizes (35×35×35 mm, 25×25×25 mm and 15×15×15 mm) from 316L stainless steel were built using Insstek MX-600 additive manufacturing machine. They were then used for evaluating residual stresses in the z-direction (σ_z), i.e. perpendicular to the cut surface. A pattern involving compressive residual stresses in the centre and tensile stresses along edges was noted. The possible reason is the variation in cooling rates between the centre and edges. Residual stresses in the samples were compared. The smallest sample had the highest values of compressive residual stress. This occurrence will be explored in more depth in a future study in which FE simulation and neutron diffraction measurement will be considered.

Acknowledgment
This paper was developed thanks to the support under project entitled of Pre-Application Research of Functionally Graduated Materials by Additive Technologies, No.: CZ.02.1.01/0.0/0.0/17_048/0007350, financed by the MEYS of the Czech Republic.

References

[1] T. D. Ngo, A. Kashani, G. Imbalzano, K. T. Nguyen and D. Hui, "Additive manufacturing (3D printing): A review of materials, methods, applications and challenges," *Composites Part B: Engineering,* pp. 172-196, 2018.

[2] EPMA, Introduction to additive manufacturing technology - a guide for designers, 2015.

[3] X. L. Jamison L. Bartlett, "An overview of residual stresses in metal powder bed fusion," *Additive Manufacturing,* pp. 131-149, 2019.

[4] T. DebRoy, H. Wei, J. Zuback, T. Mukherjee, J. Elmer, J. Milewski, A. Beese, A. Wilson-Heid, A. De and W. Zhang, "Additive manufacturing of metallic components – Process, structure and properties," *Progress in Materials Science,* pp. 112-224, 2018.

[5] F. Hosseinzadeh, J. Kowal and P. J. Bouchard, "Towards good practice guidelines for the contour method of residual stress," *Engineering and Innovation Department,* 2014.

[6] B. Vrancken, V. Cain, R. Knutsen and J. V. Humbeeck, "Residual stress via the contour method in compact tension specimens produced via selective laser melting," *Scripta Materialia,* pp. 29-32, 2014.

[7] B. Ahmad, S. O. v. d. Veen, M. E. Fitzpatrick and H. Guo, "Residual stress evaluation in selective-laser-melting additively manufactured titanium (Ti-6Al-4V) and inconel 718 using the contour method and numerical simulation," *Additive Manufacturing,* pp. 571-582, 2018.

[8] P. Rangaswamy, M. Griffith, M. Prime, T. Holden, R. Rogge, J. Edwards and R. Sebring, "Residual stresses in LENS® components using neutron diffraction and contour method," *Materials Science and Engineering: A,* pp. 72-83, 2005.

[9] M. Olson, A. DeWald, M. Prime and M. Hill, "Estimation of Uncertainty for Contour Method Residual Stress Measurements," *Exp Mech,* pp. 577-585, 03 2015.

[10] J. Beuth and N. Klingbeil, "The role of process variables in laser-based direct metal solid freeform fabrication," *The Journal of The Minerals, Metals & Materials Society (TMS),* pp. 36-39, September 2001.

Rolling of ingots of third-generation high-strength steels into sheets

Jiří Vrtáček[1,1], Hana Jirková[1,2], Michal Peković[1,3], Ladislav Tříska[1,4]

[1]UWB in Pilsen, Regional Technological Institute, Univerzitní 22, 301 00 Pilsen, Czech Republic

[1]vrtacekj@rti.zcu.cz, [2]hstankov@rti.zcu.cz, [3]pekovicm@rti.zcu.cz, [4]triska7@rti.zcu.cz

Abstract. One of the essential branches of today's engineering production involves the production of sheet stock by rolling. In the automotive industry is used the majority of steel sheet stock. In recent decades, this sector has been striving to reduce vehicle emissions. One of the available solutions involves the use of advanced high-strength steels whose chemical composition and strengthening mechanisms make it possible to build the car body with thinner sheet blanks than before. For rolling trials in which ingots were converted into 1.8 mm sheet by combined hot and cold rolling were used two advanced high-strength steels containing 0.2 wt. % carbon and additions of manganese, silicon and different levels of aluminium. This procedure was found to produce strengths in excess of 1000 MPa combined elongation of more than 15 %.

1 Introduction

Steel stock is often rolled into flat sheet or products of more or less complex shapes. Parts with a simple geometry include various shaped sections or tubes with different cross sections. In lightweight structures, such as front or rear crumple zones of vehicles, including bumpers, more intricate rolled products are often used [1, 2]. Rolled stock and products are often formed into vehicle frame structures. Their materials are high-strength steels [1, 3]. Third generation of high-strength steels combines several types of reinforcement mechanisms, mainly TRIP (Transformed induced plasticity) and TWIP (Twinning induced plasticity) and alloying with manganese, silicon and aluminum [1-3]. Thanks to mechanical properties of these steels, designers can use thinner sheet for making equal components. This translates into cost savings, in terms of both materials and energy, as the weight of parts and fuel consumption are reduced [1, 3, 4]. Because the primary use of high-strength steels is the manufacture of safety components of vehicle body, there is a need to study the entire process of sheet production and not only the manufacture of the shaped part [5]. During hot and cold rolling of ingots, a number of problems may arise, prohibiting the production of high-quality defect-free sheet with a desired microstructure. This paper discusses rolling of small ingots into 1.8 mm sheets of newly-designed experimental high-strength steels with different manganese to aluminium ratio.

2 Experimental program

Two advanced high-strength steels identified as C3Mn1.4Al and C3Mn2Al with different aluminium levels were designed for this series of experiments. This enabled the effects of the manganese-aluminium ratio on microstructure evolution and mechanical properties to be studied. The main alloying additions in C3Mn1.4Al were manganese, silicon and aluminium (table 1). The only difference in the composition of the second material, C3Mn2Al, was a higher aluminium content: 2 % (table 1). Measurement in a dilatometer was employed to find heating and forming temperatures with dilatometer SG instruments L75PT with argon atmosphere. Heating to the temperature 600 °C was 10 °C/min, then the heating rate dropped to 3 °C/min to the temperature 1000 °C. Austenitizing temperatures for C3Mn1.4Al and C3Mn2Al were 961 °C and 1025 °C. The heating temperature of 1050 °C was chosen based on the dilatometer data and empirical findings. The materials were cast into 50 kg ingots. They were rolled in a combined rolling mill at the company COMTES FHT, where a two-high stand for hot rolling and four-high stand for cold rolling are available.

Table 1. Chemical composition of the experimental materials.

	C	Si	Mn	P	S	Cr	Al
C3Mn1.4Al	0.22	0.59	3.06	0.008	0.003	0.19	1.45
C3Mn2Al	0.22	0.56	3.13	0.008	0.003	0.17	2.02

2.1 Hot rolling

Prior to hot rolling were the top and bottom of the ingots removed. The ingots were placed into a pre-heated furnace at 1050 °C for 2 hours. Then they were transferred from the furnace to the rolling mill. The rolling sequence consisted of nine successive reductions with no reheating operation. The initial diameter was 97 mm and the final thickness of the strip was 13 mm (figure 1a). The temperature at the entry was about 1035 °C and temperature did not dropped below 950 °C during all reductions. With each reduction, the rolling force increased: from the initial 680 kN to 2000 kN in the last reduction. No excessive buildup of force, irregular thickness or waviness were experienced in the rolling process. Immediately after finish rolling to 13 mm, the rolled part was cut into three identical segments 800 mm in length (figure 1b) and reheated in a furnace for 1 hour.

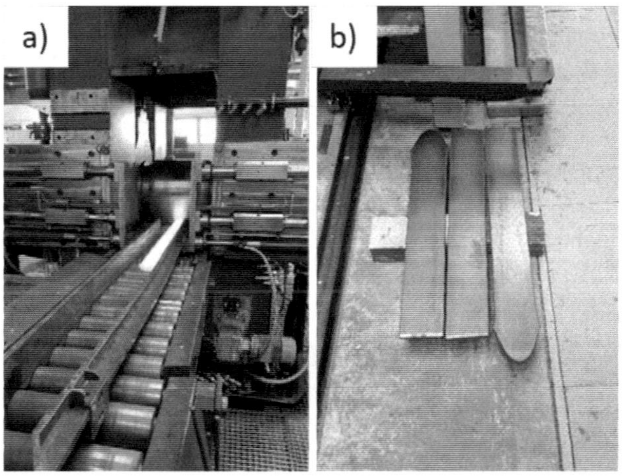

Figure 1. Rolled product passing through the rolling mill (a); rolled product 13 mm in thickness cut into three parts (b).

Subsequent hot rolling involving four reductions to the final stock thickness of 5.2 ± 0.1 mm. Rolling force varied from 1060 kN in the first reduction to 2000 kN in the last pass. Over the course of the rolling process, temperature decreased from the initial 1050 °C to 850 °C in the last reduction. After this step, no defects were found in the materials. After rolling, the rolled product was cut into three

IOP Conf. Series: Materials Science and Engineering **723** (2020) 012033 doi:10.1088/1757-899X/723/1/012033

segments approximately 480 mm in length. The segments then cooled in air to room temperature. The segments were annealed at 680 °C for one hour before cold rolling.

2.2 Cold rolling

Following the in-process annealing at 680 °C, both sides of the rolled products were ground to remove scale and poor-quality surface (figure 2a). The rolling stand was changed over to a four-high mill for cold rolling (figure 2b). The materials were rolled to the final thickness of 1.8 ± 0.1 mm. As the ability of the material to undergo plastic deformation had not been depleted by cold rolling, there was no need for additional in-process annealing. The cold-rolled products were normalized. The normalizing sequences for C3Mn1.4Al and C3Mn2Al were 1000 °C/1.5 hr and 950 °C/1.5 hr, respectively. Finally, the rolled sheets were straightened and ground to the desired thickness (1.5 mm) for subsequent processing.

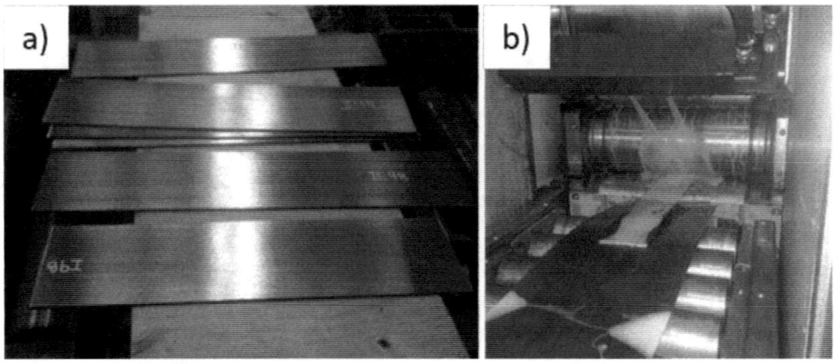

Figure 2. Ground rolled products (a), cold rolling process (b).

3 Results and Discussion

3.1 Microstucture analysis

Hot-rolled microstructures were examined under a light microscope. The structure contained distinct strain bands which are typical of rolling. The microstructure of C3Mn1.4Al consisted of martensite and bainite. Hardness was 470 HV10 (figure 3a). C3Mn2Al, which had a higher aluminium content, contained a predominantly martensitic-bainitic microstructure as well. It was coarser than in the other material. In addition, there were notable bands of free ferrite. They were the result of chemical heterogeneity and the shift of the ferrite nose towards faster cooling rates caused by the higher aluminium content. The presence of ferrite affected hardness, the value of which was 419 HV10 (figure 3b).

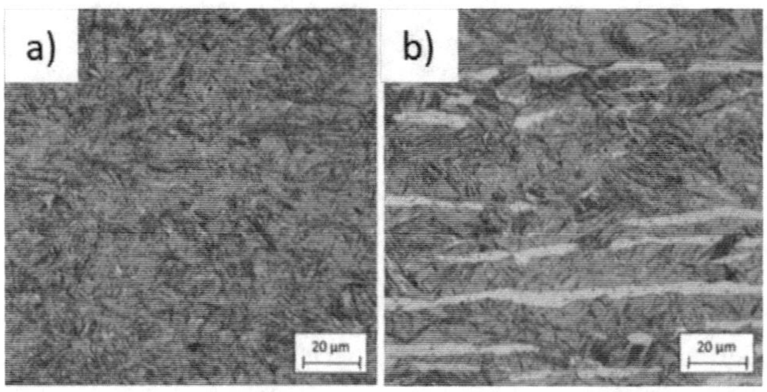

Figure 3. Micrographs of C3Mn1.4Al (a) and C3Mn2Al (b) after hot rolling.

Hot rolling was followed by in-process annealing at 680 °C for one hour. The final microstructures were heavily tempered (figure 4). Dense cementite and carbide precipitates were found along the boundaries of poorly-visible martensite laths. There was a drop in hardness of C3Mn1.4Al and C3Mn2Al to 250 HV10 and 264 HV10, respectively.

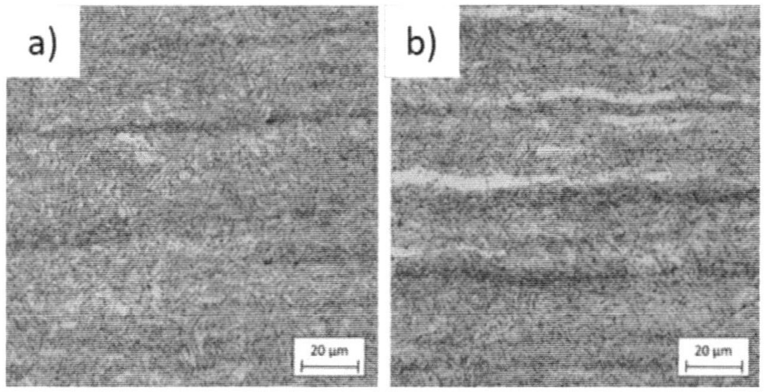

Figure 4. Micrographs of C3Mn1.4Al (a) and C3Mn2Al (b) after in-process annealing.

The resulting microstructures of C3Mn2Al and C3Mn1.4Al were consistent with the appearance of cold-rolled structures (figure 5). They contained a fine mixture of ferrite and cementite produced by tempering. C3Mn2Al also contained conspicuous deformation bands consisting of ferrite. By comparison, the width of the deformation bands was different in each material. In addition, there were more impurities in C3Mn2Al than in C3Mn1.4Al.

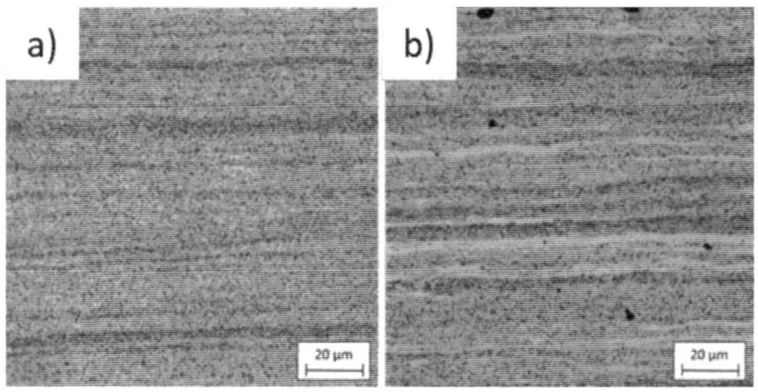

Figure 5. Micrographs of cold-rolled C3Mn1.4Al (a) and C3Mn2Al (b).

3.2 Mechanical properties

Mechanical properties were assessed by means of tensile tests. Flat test specimens were used, in which the gauge section was 10×1.5 mm in cross section and 50 mm in length (figure 6). In order to characterize anisotropy in the sheet stock, some specimens were oriented parallel to the rolling direction and the rest perpendicularly to the rolling direction. After hot rolling, the mechanical properties in both materials were favourable. C3Mn1.4Al showed an ultimate strength of just above 1500 MPa and elongation of 20 % in the rolling direction. The ultimate strengths in the perpendicular direction were nearly equal. Elongation was slightly lower, just under 18 % (table 2). In C3Mn2Al, which had a higher aluminium content, the ultimate strength was 100 MPa less: 1400 MPa. Elongations in the rolling direction and crosswise were just under 18 % and below 15 %, respectively (table 3). Higher aluminium content led to a lower ultimate strength and equal elongation.

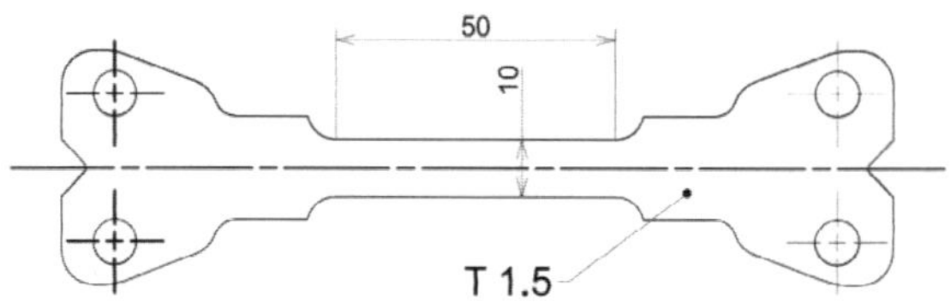

Figure 6. Drawing of the tensile test specimen.

Table 2. Results of mechanical testing of hot-rolled material.

	Offset yield strength ($R_{p0.2}$) [MPa]	$R_{p0.2}$ standard deviation	UTS (R_m) [MPa]	R_m standard deviation	A_{20mm} [%]	A_{20mm} standard deviation
C3Mn1.4Al	1089	12	1523	15	20.1	1.0
C3Mn1.4Al-90°	1119	12	1556	3	17.7	0.4
C3Mn2Al	1019	7	1423	10	18.3	0.2
C3Mn2Al – 90°	1010	3	1418	2	14.7	1.2

Annealing at 680 °C for one hour, which followed the hot rolling operation, led to a notable decrease in ultimate strength. In C3Mn1.4Al, the strength was around 810 MPa and elongation reached 28 % (table 3). In the other material, strength dropped to 770 MPa, while elongation rose to 31 % (table 3). The comparison between them shows that higher aluminium content was manifested in higher elongation and reduced yield strength and ultimate strength.

Table 3. Results of mechanical testing of hot-rolled material and in-process annealed material.

	Offset yield strength ($R_{p0.2}$) [MPa]	$R_{p0.2}$ standard deviation	UTS (R_m) [MPa]	R_m standard deviation	A_{20mm} [%]	A_{20mm} standard deviation
C3Mn1.4Al	704	15	806	0	31.0	0.8
C3Mn1.4Al-90°	717	3	811	2	30.0	1.0
C3Mn2Al	630	3	773	2	34.1	0.6
C3Mn2Al – 90°	620	2	779	4	30.0	0.3

In both cold-rolled materials, ultimate strength in the rolling direction was approximately 1100 MPa and elongations were low, 3-5 % (table 4). Perpendicular to the rolling direction, ultimate strengths were approx. 100 MPa higher: 1200 MPa. After cold rolling, both materials showed similar properties. C3Mn1.4Al exhibited higher ultimate strength and yield strength, by mere units of MPa.

Table 4. Results of mechanical testing of cold-rolled material.

	Offset yield strength ($R_{p0.2}$) [MPa]	$R_{p0.2}$ standard deviation	UTS (R_m) [MPa]	R_m standard deviation	A_{20mm} [%]	A_{20mm} standard deviation
C3Mn1.4Al	1022	8	1119	14	5.2	0.05
C3Mn2Al	999	4	1100	3	4.9	0.30
C3Mn1.4Al – 90°	1036	0.2	1203	11	4.5	0.20
C3Mn2Al – 90°	1025	3	1193	2	3.2	0.14

For the purpose of restoring the plastic deformation ability of the material, normalizing annealing was performed. The normalizing sequences for C3Mn1.4Al and C3Mn2Al were performed at 1000 °C and 950 °C/1.5 hr, respectively. The annealing operation greatly improved elongation values without altering ultimate strength in both materials. In the rolling direction, elongations in excess of 15 % were obtained, and those perpendicular to the rolling direction were 10 % (table 5).

Table 5. Mechanical properties after rolling to final dimension and normalizing.

	Offset yield strength ($R_{p0.2}$) [MPa]	$R_{p0.2}$ standard deviation	UTS (R_m) [MPa]	R_m standard deviation	A_{20mm} [%]	A_{20mm} standard deviation
C3Mn1.4Al	1041	12	1099	4	16.3	1.4
C3Mn1.4Al-90°	1026	0.2	1174	2	11.2	0.8
C3Mn2Al	1004	15	1064	9	15.0	0.6
C3Mn2Al – 90°	1017	5	1144	4	8.6	0.1

4 Conclusion

Newly-designed third-generation high-strength steels with higher manganese levels than classical TRIP steels were used for an experimental procedure, in which ingots were rolled into 1.8-mm sheet. Based on known phase transformations and results of earlier research, appropriate soaking temperatures and reductions in hot and cold rolling were proposed. Using this manufacturing procedure, sheet stock free from internal defects was obtained. Cold rolling and normalizing led to ultimate strengths between 1000 and 1100 MPa and elongations up to 16 %. The higher aluminium level in C3Mn2Al was reflected in a slightly lower ultimate strength. In this steel, distinctive ferrite bands and higher levels of impurities were found in final state. The data collected in the process of rolling of ingots into sheet stock was crucial for mapping the initial condition of the multiphase high-strength steels.

Acknowledgement
The present contribution has been prepared with the support of the student grant competition of University of West Bohemia in Pilsen, SGS 2018-045 "Design of heat and thermo-mechanical treatment of high strength steels of the third generation". The project is subsidized from specific resources of the state budget for research and development.

IOP Conf. Series: Materials Science and Engineering **723** (2020) 012033 doi:10.1088/1757-899X/723/1/012033

References

[1] Demeri M Y 2013 *Advanced high-strength steels: Science, Technology and Applications,* Materials Park, Ohio ASM International.

[2] Tamarelli C M 2011 *AHSS 101: The Evolving Use of Advanced High-Strength Steels for Automotive Applications,* Southfield, Michigan.

[3] Rowe J 2012 *Advanced Materials in Automotive Engineering,* ed. Rowe J, Elsevier Amsterdam, Netherlands.

[4] Chen S C et al., 2019 Microstructure and mechanical bevaiors of GPa-grade TRIP Steels Enabled by hot-rolling processes *Materials Science & Engineering,* A **761.**

[5] Kučerová L, Jirková H and Mašek B 2014 The Effect of Alloying on Mechanical Properties of Advanced High Strength Steels *Archives of Metallurgy and Materials* **59** 1189-92.

AUTHOR INDEX